CRASH COURSE

Third Edition

Cell Biology and Genetics

First and second edition authors:

Emma Jones

Anna Morris

Ania L Manson

CRASH COURSE

Third Edition

Cell Biology and Genetics

Series editor
Daniel Horton-Szar,
BSc (Hons), MBBS (Hons), MRCGP
Northgate Medical Practice,
Canterbury, Kent, UK

Faculty advisor
Melanie Newport,
FRCP PhD
Reader in Infectious Diseases and International Health,
Brighton and Sussex Medical School, University of Sussex, Brighton, UK

Joanne Evans,
DPhil
Medical Student, Barts and The London School of Medicine and Dentistry, London, UK

MOSBY

Edinburgh • London • New York • Oxford • Philadelphia • St Louis • Sydney • Toronto 2008

MOSBY
ELSEVIER

Commissioning Editor: *Alison Taylor*
Development Editor: *Kim Benson*
Project Manager: *Elouise Ball*
Page design: *Sarah Russell*
Icon illustrations: *Geo Parkin*
Cover design: *Stewart Larking*
Illustration Manager: *Merlyn Harvey*
Illustrator: *Cactus*

First edition 1998
Second edition 2002
Reprinted 2003 (twice), 2005
Third edition 2008

ISBN: 978-0-7234-3421-4

British Library Cataloguing in Publication Data
A catalogue record for this book is available from the British Library

Library of Congress Cataloging in Publication Data
A catalog record for this book is available from the Library of Congress

Note
Knowledge and best practice in this field are constantly changing. As new research and experience broaden our knowledge, changes in practice, treatment and drug therapy may become necessary or appropriate. Readers are advised to check the most current information provided (i) on procedures featured or (ii) by the manufacturer of each product to be administered, to verify the recommended dose or formula, the method and duration of administration, and contraindications. It is the responsibility of the practitioner, relying on their own experience and knowledge of the patient, to make diagnoses, to determine dosages and the best treatment for each individual patient, and to take all appropriate safety precautions. To the fullest extent of the law, neither the Publisher nor the Authors assumes any liability for any injury and/or damage to persons or property arising out of or related to any use of the material contained in this book.

The Publisher

The publisher's policy is to use **paper manufactured from sustainable forests**

Printed in China

Cell biology and genetics are subjects often much feared by medical students. They are central to all disease processes and yet, because of the current educational emphasis on 'soft skills' on top of academics, are receiving less and less attention in the undergraduate curriculum. Neither subject need be difficult (honest!), and this book aims to present material in a clear and concise manner, to support lecture and PBL notes and to break down some of those dark walls of fear!

In the 5 years that has passed since the second edition was published, much has changed in the world of genetics, molecular and cellular biology. Many chapters have been reordered and rewritten in order to take into account the many changes that have occurred in this fast paced and exciting field. The human genome project was completed in 2003, and the genomes of many other organisms have also been unlocked, leading to the new fields of genomics, proteomics, metabolomics and bioinformatics. Powerful new technologies, such as RNAi, hold much promise not only as research tools, but also as potential treatments to many genetic conditions.

Drawing upon the first two editions, this book aims not only to give you essential information that will help you ace those exams, but also to highlight areas of extra study for the diligent student. So, whether you have bought this book to pass an exam or as an introduction to the most fascinating subjects in medicine, I hope it does what it says on the tin! Enjoy!

Joanne Evans

Progress in the basic sciences, especially genetics, continues to advance at an exponential rate. Exciting new developments resulting from the human genome project coupled with parallel advances in technology have led to growth in the identification of susceptibility genes for common, multifactorial diseases. In addition to enhancing our understanding of the basic mechanisms of disease, this knowledge will translate into benefit for patients; for example, through the development of novel therapies, or prevention of disease through better understanding of risk factors and gene–environment interactions. As the current generation of medical students, you are in the exciting position of being among the first generation of doctors applying the fruits of these advances in your clinical practice. A good understanding of the basics of genetics and how cells work is, therefore, critical, whichever specialty of medicine (in its broadest sense) you ultimately practice. We have updated this book to include these latest advances, as well as covering the basics, in what we hope is a readable and useful format. We hope that the strong emphasis on the clinical relevance of what you are learning inspires a lifelong enthusiasm for what is an exciting and fundamental subject.

Melanie Newport
Faculty Advisor

More than a decade has now passed since work began on the first editions of the *Crash Course* series, and over 4 years since the publication of the second editions. Medicine never stands still, and the work of keeping this series relevant for today's students is an ongoing process. These third editions build upon the success of the preceding books and incorporate a great deal of new and revised material, keeping the series up to date with the latest medical research and developments in pharmacology and current best practice.

As always, we listen to feedback from the thousands of students who use *Crash Course* and have made further improvements to the layout and structure of the books. Each chapter now starts with a set of learning objectives, and the self-assessment sections have been enhanced and brought up to date with modern exam formats. We have also worked to integrate points of clinical relevance into the basic medical science material, which will not only add to the interest of the text, but will also reinforce the principles being described.

Despite fully revising the books, we hold fast to the principles on which we first developed the series: *Crash Course* will always bring you all the information you need to revise in compact, manageable volumes that integrate basic medical science and clinical practice. The books still maintain the balance between clarity and conciseness, and providing sufficient depth for those aiming at distinction. The authors are medical students and junior doctors who have recent experience of the exams you are now facing, and the accuracy of the material is checked by senior faculty members from across the UK.

I wish you all the best for your future careers!

Dr Dan Horton-Szar
Series Editor

Acknowledgements

I would like to thank all at Elsevier for their constant encouragement and many, many reminder e-mails! I am grateful to Munira Kadhim at the MRC Radiation and Genome Stability Unit for some amazing karyotypes and FISH images, not all of which could be used due to monochrome! A big thanks must also go to Mel Newport for all her help, guidance, patience, wine and croissants! Finally, I wish to acknowledge the support of The Worshipful Company of Barbers and the Sir Richard Stapley Educational Trust, without whom affording study would be even more of a struggle.

Figure acknowledgements

Figure 1.1 adapted from Medical Microbiology, 3rd edn., by C. Mims et al., Mosby, 2004, Fig. 2.1

Figure 1.17 adapted from Medical Microbiology, 5th edn., by P.R. Murray, M.A. Pfaller and K.S. Rosenthal, Mosby, 2005, Fig. 6.9

Figures 2.1, 4.4, 4.6, 4.13–4.16, 4.27, 5.3, 6.2, 6.37 adapted from Human Histology, 2nd edn., by A. Stevens and J. Lowe, Mosby, 1997

Figure 2.3 electron micrographs reproduced courtesy of Dr Trevor Gray

Figures 3.10, 3.14, 5.14, 6.11 adapted from Medical Biochemistry, 1st edn., by J. Baynes and M. Dominiczak, Mosby, 1999

Figure 3.17 adapted from Physiology, 4th edn., by R. Berne et al., Mosby, 1998

Figures 4.2, 4.5, 4.9, 4.12, 6.4 adapted from Medical Cell Biology Made Memorable by R. Norman and D. Lodwick, Churchill Livingstone, 1999

Figures 6.17, 6.28, 7.18, 8.6, 8.7 from Thompson & Thompson Genetics in Medicine by R.L. Nussbaum, R.R. McInnes and H.F. Willard, WB Saunders, 2001

Figures 6.18A, 8.20B adapted from Clinical Medicine, 6th edn., by P. Kumar and M. Clark, WB Saunders, 2005, Figs. 4.8, 6.40

Figures 7.3A, 7.7, 7.9B reproduced courtesy of Dr Steve Howe

Figure 7.4 B reproduced courtesy of Dr Ajay Mistry

Figures 7.1, 7.11, 7.12, 7.19, 7.20, 7.25, 8.26 adapted from Emery's Elements of Medical Genetics, 11th edn., by R. Mueller and I. Young, Churchill Livingstone, 2001

Figure 7.13 reproduced courtesy of Linda E. Ritter

Figure 7.14 reproduced courtesy of Dr Paul Scriven, GSTT

Figure 7.21 reproduced courtesy of Dr Kathy Mann

Figure 8.24 reproduced courtesy of Dr A. Stevens and Professor J. Lowe

Figures 8.30B, 8.32 adapted from Robbins & Cotran Pathologic Basis of Disease, 7th edn., by V. Kumar, A. Abbas and N. Fausto, WB Saunders, 2004, Fig. 5.26

Figure 8.14 and 9.6 adapted from Emery's Elements of Medical Genetics, 12th edn., P. Turnpenny and S. Ellard, Churchill Livingstone, 2005, Fig. 21.4A

Figure 6.19 and 9.10 adapted from Medical Genetics, 3rd edn., by L. Jorde, J. Carey and M. Bamshad, Mosby, 2003, Fig. 13.9

To my family (Mum, Dad, Nan, Darren, Jonathan and Daniel), to my wonderful boyfriend Barnaby, and to all those who have helped me get to where I am today, especially Prof. Jim Parry (Swansea University) and the late Prof. Sir Gareth Roberts (Oxford University) – both of whom saw something in me and my ability that I could not see myself, and to whom I will always be eternally grateful.

Allele An allele is one of a series of possible alternative forms of a given gene or DNA sequence at a given locus.

Aneuploidy The condition in which the chromosome number of the cell is not an exact multiple of the haploid number. Monosomies and trisomies are examples of aneuploidy.

Antisense A piece of nucleic acid, typically created in the lab, which has a sequence exactly opposite to an mRNA molecule made by the body. Antisense can bind tightly to its mirror image mRNA, preventing a particular protein from being made.

Autosomal dominant A trait or disease that is produced when only one copy of a polymorphism or mutation is present on an autosome.

Autosomal recessive A trait or disease that is produced when two copies of a polymorphism or mutation are present on an autosome.

Autosomes These are any chromosomes other than the sex chromosomes.

Cell The cell is the basic unit of life. If it is to survive, each cell must maintain an internal environment that supports its essential biochemical reactions, despite changes in the external environment. Therefore, a selectively permeable plasma membrane surrounding a concentrated aqueous solution of chemicals is a feature of all cells.

Clone A member of a group of cells that all carry the same genetic information, and which are all derived from a single ancestor by repeated mitoses.

Compound heterozygote This is an individual with two different mutant alleles at the same locus.

Consultand An individual seeking, or referred for, genetic counselling.

Exon Region of a gene containing DNA that codes for a protein.

Gamete The reproductive cell formed by meiosis containing half the normal chromosome number.

Genome This is the entire genetic complement of a cell.

Genotype This is the genetic constitution of an individual, and it is also used to refer to the alleles present at one locus.

Heritability The degree to which a characteristic is determined by our genes.

Heterozygote This is an individual or genotype with two different alleles at a given locus on a pair of homologous chromosomes.

Holoenzyme The complete enzyme including all subunits and co-factors.

Homozygote This is an individual or genotype with identical alleles at a given locus on a pair of homologous chromosomes.

Hormone A molecule produced by an endocrine cell, which is released into the bloodstream and acts on specific receptors to elicit its effect.

Host The organism used to propagate a recombinant DNA molecule (usually *E. coli* or *S. cerevisiae*).

Insert The fragment of foreign DNA cloned into a particular vector.

Intron A section of a gene that does not contain any instructions for making a protein.

Karyotype This is the chromosome complement of a cell. In a standard karyotype the chromosomes are conventionally arranged in an order depending upon size. Chromosomes are distinguished individually by their size, centromere position and banding pattern. The normal human karyotype is 46, XY (male) or 46, XX (female).

Library A collection of cloned DNA fragments which, taken together, represent the entire genome of a specific organism. Using traditional techniques they allow the isolation and study of individual genes.

Ligand A molecule, such as a hormone or neurotransmitter, which binds the receptor, and is termed the first messenger.

Locus The position of a gene on a chromosome.

Microarray A large set of cloned nucleic acid molecules or proteins spotted onto a solid matrix, usually a microscope slide, and used to profile gene and protein expression in cells and tissues.

Monosomy A chromosome constitution in which one member of a chromosome pair is missing (e.g. Turner's syndrome (45, XO)).

Multifactorial disorder This is a term used to describe disorders in which both environmental and genetic factors are important.

Mutation A mutation is a permanent heritable change in the sequence of DNA.

Neurotransmitter A molecule that is used to transmit nerve impulses across a synapse.

Operon A prokaryotic locus consisting of two or more genes that are transcribed as a unit and are expressed in a coordinated manner.

Organism An organism is a system capable of self-replication and self-repair, which may be unicellular or multicellular. Unicellular organisms consist of a solitary cell able independently to perform all the functions of life. Multicellular organisms contain several different cell types that are specialized to perform specific functions.

Pedigree charts These are used to illustrate inheritance.

Penetrance The proportion of individuals with a specified genotype that show the expected phenotype under defined environmental conditions. The term is usually used in association with dominant disorders.

Phenocopy This is the alteration of the phenotype by environmental factors during development to produce a phenotype that is characteristically produced by a specific gene (e.g. rickets due to a lack of vitamin D would be a phenocopy of vitamin D-resistant rickets).

Phenotype This is the observed biochemical, physiological or morphological characteristics of an individual that are determined by the genotype and the environment in which it is expressed.

Plasmid Autonomously replicating, extrachromosomal circular DNA molecules. Often expoited in the laboratory as a vector for gene cloning.

Pleiotrophy Pleiotrophy is the phenomenon in which a gene is responsible for several distinct and apparently unrelated phenotypic effects, which may concern the organ systems involved and the signs and symptoms that occur.

Ploidy This refers to the number of complete sets of chromosomes in a cell. A haploid cell contains a single set of chromosomes (e.g. gametes); a diploid cell contains two copies of each chromosome (e.g. somatic cells); a polyploid cell contains more than two sets of chromosomes (sometimes this occurs normally in plants and in some animal cells, such as megakaryocytes).

Polygenic inheritance This is a term used to describe the inheritance of traits that are influenced by many genes at different loci.

Polymorphism Polymorphism is the occurrence in a population of two or more alternative genotypes, each at a frequency greater than that which could be maintained by recurrent mutation alone. A locus is arbitrarily considered to be polymorphic if the rarer allele has a frequency of at least 0.01. Any allele rarer than this is a 'rare variant'.

Proband The first person in a pedigree to be clinically identified as having a disease in question.

Probe A probe is a radioactively or fluorescently labelled piece of single-stranded DNA of defined sequence.

Recombinant DNA A molecule of DNA created *in vitro* that contains elements of more than one original sequence, e.g. vector and insert.

Rolling circle replication A process of nucleic acid replication that can rapidly synthesize multiple copies of circular molecules of DNA such as bacterial chromosomes and plasmids.

Translocation The transfer of one segment of a chromosome to another.

Trisomy The state of having three representatives of a given chromosome instead of the usual pair (e.g. as in trisomy 21 – Down syndrome).

Variable expressivity This occurs when a genetic lesion produces a range of phenotypes:

for example, tuberous sclerosis can be asymptomatic with harmless kidney cysts, but in the next generation it may be fatal, owing to the development of brain malformations.

Vector A DNA molecule capable of replicating within a particular host, into which foreign DNA may be inserted.

X-linked dominant A trait (or a disease) that is produced when only one copy of a polymorphism or mutation is present on an X chromosome. This means that both males and females can display the trait or disorder, by only having one copy of the gene.

X-linked recessive A trait or disease that is produced when a polymorphism or mutation in a gene on the X chromosome causes the phenotype to be expressed. Remember, females have two X chromosomes, while males have one X and one Y.

PRINCIPLES OF CELL BIOLOGY AND MOLECULAR GENETICS

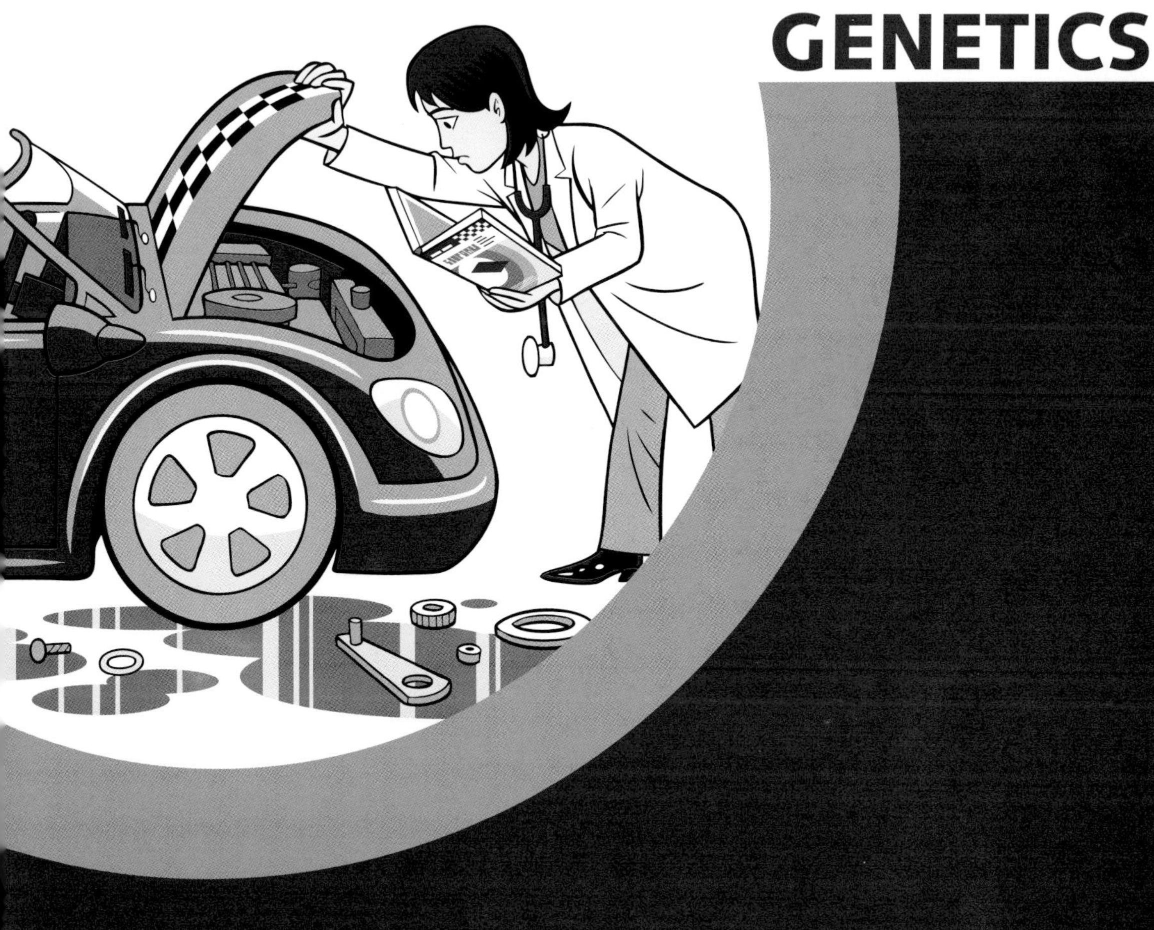

Cell biology and genetics of prokaryotes

1

Objectives

By the end of this chapter you should be able to:

- Explain why prokaryotic cells are smaller than eukaryotic cells.
- Describe the key features of a prokaryotic cell.
- Understand the differences between bacterial transformation, transduction and conjugation.
- Describe the processes of DNA replication, transcription and translation in prokaryotic cells.
- Explain how prokaryotic genes are regulated.
- Appreciate the main sites of action of antimicrobial agents.
- Recognize the main classes of virus based on nucleic acid composition.
- Understand key stages in the viral life-cycle.
- Understand the key targets of antiviral chemotherapy.

PROKARYOTIC CELL

Prokaryotes are the simplest unicellular organisms. It is generally accepted that all living organisms evolved from a common prokaryotic ancestor.

The basic molecular machinery of life has been conserved in all species. The enzymes that perform common reactions, such as glycolysis, in bacterial and human cells show significant homology at both the DNA and protein level.

All microorganisms lacking a membrane-bound nucleus (i.e. the various types of bacteria) are classified in the prokaryote superkingdom. Classically prokaryotic cells (Fig. 1.1) show the following features:

- A single cytoplasmic compartment containing all the cellular components.
- Cell division usually by binary fission.

In order for the prokaryotic cell to survive, molecules required for energy and biosynthesis must diffuse into the cell, and waste products must diffuse out of the cell across the plasma membrane. The rate of diffusion is related to membrane surface area. When the diameter of a cell increases:

- the cell volume expands to the cube of the linear increase
- the surface area only expands to the square of the linear increase.

Thus, small cells have a larger surface area to volume ratio than large cells. If prokaryotes expand above a certain size, the rate of diffusion of nutrients across the plasma membrane will not be sufficient to sustain the increased needs of its larger cell volume. Hence, prokaryotic cells are small.

Remember: **P**rokaryotes are **p**rimitive. ***You*** are a ***eu***karyote.

Prokaryotic cell structure and organelles

Plasma membrane

The prokaryotic cell membrane is a fluid phospholipid bilayer. With the exception of the mycoplasmas, they lack sterols, instead containing sterol-like molecules called hopanoids. The main function of the plasma membrane is to act as a selectively permeable membrane.

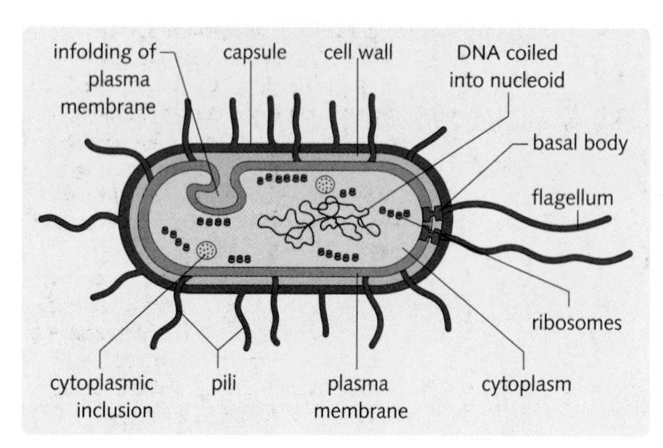

Fig. 1.1 Structure of a typical prokaryotic cell. The DNA molecule is free in the cytoplasm. Bacteria contain some sub-cellular structures, such as ribosomes, but do not contain membrane bound organelles.

Cell wall

All bacteria (except mycoplasmas) possess a complex cell wall. In the true bacteria, this is usually a peptidoglycan structure, a polymer of amino sugars. As bacteria concentrate dissolved nutrients through active transport, their cytoplasm is usually hypertonic. The cell wall acts to prevent osmotic lysis of the cell and is, therefore, a common antibiotic target (see p. 16) (Fig. 1.2).

Mycoplasmas are degenerate bacteria that do not have cell walls. They are the smallest known free-living life forms. Several species are pathogenic in humans, including *M. pneumoniae*, which is an important cause of atypical pneumonia and other respiratory disorders.

Bacterial cell walls can be described according to their staining ability, primarily being *Gram-positive* (retain the crystal violet dye during the Gram-stain procedure), *Gram-negative* (decolorize during the Gram-stain procedure) and *acid fast*. Gram-positive bacteria have thick (20–80 nm) peptidoglycan walls external to the cell membrane, while Gram-negative species have a protected thin (5–10 nm) peptidoglycan layer, overlaid by a lipopolysaccharide rich outer membrane. When coupled with cell shape, such properties aid organism identification; for example Gram-negative diplococci in cerebrospinal fluid are typical of meningococcal meningitis.

Along with the cytoskeleton, the cell wall maintains the overall shape of the bacterial cell. The most common shapes held by bacteria are:

- coccus (spherical or oval)
- bacillus (rod-shaped)
- helical.

Ribosomes

Ribosomes are small cellular components composed of ribosomal RNA (rRNA) and ribosomal protein (ribonucleoprotein). They are the site of protein synthesis (translation). Bacterial ribosomes are composed of two subunits with densities of 50S and 30S. The two subunits combine during protein synthesis to form a complete 70S ribosome, about 25 nm in diameter. The smaller size of the bacterial ribosome, compared to the 80S eukaryotic ribosome, makes it an ideal antimicrobial target.

Nucleiod

The bacterial genome is encoded by a single chromosome also known as the nucleoid. A structure measuring approximately 0.2 μm in diameter, it is formed from supercoiled DNA and histone-like proteins. It is not membrane bound and, therefore, is free in the cytoplasm. Bacterial DNA lacks introns, instead comprising a continuous coding sequence of genes, usually clustered into functional units called operons (see p. 14).

Cytoskeleton

It was long assumed that prokaryotic cells did not contain a cytoskeleton; however, recently at least two major components of the bacterial cytoskeleton have been identified: the bacterial tubulin and actin homologues FtsZ and MreB. FtsZ is thought to be involved in cell division, while MreB proteins are thought to be involved in the regulation of cell shape and the segregation of some bacterial plasmids.

Cell specializations

Glycocalyx

The glycocalyx, forming the capsule or slime layer, is a mucopolysaccharide sheet external to the cell wall. Where present, the main functions of the glycocalyx are to protect against phagocytic engulfment by the host cell and to enable bacteria to adhere to and colonize surfaces.

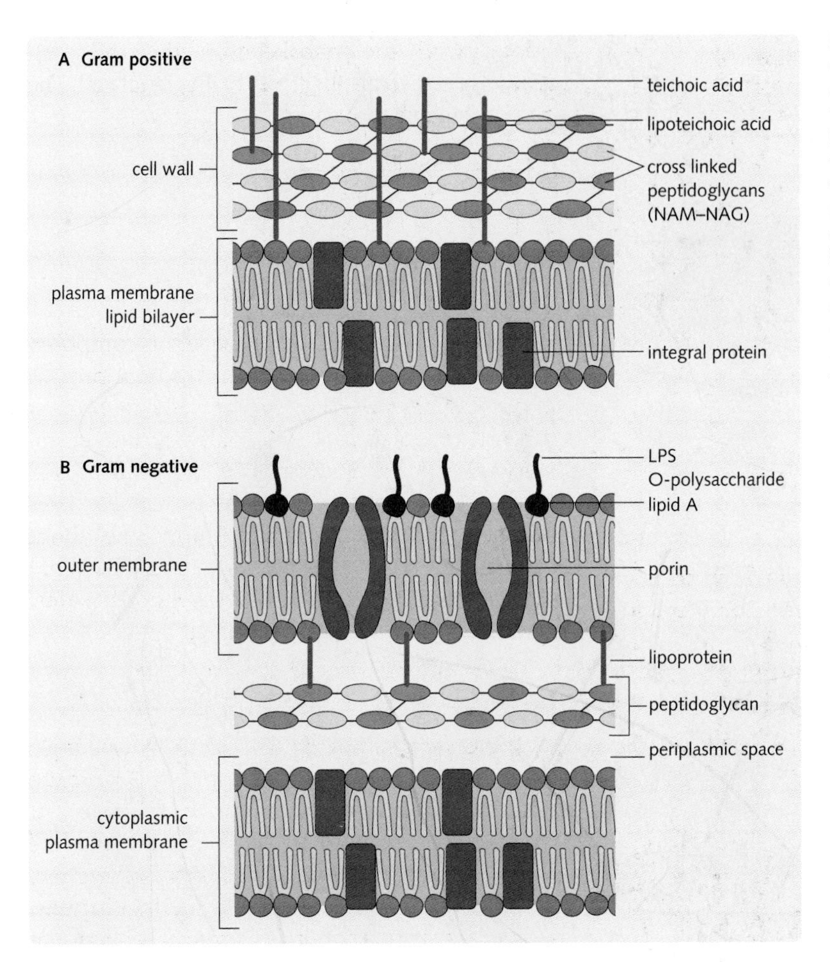

Fig. 1.2 Bacterial cell wall structure. (A) The Gram-positive cell wall is made mostly of peptidoglycan, interspersed with teichoic acid, which chelate the different layers together. (B) The Gram-negative cell wall contains a thin peptidoglycan layer with much less crosslinking between the peptidoglycan than seen in Gram-positive cell walls. The peptidoglycan layer is surrounded by an outer membrane, the outer leaflet of which contains antigenic lipopolysaccharide (LPS). In most cases this membrane structure is anchored non-covalently to lipoprotein molecules, which are covalently linked to the peptidoglycan.

Streptococcus pneumoniae is a spherical, Gram-positive human pathogen. Infection can lead to pneumonia, meningitis, septic arthritis, endocarditis and otitis media, among others. The virulence of these organisms is associated with the presence of a polysaccharide capsule. Mutants unable to synthesis a capsule are not pathogenic to humans; loss of the capsule is accompanied by a 10^5-fold reduction in virulence. As asplenic patients are particularly at risk from infection with capsulated organisms such as pneumococcus, with an associated mortality of around 60%, they should be offered pneumococcal vaccination.

Flagella

Flagella are long structures extending from the cell surface, enabling bacteria to move in their environment. The bacterial flagellum is made of the protein flagellin and is a hollow tube 20 nm in diameter. It has three parts:

- a basal body
- a hook
- a filament.

The basal body consists of a reversible rotary motor embedded in the cell wall, beginning within the cytoplasm and ending at the outer membrane. The hook is a flexible coupling or universal joint, and the long helical filament acts as a propeller (Fig. 1.3).

Pili

Pili are thin, protein tubes originating from the cytoplasmic membrane. They are found almost exclusively in Gram-negative bacteria. Pili are more rigid than flagella and have a role in bacterial attachment, either to another bacterium via 'sex' pili or to a host cell via the 'common' pili. The presence of many pili is thought to prevent

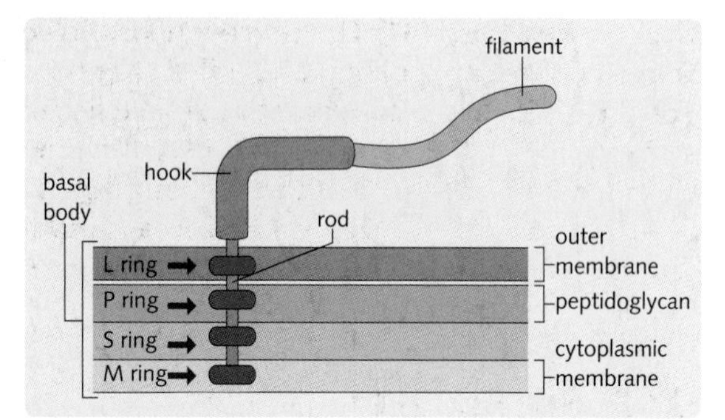

Fig. 1.3 Schematic of a flagellum of a Gram-negative bacteria. The basal body acts as a molecular motor, enabling the flagellum to be rotated, and consists of a rod and a series of rings that anchor the flagellum to the cell wall and the cytoplasmic membrane. While Gram-negative organisms have four basal rings, Gram-positive organisms only have two: one in the peptidoglycan layer and one in the plasma membrane. The hook provides a flexible coupling between the filament and the basal body. The filament is a hollow tube composed of the protein flagellin, ending with a capping protein.

phagocytosis, reducing host resistance to bacterial infection.

TRANSFER OF GENETIC MATERIAL

Although bacteria are not capable of true sexual reproduction, they may exchange genetic material by three mechanisms:

- Transformation. Certain bacteria release DNA into the environment that can be taken up by other bacteria via specific receptors. If the DNA is compatible, it is incorporated into the bacterial genome; otherwise it is degraded by exonucleases.
- Transduction. A fragment of bacterial DNA may become incorporated into a bacteriophage (a virus that infects bacteria) during its assembly. One in 10^6 bacteriophages contains such bacterial DNA. This may be introduced into the host bacterial cell along with viral genes during the infection process. Again, if the bacterial DNA is compatible it can be integrated into the host's genome.
- Conjugation. This is sometimes loosely called bacterial sexual reproduction (Fig. 1.4). Bacteria may be designated F positive (F^+) or F negative (F^-). F^+ cells possess a plasmid called the F factor, which includes genes for a 'sex' pilus, giving F^+ cells the ability to attach to other

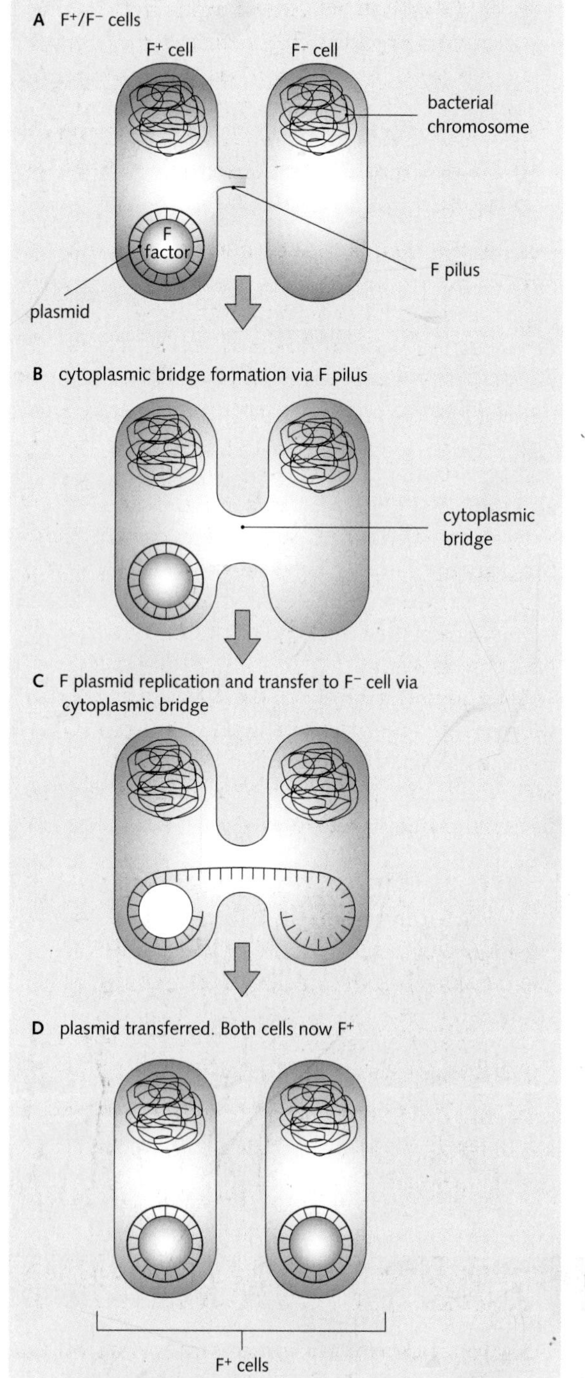

Fig. 1.4 Bacterial conjugation. (A) The ability to conjugate is conferred by a plasmid called the F factor. Bacteria with the plasmid are F^+. (B) F^+ cells contain thread-like projections called F pili, which can attach to F^- cells to form a cytoplasmic bridge. (C) F plasmid is replicated and a single-stranded replica is transferred along the bridge. (D) Within the recipient, the transferred material is replicated to form a new plasmid.

bacterial cells. In this way it forms cytoplasmic bridges through which genetic material may then be transferred after it has been replicated by 'rolling circle replication'. Other plasmids present, for example those conferring antibiotic resistance, may also be transferred during this process.

Usually F plasmids exist extra-chromosomally. On rare occasions, the F plasmid may become integrated into the bacterial genome, resulting in Hfr (high frequency of recombination) cells. When this happens, the whole bacterial chromosome may be replicated by rolling circle replication, starting at an origin in the F factor. The replicated DNA, that now includes bacterial chromosomal material, is passed along the cytoplasmic bridge into the recipient cell. Variable amounts of DNA are transferred because the bridge invariably breaks down before the whole chromosome has been transmitted. The transmitted material may then undergo recombination with the host chromosome.

Infrequently, an F plasmid that has integrated into the genome may 'pop-out' again, taking part of the bacterial chromosome with it. These are called F′ plasmids and they may act as a vehicle to transmit bacterial genes to new hosts.

> The conjugation process provides bacteria with a means of acquiring genes that, although beneficial to the organism, are not to their hosts. During conjugation, the F^+ cell can also pass an 'R plasmid', containing several antibiotic resistance genes, to the recipient F^- bacterium. The recipient bacterium is now not only antibiotic resistant, but also capable of producing a sex pilus (F^+) and passing on antibiotic resistance to surrounding F^- cells.

DNA REPLICATION

DNA replication is the process by which double-stranded DNA molecules are divided longitudinally, such that each strand is conserved to act as a template for the formation of a new strand. It is said to be semi-conservative, since only one strand is newly synthesized in each daughter molecule.

The bacterial nucleoid does not divide by mitosis. With its single chromosome and a division time of 20 min at 37 °C, replication has been studied extensively in *Escherichia coli*. It has been possible to isolate a range of replication deficient mutants, which have been used to identify and characterize the corresponding replication proteins. Such studies suggest that even prokaryotic replication is a complex process that requires about 30 proteins.

DNA polymerases

DNA polymerases are the enzymes responsible for DNA-chain synthesis. They couple nucleoside triphosphates onto a growing DNA strand by adding a phosphate group onto the free 3′-OH group. New DNA molecules are thus synthesized in a 5′–3′ direction. Polymerization is driven thermodynamically by the elimination of a pyrophosphate (PP_i) and its subsequent hydrolysis:

$$(DNA)_n + dNTP \rightarrow (DNA)_{n+1} + PP_i$$

Three DNA polymerases have been characterized in *E. coli*, of which two are important in replication (Fig. 1.5). RNA polymerase can begin a polynucleotide chain by linking two nucleoside triphosphates together directly (see pp. 10–11). In contrast, DNA polymerase has an absolute requirement for a perfectly base-paired nucleotide, onto which it can then add nucleotides at the 3′-OH end. This has important consequences:

- DNA polymerase requires a primer on which to initiate extension.
- It will pause if an incorrect base is inserted.

The primers that are required for DNA polymerase activity are synthesized by RNA polymerase, since this enzyme does not itself require priming oligonucleotides. It synthesizes short stretches (10–20 nucleotides) of RNA primer sequence.

If Pol III, the main replication enzyme, inserts an incorrect base into the extending DNA chain, it cannot proceed until the erroneous nucleotide is excised. This is achieved by its own 3′–5′ exonuclease activity (Fig. 1.5). In this manner, the enzyme corrects its own mistakes as it goes along. This function is called 'proofreading', and it maintains the fidelity of DNA sequence after replication. Errors in replication occur at a rate of about 1 in 10^5 base pairs, reduced to about 1 in 10^8 by proofreading mechanisms.

Fig. 1.5 DNA polymerases in *Escherichia coli* (prokaryote)

	DNA Pol I	DNA Pol III
Notes	First polymerase to be discovered by Kornberg in 1957	Discovered when a mutant strain of *E. coli* with very low Pol I activity was shown to have a normal rate of reproduction
Structure	Single polypeptide with 928 residues, 109 kDa mass; forms one large ('Klenow') fragment and one small fragment	Three subunits with total 140 kDa; subunits—a, e, q
Functions		Polymerase (a subunit), 3'–5' exonuclease (e subunit) 5'–3' exonuclease
Associated proteins	Nil	At least seven other proteins associate to form a complex called the Pol III holoenzyme
Processivity (the number of consecutive reactions the enzyme is capable of performing)	At least 20 consecutive polymerization steps can occur before Pol I becomes dissociated from the DNA	In the holoenzyme the extra subunits interact with DNA and other proteins to clamp the polymerase onto the DNA, creating very high processivity (> 5000 residues); Pol III alone has a processivity of 10–15 residues
Main biological function	Proofreading and error correction	DNA replication, proofreading

Fig. 1.5 Prokaryotic DNA polymerases. *Eschericheria coli* also has a Pol II. Its main physiological function is unknown.

DNA replication fork

Replication is initiated at the 'origin of replication', giving rise to two replication forks. Replication complexes assemble at both of these and proceed in opposite directions (Fig. 1.6).

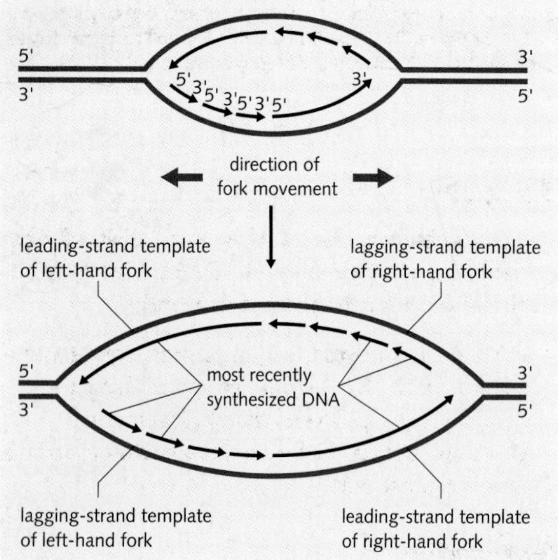

Fig. 1.6 The replication bubble. Two replication complexes form at each origin of replication and proceed in opposite directions. Note that the strand that is leading and the strand that is lagging depends upon the direction the replication complex is migrating.

Since DNA-dependent polymerases can only add nucleotides to the 3′ end, the leading strand can be synthesized in a continuous process from a single RNA primer. However, the lagging strand must be synthesized in pieces (called Okazaki fragments), which are subsequently joined together by a DNA ligase (Fig. 1.6).

Prokaryotic replication

Initiation

E. coli has a single origin of replication (*OriC*), which is a 245bp DNA sequence. Initially, sequence-specific protein binding around the origin causes the DNA to denature and unwind (Fig. 1.7). This produces a pre-priming complex, which facilitates further unwinding and entry of the other components of the replication complex (Fig. 1.8):

- DNA helicase clamps to the lagging strand and unwinds the DNA helix.
- Primase binds to the helicase to form the primosome, which synthesizes the RNA primer.
- Single strand binding proteins stabilize ssDNA on the lagging strand.
- DNA polymerase III clamps to the leading strand and synthesizes DNA.
- A regulated sliding clamp protein holds DNA polymerase on the DNA.

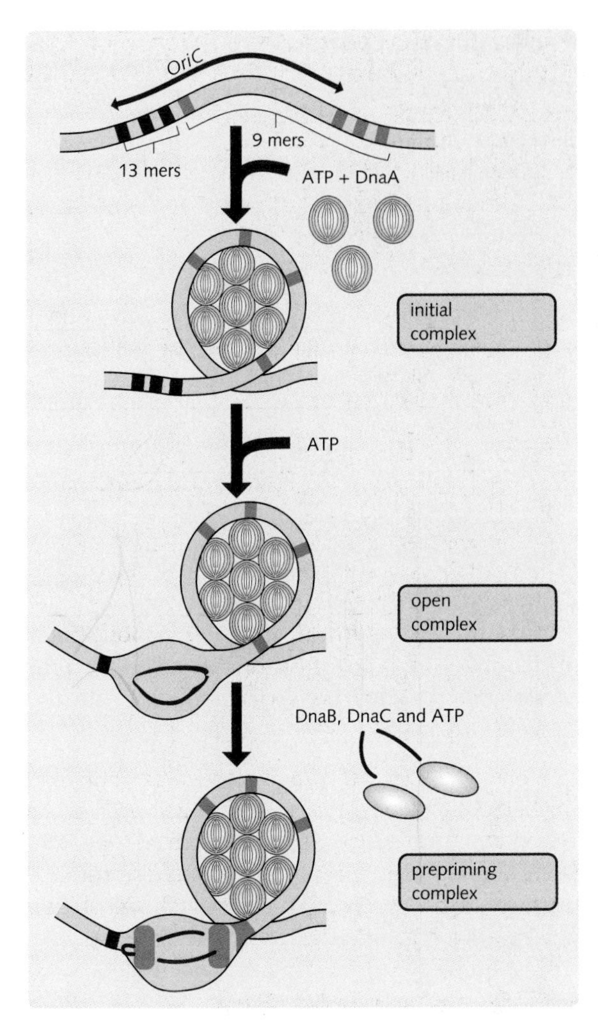

Fig. 1.7 Formation of pre-priming complex for prokaryotic DNA replication. DnaA is a protein that recognizes and binds to 9bp segments of *OriC*. The process is facilitated by Hu protein. The DnaA–DNA complex becomes negatively supercoiled. The DnaA–DNA complex guides the binding of DnaB–DnaC complex onto the adjacent region of *OriC*. DnaB has helicase activity and unwinds the DNA in the prepriming complex.

- DNA topoisomerase relieves helical winding and tangling problems.

Two replication complexes (replisomes), containing two polymerases, are formed at each origin of replication (Fig. 1.8).

When answering questions on replication, transcription or translation, consider the processes in three stages: initiation, elongation and termination.

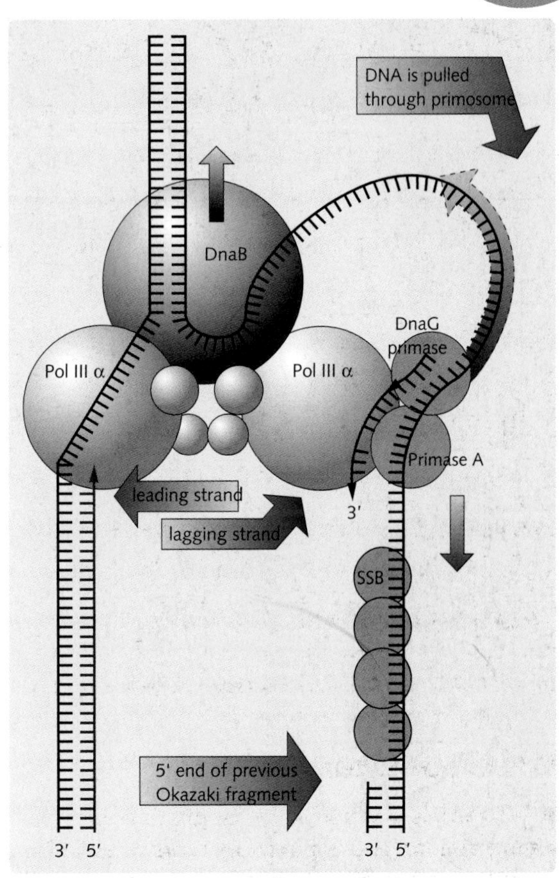

Fig. 1.8 Replication of DNA in *Escherichia coli*. The replisome consists of two polymerase enzymes, one that binds to the leading strand and one that binds to the lagging strand. The holoenzyme moves continuously along the template for the leading strand while the lagging strand is 'pulled through' by the primosome creating a loop in the DNA. Helicase (DnaB) creates the unwinding point as it translocates along the DNA. SSB, single strand binding protein.

Elongation

The DNA chain is elongated by the action of DNA polymerase III (see above). Other enzymes are required to deal with the lagging strand:

- DNA Pol I removes the RNA primer sequence and completes the DNA strand by space filling the resulting gaps between Okazaki fragments.
- DNA ligase joins the fragments together to form a continuous strand.

Termination

Replication is complete when the replisomes, which set off around the circular chromosome in opposite directions, meet in the middle.

TRANSCRIPTION AND TRANSLATION

Escherichia coli RNA polymerase

E. coli RNA polymerase (RNAP) is a large holoenzyme containing two Zn^{2+} ions, required for catalytic activity. It has a core subunit composition of α-2-β-β'-ω, in which:

- the two α subunits are required to assemble the enzyme and recognize regulatory factors
- the β subunit contains polymerase activity (catalyses the synthesis of RNA).
- the β' subunit binds nonspecifically to DNA
- the ω subunit is of unknown function, but has been shown to restore denatured RNA polymerase to its fully functional form *in vitro*.

The core enzyme associates with another subunit, σ, to form the initiation complex, α-2-β-β'-ω-σ. The σ factor:

- reduces the affinity of the enzyme for non-specific DNA
- increases the affinity of the enzyme for promoters.

E. coli has a number of σ factors that recognize the promoter regions of specific, coordinated sets of genes.

Promoters

The promoter is a DNA sequence to which RNA polymerase binds in order to initiate transcription. In prokaryotes, the consensus promoter features two sequences upstream (5′) of the gene under control (transcription starts at 0):

- One at –10, also known as the TATA box. It usually consists of the six nucleotides TATAAT, and is absolutely essential to start transcription in prokaryotes.
- One at –35, which usually consists of the six nucleotides TTGACA. It appears that the –35 sequence is the sequence recognized by the σ factor.

In RNA synthesis, promoters are a means to control which genes should be transcriptionally active and, thus, which proteins the cell manufactures.

Prokaryotic transcription

Transcription is divided into three phases:

1. Initiation
2. Elongation
3. Termination.

Initiation

Subunit σ binds to the promoter region, which induces a conformational change in RNAP that allows nucleotides to associate with the β-subunit. This is followed by chemical initiation, which involves coupling of two nucleotide triphosphates. The first is almost always a purine, usually adenine:

$$\text{pppA} + \text{pppN} \rightarrow \text{pppApN} + \text{PP}_i$$

(pppA, adenosine triphosphate; pppN, nucleoside triphosphate; PP_i, pyrophosphate).

After the enzyme has synthesized about eight nucleotides, the σ factor dissociates and a number of elongation factors become associated with the enzyme instead.

Elongation

The RNA molecule is synthesized 5′–3′. The DNA template is gradually unwound by RNAP pushing the DNA's coils ahead of it, creating positive superhelicity ahead of the complex and correspondingly unwinding behind it (Fig. 1.9). Only about 17 bases are exposed at a time. The RNA quickly leaves the DNA template and the DNA re-forms its double helix.

The overall equation for the polymerization process is:

$$(\text{RNA})_n + \text{XTP} \rightarrow (\text{RNA})_{n+1} + \text{PP}_i \rightarrow 2\text{P}_i$$

(XTP, nucleoside triphosphate; PP_i, pyrophosphate; P_i, inorganic phosphate).

Termination

Elongation reactions continue until the core RNA polymerase encounters a transcription termination signal, the RNA is released from DNA template and from the enzyme, and the RNA polymerase dissociates from the DNA helix. In prokaryotes there are two different mechanisms of termination:

1. Intrinsic (also known as rho-independent)
2. Rho-dependent.

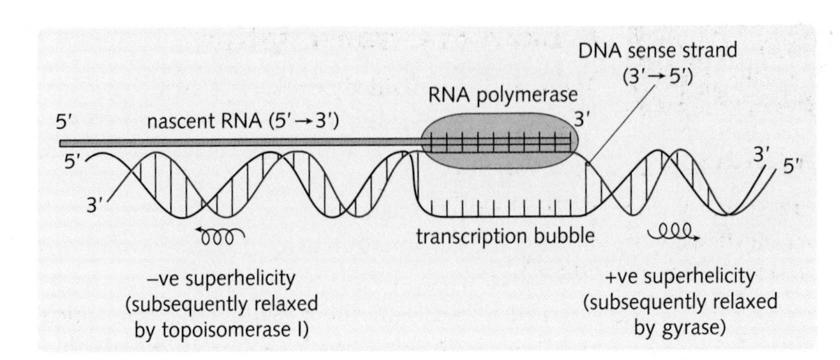

Fig. 1.9 RNA chain elongation. The RNA molecule develops in a straight line while DNA rotates. (Adapted with permission from cell 52: 752, by Branhill and Kornberg, 1988.)

Intrinsic termination

The intrinsic terminator sequence is an inverted repeat of a GC-rich sequence followed by six or more adenines. The transcribed RNA forms stem-loop hairpin structure at inverted repeats via internal base pairing, the formation of which causes the RNAP to pause. Formation of the hairpin in the nascent RNA limits the area of RNA/DNA hybrid to the short segment of uracil in the RNA strand and adenine in the DNA strand. The uracil–adenine bonds are unstable, and are not strong enough to maintain the RNA–DNA association. The RNA polymerase is released from the DNA template, resulting in termination of transcription (Fig. 1.10).

Rho-dependent termination

Rho-dependent termination requires the action of a protein factor called Rho, an ATP-dependent hexamer of six identical 60 kDa units, to stop RNA synthesis at specific sites. Rho initially binds to the RNA transcript at the upstream site, which is 70–80 nucleotides long and rich in C residues. After binding, Rho chases the RNA polymerase, unwinding the RNA–DNA hybrid by helicase activity. The formation of a hairpin loop in the RNA structure causes the RNA polymerase to pause, allowing the Rho protein to catch up with and displace it from the template, terminating transcription (Fig. 1.11).

Prokaryotic post-transcriptional modification

In prokaryotes there is very little or no post-transcriptional modification of mRNA and translation begins while the RNA transcript is still being produced. However, rRNA (ribosomal RNA) and tRNA (transfer RNA) transcripts do show a degree of post-transcriptional modification. They are synthesized as a continuous strand that undergoes post-transcriptional cleavage with a nuclease, which may be followed by base modification:

- CCA is added to the 3′ end of each tRNA.
- Bases in rRNA may be methylated.
- Bases in tRNA may be modified to produce inosine, pseudouridine and dihydrouridine.

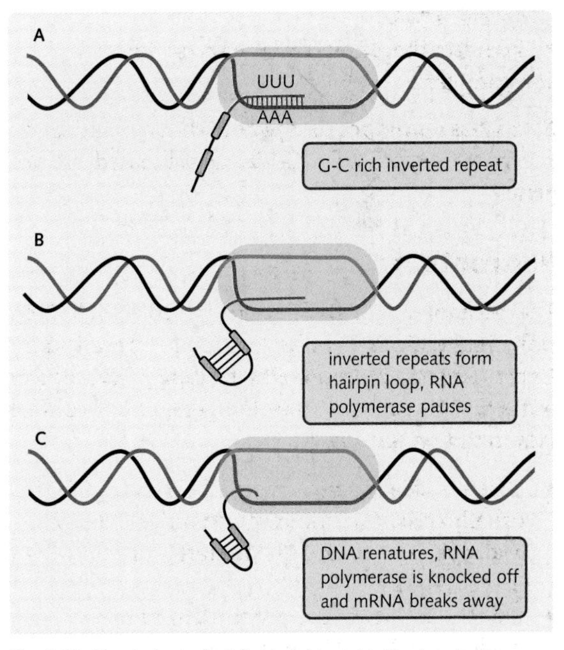

Fig. 1.10 Rho-independent (intrinsic) transcription termination. (A) The inverted CG rich repeat regions self-hybridize to form a 'stem-loop' structure. (B) The GC rich stem loop structure interacts with RNA polymerase to cause it to pause. (C) The short UUU region, which is base pairing with AAA sequence on the anti-sense DNA strand, has low thermal stability and melts, releasing the nascent RNA transcript. The DNA renatures with the removal of RNA polymerase.

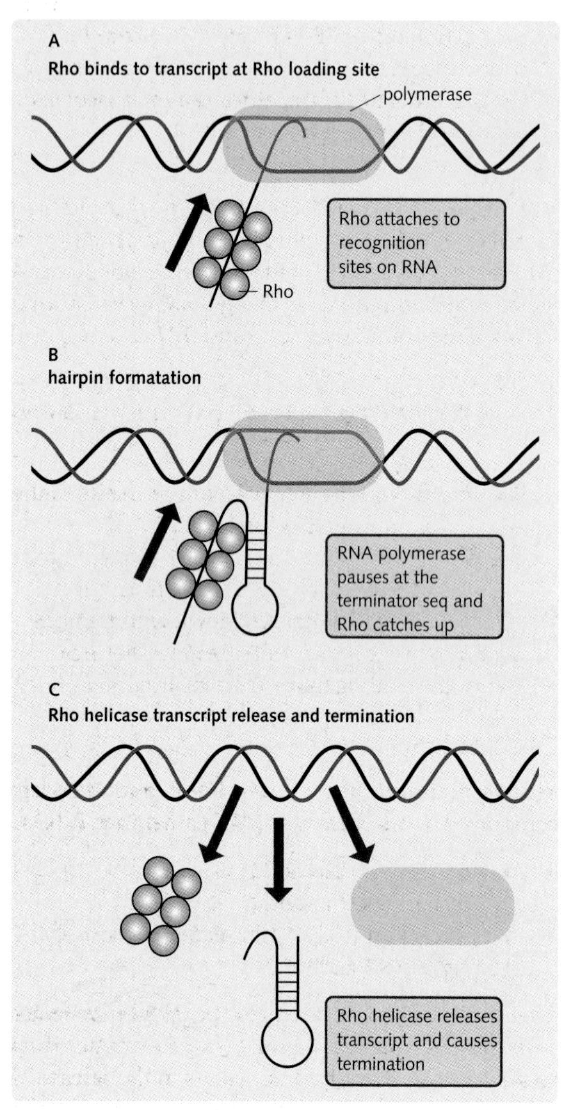

Fig. 1.11 Rho-dependent transcription termination. The Rho protein hexamer is required to terminate transcription in about 50% of cases in *Escherichia coli*, but the precise mechanism is not clear. Experiments suggest that: (A) Rho associates with the nascent RNA transcript at a C rich binding domain. The interaction of Rho with nascent RNA activates an ATPase activity, which appears to allow Rho to translocate along the mRNA in the 3′ direction. (B) Hairpin formation stalls the RNA polymerase, allowing the Rho hexamer to catch up. (C) Rho helicase activity releases the transcript and leads to transcriptional termination.

Prokaryotic translation

The components needed for translation are mRNA, tRNA, ribosome, GTP, initiation factors and elongation factors.

Amino acids combine with their corresponding tRNA in a reaction catalysed by a specific aminoacyl transferase.

$$\text{Amino acid} + \text{tRNA} \xrightarrow{\text{Aminoacyl transferase}} \text{Aminoacyl-tRNA}$$

This incorporates a high-energy ester bond between the aminoacyl group and the 3′ CCA group of the tRNA. This is known as 'charging' the tRNA and the energy released when this bond is broken drives the peptide bond formation step of chain elongation. Like transcription, translation consists of three stages: initiation, elongation and termination.

In prokaryotic cells both transcription and translation take place in the cytoplasm.
In eukaryotic cells, transcription takes place in the nucleus and translation occurs in the cytoplasm.

Initiation

The initiation complex is composed of a ribosome, mRNA and initiator tRNA. The process of initiation requires:

- three initiation factors, IF-1, IF-2, and IF-3
- a molecule of GTP.

Initiator tRNA is f-met-tRNA. It carries a formylated methionine residue and recognizes an AUG codon on the mRNA. Each mRNA contains many AUG codons in its various reading frames, and the AUG corresponding to the start of translation is preceded by a purine rich tract of nucleotides called the Shine–Dalgarno sequence. This binds to a corresponding pyrimidine rich sequence in the ribosomal S unit (Fig. 1.12).

Elongation

Chain elongation involves the addition of aminoacyl residues to the growing polypeptide. It is a three-stage process.

Step one

Aminoacyl-tRNA binds to the ribosomal A site as follows:

- A complex of aminoacyl-tRNA, GTP and elongation factor (EF)-Tu is formed.
- Codon–anticodon binding occurs with concomitant hydrolysis of GTP.

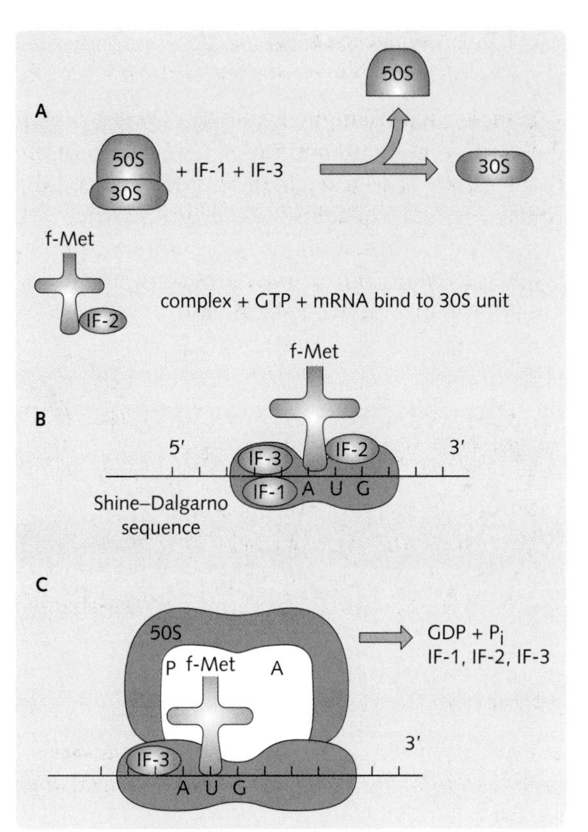

Fig. 1.12 Initiation factors of prokaryotic translation. (A) IF-1 and IF-3 bind to 30S subunit of ribosome to promote dissociation of 50S subunit. (B) f-Met-IF-2 complex and GTP and mRNA bind to 30S unit.IF-2 is required for f-Met-tRNA binding to mRNA. f-Met-tRNA–mRNA binding is not, therefore, dependent upon codon–anti-codon association. IF-3 assists ribosomal binding to the Shine–Dalgarno sequence on mRNA. (C) IF-3 is released and GTP hydrolysed. The 50S subunit associates and IF-1 and IF-2 are released. f-Met-tRNA lies in the P site so the A site is ready to accept incoming tRNA. (Note that f-Met-tRNA is the only tRNA that does not enter the ribosome via the A site.)

- Elongation factor EF-Ts interacts in a recycling process to release GDP and inorganic phosphate (P_i) and liberate EF-Tu, which can then associate with another free aminoacyl-tRNA.

Step two

Peptide bond formation is catalysed by a peptidyl transferase in the 50S subunit. The peptidyl group in the P site is added onto the aminoacyl group in the A site in a reaction driven by a high-energy ester bond between the aminoacyl group and tRNA.

Step three

The translocation (Fig. 1.13) process occurs as follows:

- The uncharged tRNA is expelled.
- Peptidyl-tRNA is transferred from the A site to the P site. This requires EF-G and GTP.
- The tRNA is still bound to the mRNA codon, so, as the peptidyl-tRNA moves across from the A site to the P site, the mRNA moves with it. A new codon now lies in the A site. This mechanism allows the reading frame to be maintained.

Termination

The termination codons, UAA, UAG and UGA are recognized by release factors (RFs) rather than tRNAs:

- RF-1 recognizes UAA and UAG.
- RF-2 recognizes UAA and UGA.
- RF-3, which binds GTP, stimulates ribosomal binding of RF-1 and RF-2.

The binding of an RF causes the peptidyl transferase to transfer the peptidyl group to water rather than to an aminoacyl group. This results in the release of a free polypeptide. The uncharged tRNA is released from the ribosome and the RFs are expelled.

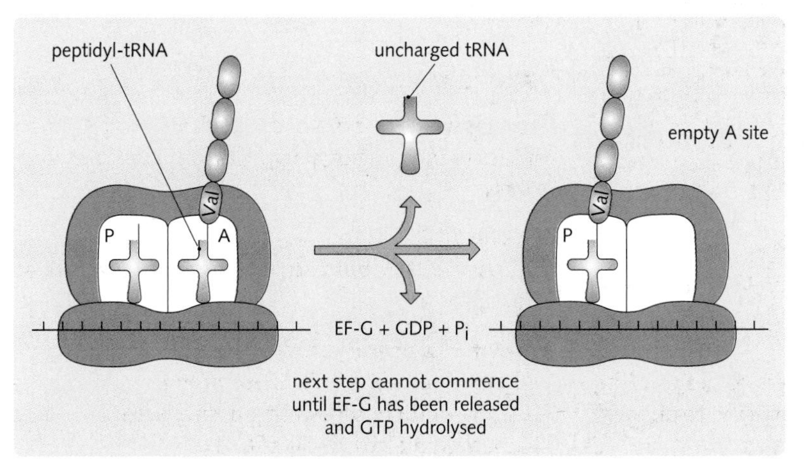

Fig. 1.13 Step three of chain elongation: translocation. Unchanged tRNA is expelled from the A site and the peptidyl-tRNA is transferred from the A site to the P site.

GENE REGULATION

As prokaryotes live in environments where competition for the available nutrients is fierce, the ability to regulate gene expression is of paramount importance, allowing bacteria to adapt to their environment. It is clearly advantageous for bacteria to have the ability to metabolize a wide variety of nutrients. Synthesizing proteins consumes energy and so bacteria regulate the proteins they are expressing at any one time. In bacteria, the genes that encode the enzymes of a metabolic pathway are usually clustered together on the chromosome in a functional complex called an operon.

Operons

A typical bacterial operon consists of:

- structural genes, which code for the enzymes themselves
- a promoter region
- an operator region, which serves as the binding site for the protein called the repressor
- a regulatory gene, which encodes the repressor protein.

Some enzymes, such as those required for lactose catabolism, are only synthesized when the appropriate substrate is present. The control of expression of these genes is said to be by induction (Fig. 1.14).

Other enzymes, such as those required for the synthesis of tryptophan, are controlled by repression. Tryptophan is essential for cell survival, so the enzymes that synthesize this amino acid are expressed constitutively. However, if tryptophan happens to be present in the growth media, expression of these genes is turned off (Fig. 1.15).

ANTIMICROBIAL AGENTS

There are fundamental differences in the cellular machinery of bacterial and mammalian cells, and mammals can tolerate some chemicals that are toxic to bacteria. Therefore, humans can generally take antimicrobial agents in appropriate amounts to treat

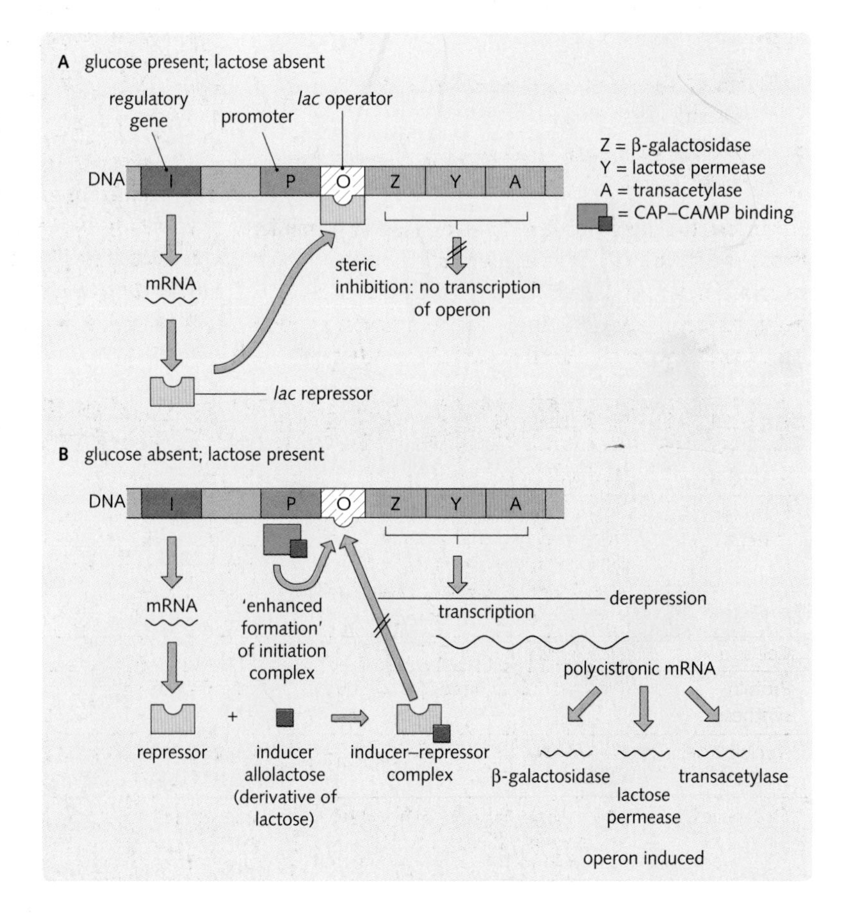

Fig. 1.14 The *lac* operon. (A) The '*lac* repressor' binds to the '*lac* operator' with great affinity. The *lac* operator is formed by a stretch of DNA that overlaps with the operon's initiation site. *lac* repressor binding prevents the formation of a transcription initiation complex so transcription cannot occur. (B) Lactose binds to the *lac* repressor and changes its shape so it has a much lower affinity for the *lac* operator, and derepression occurs. An initiation complex forms and transcription begins. At the same time, low levels of glucose cause an increase in intracellular cAMP, which binds to catabolite activator protein (CAP). The CAP–cAMP complex binds to the *lac* operon promoter region and encourages the formation of an RNA initiation complex. (*lacA*, *lacY* and *lacZ* are genes for transacetylase, lactose permease and β-galactosidase respectively.)

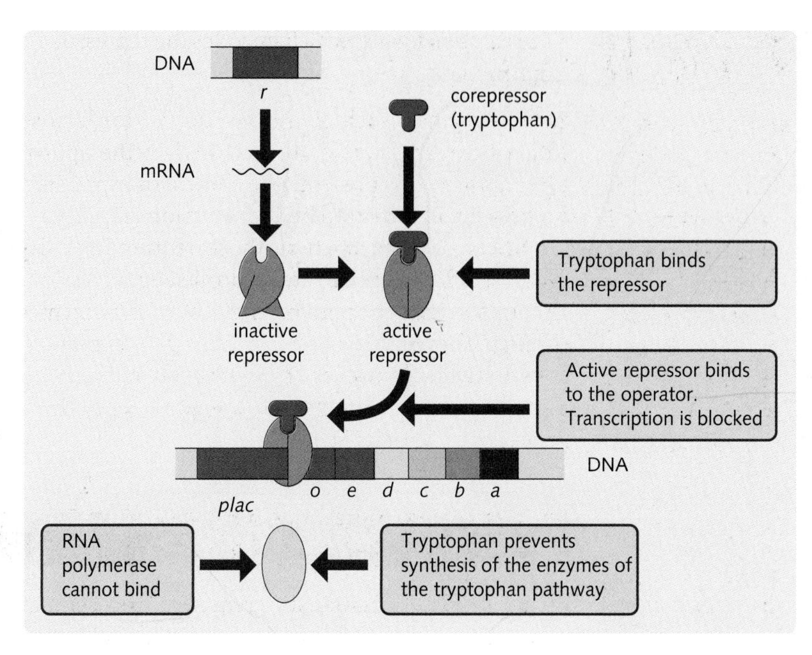

Fig. 1.15 The *trp* operon. When tryptophan is in abundance in the environment, tryptophan-synthesizing genes are turned off. This is achieved by the binding of tryptophan to the repressor protein, activating it. As the *trp* repressor will not bind to the DNA operator sequence unless it is activated by binding with tryptophan, tryptophan is, therefore, a co-repressor. When tryptophan levels are low, the tryptophan dissociates from the repressor, and the repressor loses its ability to bind to DNA. The operator is now free for RNA polymerase and transcription proceeds, making tryptophan biosynthetic genes and replenishing the cell's supply of tryptophan. (*trpE* and *trpD* encode anthranilate synthetase, *trpC* encodes indole-3-glycerol phosphate synthase and *trpB* and *trpA* encode tryptophan synthase.)

bacterial infections without harming themselves. A summary of the differences between bacterial and human cells is given in Fig. 1.16.

There are three ways of classifying antimicrobial agents:

1. By target site, which is the most practical
2. By chemical structure
3. According to whether they are bactericidal (kill) or bacteriostatic (inhibit growth).

Antimicrobial agents that kill bacteria are irreversible; those that inhibit growth are reversible. Some agents may act in different ways depending on the organism being treated, so what may be bactericidal in one organism, may only be bacteriostatic in another.

Bacteriostatic agents include sulphonamides, tetracyclines, chloramphenicol and erythromycin. In order to be effective, they require a competent immune system. The bacteriostatic agent prevents the bacterial population from increasing, allowing the host immune system to eliminate the remaining pathogens. Therefore, the duration of therapy must be sufficient to allow cellular and humoral defence mechanisms to become active. Unsurprisingly, in the case of immunocompromised patients, antibiotics with bactericidal activity are, therefore, preferred.

Fig. 1.16 Differences between bacterial and human cells

	Bacterial	**Human**
Form of genetic material	Prokaryotic, no nucleus or nucleolus, single DNA thread tightly coiled by topoisomerases, extra-chromosomal elements called plasmids, double stranded	Eukaryotic, linear DNA, associated with proteins to form chromosomes, double stranded
Cell size	Average diameter 0.5–5 μm	Up to 40 μm
Protein synthesis	Ribosomal subunits 50S and 30S	Subunits 60S and 40S
Nucleic acid synthesis	Folate must be synthesized *de novo* for base synthesis	Folate can be obtained from the diet
Organelles	Few, associated with respiration and photosynthesis	Many and diverse

Fig. 1.16 Summary of the differences between bacterial and human cells.

There are five main target sites for antibacterial agents:

1. Nucleic acid synthesis
2. Cell wall synthesis
3. Protein synthesis
4. Metabolic pathways
5. Cell membrane function.

Inhibitors of nucleic acid synthesis

Significant differences between eukaryotic and prokaryotic replication allow for the differential targeting of prokaryotic replication. Nucleic acid synthesis can be inhibited at the level of:

- DNA replication
- RNA polymerase activity
- inhibition of nucleic acid precursors (see pp. 92–94).

DNA replication

Quinolones (e.g. ciprofloxacin) inhibit bacterial DNA gyrase and topoisomerase IV, which are the enzymes responsible for unwinding and rewinding supercoiled DNA either side of the replication fork, thus inhibiting DNA replication. They are bactericidal and are selective for bacteria, as they do not affect the mammalian versions of these enzymes.

RNA polymerase activity

Rifampicin inhibits RNA synthesis by inhibiting DNA-dependent RNA polymerase, blocking the synthesis of mRNA. All members of the rifampicin family are bactericidal and show selective toxicity with a greater affinity for bacterial polymerases than for the equivalent human enzymes.

Inhibitors of cell-wall synthesis

The bacterial cell wall is rigid, containing linear peptidoglycans that are crosslinked by peptides (see Fig. 1.2). The absence of such molecules in mammalian cell membranes allows them to be targeted. Disrupting the cell wall leaves the bacteria susceptible to osmotic lysis and death.

β-lactams

Penicillins were the first group of antibiotics to be discovered and they are still important clinically. Penicillin itself is active mainly, but not exclusively, against Gram-positive organisms, whereas synthetic penicillins have been developed for their activity against Gram-negative rods.

Benzylpenicillin:

- is used in the treatment of pneumococcal, other streptococcal and meningococcal infections. However, there have been reports suggesting the emergence of benzylpenicillin-resistant pneumococci
- is not effective orally, so it is given by injection.

Cephalosporins are also β-lactams and have the same mode of action as penicillin. First-generation cephalosporins were most active against Gram-positive bacteria, but modified second- and third-generation drugs have a much broader spectrum of activity, which includes Gram-negative bacteria. They are also resistant to β-lactamase. Thus, cefuroxime (second-generation) is active against *S. aureus*, and ceftazidime (third-generation) has activity against the *Pseudomonas* species that can cause infection in immunocompromised individuals.

Glycopeptides

The glycopeptides vancomycin and teicoplanin inhibit peptidoglycan synthesis by acting at an earlier stage than β-lactams. They covalently bind to terminal D-alanine–D-alanine at the end of pentapeptide chains, thus sterically inhibiting the elongation of the peptidoglycan backbone by preventing the incorporation of new subunits into the growing cell wall. They are only effective against Gram-positive organisms, as their large size means they cannot penetrate easily into Gram-negative cells. As they are expensive and potentially toxic, glycopeptides are reserved for severe infections, for infections with organisms that are resistant to other antibiotics, or in cases where a patient has displayed hypersensitivity to β-lactams.

Inhibitors of protein synthesis

Many antibiotics block protein synthesis, either by blocking translation or by other means. Subtle difference between prokaryotes and eukaryotes, for example in ribosome size, mean that selectivity is possible. For example, protein synthesis inhibitors, such as tetracycline, kanamycin and erythromycin, target prokaryotic ribosomes, but they do not affect mammalian ribosomes. Inhibition can be effected at all stages of translation from initiation to elongation to termination.

To recall antibiotics that interact with the different subunits of the bacterial ribosome remember 'Buy ***AT*** 30, ***CELL*** at 50'. A= **A**minoglycosides; *T*= **T**etracycline; C= **C**hloramphenicol; E= ***E***rythromycin; L=***L***inezolid; L= c***L***indamycin.

Antibiotics and mitochondria

Prokaryotic cells do not possess membrane-bound organelles and, as such, do not possess mitochondria. However, it is generally accepted that eukaryotic mitochondria originated as as bacterial endosymbionts, evolving from prokaryotic cells and, as such, the translation process in eukaryotic mitochondria is very similar to that in prokaryotic cells. Antibiotics that inhibit prokaryotic protein synthesis can also affect mitochondrial protein synthesis. Antibiotics, however, do not harm their mammalian host because:

- some antibiotics are unable to cross the inner mitochondrial membrane
- mitochondria are replaced at cell division. This occurs relatively slowly in most cells, so mitochondria are depleted only with long-term antibiotic use
- in rapidly dividing cells, the local environment can sometimes prevent uptake of antibiotic (e.g. in bone, high calcium levels cause the formation of calcium-tetracycline, so the drug cannot be taken up by cells in the bone marrow).

Inhibitors of metabolic pathways: anti-metabolites

These agents:

- target the folic acid synthesis pathway (this pathway produces tetrahydrofolate, which is essential for nucleotide synthesis)
- do not affect mammalian cells because mammals obtain folic acid from their diet
- are bacteriostatic.

Sulphonamides (e.g. sulphadiazine) are analogues of γ-aminobenzoic acid. Trimethoprim inhibits bacterial, but not eukaryotic, dihydrofolate reductase. It is used in the treatment of urinary-tract infections.

Inhibitors of cell-membrane function

The cell membrane controls the internal composition of the cell, and the disruption of the cell membrane can lead to changes in membrane function and permeability, leading to cell damage or death. The polymyxins are active against all Gram-negative organisms, except *Proteus* species. They act as cationic detergents, disrupting the phospholipid structure of the cell membrane. They can only be given systemically and, because they are toxic, have few indications.

Antibiotic resistance

Antibiotic resistance in bacteria may be a natural characteristic of the organism (e.g. it may lack the target of the antibiotic molecule) or may be acquired. Acquired resistance is a major problem clinically and is brought about by mechanisms that can be broadly classified into three main types:

- Alteration of the antibiotic. Some bacteria can produce an enzyme that can chemically alter the structure of the antibiotic. As the structure of the antibiotic is essential to its interaction with its target molecule, this neutralizes the antibiotic before it can have an effect. Examples include the β-lactamases, aminoglycoside-modifying enzymes and chloramphenicol acetyl transferases.
- Target-mediated resistance. The target, for example a receptor, may be altered so that it has a lowered affinity for the antibacterial agent in question.
- Impermeability. Bacteria can reduce the amount of drug that reaches the target either by decreasing the permeability of the cell wall to the antibiotic or by pumping the drug out of the cell (known as an efflux mechanism).

Bacteria can acquire resistance by a number of mechanisms, including chromosomal mutation leading to class resistance; the horizontal transfer of resistance genes by conjugation, transformation, transduction; or by the acquisition of 'jumping genes' (transposable elements) or 'cassettes' of resistance genes (integrons).

VIRUSES

Viruses are infectious particles consisting of nuclear material enclosed in a protein coat called a capsid, which may be surrounded by a phospholipid envelope. The complete virus particle is known as a virion. Viruses are obligate intracellular parasites, being totally dependent on the cells that they infect to provide metabolic intermediates, energy and many, if not all, of the enzymes they require to replicate. Generally they infect their host by means of receptor-mediated endocytosis or fusion with the plasma membrane. The major stages in viral replication are generally the same for all viruses (Fig 1.17).

One of the most significant challenges in gene therapy is getting the therapeutic piece of DNA through the plasma membrane to the nucleus, while avoiding lysosomal degradation. Viruses have evolved very efficient mechanisms for doing this and, therefore, many gene therapy protocols exploit viral vectors.

Replication of RNA viral genomes is error prone and leads to genome diversity. This is especially relevant for the development of antiviral resistance by HIV. In contrast, replication of DNA genomes is relatively error free due to the proofreading activity of viral DNA polymerase.

Viral genomes

Viral genomes vary greatly in size, from approximately 3200 nucleotides (hepadnavirus) to approximately 1.2

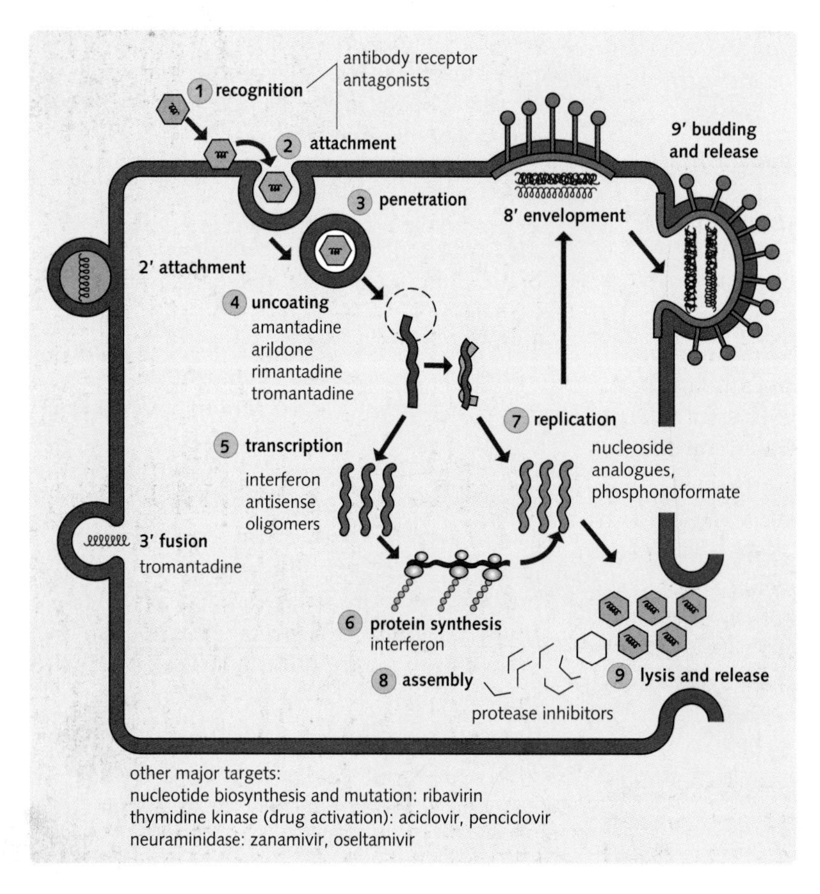

Fig. 1.17 Life-cycle of a virus. In order to replicate, each virus must infect a host cell and 'hijack' its cellular machinery. For the cycle to continue, the newly assembled virions must escape from the original host cell so that they can go on to infect new host cells. The ten stages that the virus must successfully pass through to complete this cycle are shown.

million base pairs (mimivirus). They contain either deoxyribonucleic acid or ribonucleic acid, but not usually both. The exception to this is cytomegalovirus (CMV), which contains both a DNA core and mRNA.

DNA viruses

Double-stranded DNA viruses

The genome is a molecule of double-stranded (ds) DNA that consists of a 'plus' strand and a 'minus' strand. The 'minus' DNA strand is directly transcribed into viral mRNA. Important examples include the herpes family of viruses and adenoviruses.

Single-stranded DNA viruses

The genome is a molecule of single-stranded (ss) DNA. Once inside the host cell, the ssDNA is converted into dsDNA, and the minus DNA strand is transcribed into viral mRNA. Examples of ssDNA virus include the adeno-associated virus and parvovirus B19.

Reverse transcribing DNA viruses

Although the genome is double-stranded DNA, its replication involves the generation of an intermediate mRNA molecule, known as the pre-genome, from which DNA is reverse transcribed. Hepatitis B can be found in this group.

RNA viruses

Double-stranded RNA viruses

The genome is a molecule of double-stranded RNA. The 'plus' RNA strand functions as viral mRNA. An example is the reovirus family, which includes rotavirus.

Positive-sense, single-stranded RNA viruses

Positive-sense viral RNA is identical to viral mRNA, and as such can be immediately translated into viral proteins. Examples include polioviruses, rubella and the hepatitis A and C viruses.

Negative-sense, single-stranded RNA viruses

Negative-sense viral RNA is a mirror image of mRNA and must be converted to positive-sense RNA by a virally encoded RNA-dependent polymerase before viral proteins can be translated. Examples include the influenza, measles, mumps and Ebola viruses.

Reverse transcribing RNA viruses

Also known as retroviruses, these are positive-sense ssRNA viruses that produce a DNA intermediate with a unique enzyme called reverse transcriptase

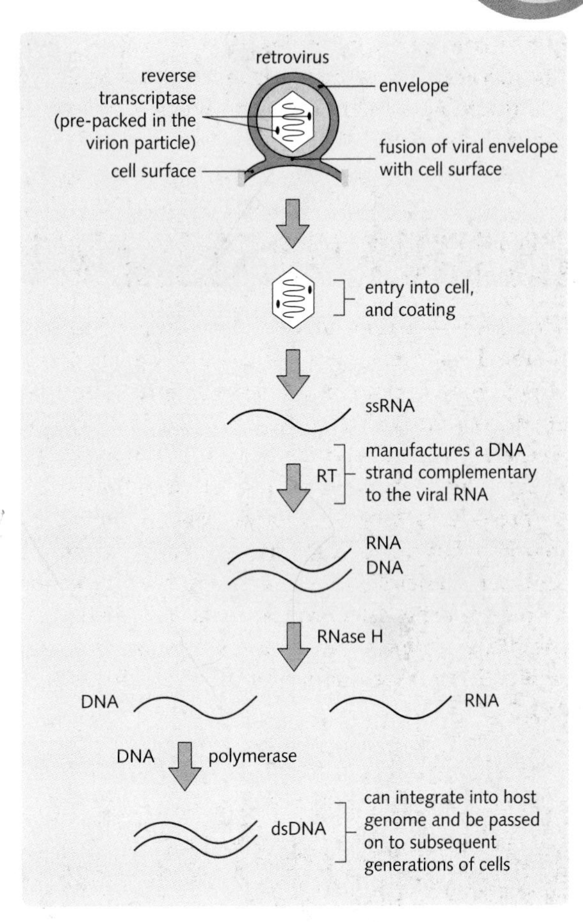

Fig. 1.18 Replication of retroviruses.

(Figs 1.18 and 1.19). Examples include the human immunodeficiency viruses (HIV) and human T lymphocyte viruses 1 and 2.

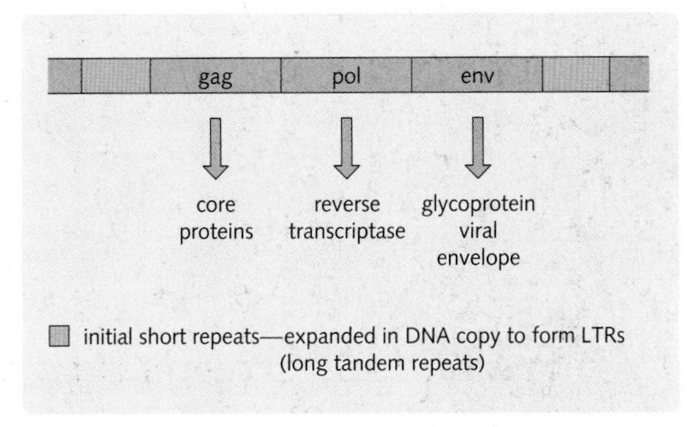

Fig. 1.19 Retrovirus genome – codes for three genes. These are preceded by a non-coding region containing enhancer and promoter regions that facilitate expression of the viral genome by host 'machinery'.

Human immunodeficiency viruses HIV-1 and HIV-2 are retroviruses and are responsible for the worldwide pandemic of acquired immune deficiency syndrome (AIDS). HIV is transmitted:

- in blood (needlestick injury, sharing needles, contaminated blood products)
- by sexual contact (heterosexual and homosexual, and from unscreened sperm donors)
- from mother to baby (during pregnancy, childbirth or breastfeeding).

HIV attacks the immune system and progressively destroys it. It does this by specifically targeting CD4-positive cells (T helper cells, dendritic cells and macrophages) by binding onto the CD4 receptor. The viral DNA integrates into the host genome and new virus particles are manufactured and mature HIV virions released.

Pathogenesis of viral infection

Following infection of a cell, virus assembly, maturation and release lead to a wide range of virus–cell interactions, which can be grouped into three main categories:

1. Lytic. Following infection, the host cell is lysed in order to release the new viral progeny.
2. Persistent. The virus infects the host cell, but does not complete the replication cycle
3. Lysogenic. Viral DNA becomes integrated with that of the host and, as such, can be transmitted to daughter cells at each subsequent cell division. This can result in latency or transformation.

Although useful for classifying infection, these categories are not mutually exclusive, and viruses can be capable of more than one life-cycle. For example, lytic infections by human papillomavirus (HPV) cause genital warts, while latent infections by some strains of HPV lead to cervical and anogenital cancer.

Antiviral chemotherapy

Antiviral drugs work by interfering with viral replication. Since viruses utilize the host cells' own metabolic pathways for replication, it is difficult to find drugs that are virus-specific. Nevertheless, there are enzymes that are only encoded by the virus, which offer potential virus-specific targets. As such, most antiviral agents are only effective while the virus is replicating:

- viral attachment to the host cell
- viral penetration and uncoating
- viral nucleic acid synthesis
- virus particle maturation
- virus release.

Viral attachment to the host cell

By interfering with the binding and adsorption of viral particles into the host cell (see Fig. 1.17), the host cell is protected from viral infection. Although largely an experimental target, one fusion inhibitor, T-20, has been approved in the US for the treatment of HIV. T-20 prevents viral attachment to the T cell by binding to the HIV transmembrane glycoprotein site, gp41.

Viral penetration and uncoating

After attachment, the virus penetrates the host cell within an endocytic vesicle. This is followed by 'uncoating', during which the endocytic vesicle and the viral capsid are enzymatically degraded and the viral genome is released (see Fig. 1.17). By inhibiting this stage, the viral genome is not released and can not take over the host's translational machinery. Amantadine prevents the influenza A virion uncoating by interacting with the viral membrane ion channel, M2, sterically blocking it.

Viral nucleic-acid synthesis

Viruses encode specific enzymes of their own. For example, polymerases are often virally encoded. These include:

- DNA-dependent DNA polymerase – DNA viruses
- RNA-dependent RNA polymerase – RNA viruses
- RNA-dependent DNA polymerase – retroviruses.

Other targets include enzymes required for nucleic acid synthesis, for example herpes simplex thymidine kinase.

DNA polymerase inhibitors

Aciclovir is an acyclic nucleoside analogue that acts as a chain terminator of herpesvirus DNA synthesis. It is a prodrug, converted to aciclovir monophosphate by α herpesvirus encoded thymidine kinase. Subsequent conversion to aciclovir triphosphate is achieved by host cellular enzymes. Aciclovir triphosphate then competes with deoxyguanosine triphosphate for incorporation into viral DNA, resulting in DNA synthesis termination.

Foscarnet is a pyrophosphate analogue that prevents nucleotide binding, thus inhibiting viral replication. It is active against herpesvirus and its main application is in the treatment of severe cytomegalovirus disease and for the treatment of aciclovir-resistant herpes simplex infection.

Reverse transcriptase inhibitors

Reverse transcriptase inhibitors (RTIs) have revolutionized the treatment of HIV. The main classes of RTIs are:

- nucleoside analogue reverse transcriptase inhibitors (NRTIs)
- nucleotide analogue reverse transcriptase inhibitors (NtRTIs)
- non-nucleoside reverse transcriptase inhibitors (NNRTIs).

NRTIs inhibit viral RNA-dependent DNA polymerase (reverse transcriptase) and are incorporated into viral DNA leading to chain termination. They require conversion into their active 'triphosphate' form to become active. Examples include zidovudine (also called azidothymidine, AZT), an analogue of thymidine, and didanosine (ddI), an analogue of deoxyadenosine, which are phosphorylated in the host cell and compete with cellular nucleotide triphosphates in the reverse transcription process. Host mitochondrial, γ DNA polymerase is also susceptible to NRTIs and this probably accounts for some of the common side effects. These include anaemia and neutropenia, gastrointestinal disturbance, alterations of liver function, headache, fever and skin rash.

NtRTIs differ from the NRTIs in that they require only one phosphorylation step within the cell to become activated, thus resulting in less toxicity. An example is tenofovir, the only NtRTI currently licensed for HIV treatment.

The NNRTIs have a different mode of action. They are not incorporated into viral DNA, but inhibit replication directly by binding non-competitively to reverse transcriptase. Examples include nevirapine and efavirenz.

Virus particle maturation

In order to be able to infect a new host cell, newly synthesized virions need to mature (see Fig. 1.17). The HIV protease is an essential enzyme in the maturation of the HIV virus. Protease inhibitors, such as saquinavir, nelfinavir and ritonavir, prevent viral assembly by inhibiting the activity of viral protease, the enzyme required to cleave HIV polyproteins into functionally active proteins. As a result, HIV cannot mature and noninfectious viruses are produced.

Virus release

The final stage in the life-cycle of a virus is the release of completed viruses from the host cell (see Fig. 1.17). Neuraminidase, one of two influenza surface glycoproteins, functions to cleave sialic acid residues from the host cell, releasing new virions. The neuraminidase inhibitors zanamivir and oseltamivir, which are active against both influenza A and B viruses, are competitive reversible inhibitors of the neuraminidase active site.

Eukaryotic organelles

2

Objectives

By the end of this chapter you should be able to:

- Draw and label a typical prokaryotic and a typical eukaryotic cell.
- Explain why eukaryotic cells are larger than prokaryotic cells.
- Outline the differences between prokaryotes and eukaryotes.
- Describe epithelial cell membrane specializations.
- Understand how the structure of membranous organelles relates to function.
- List the three components of the cytoskeleton.
- Outline the structure and function of cilia.
- Describe two types of specialized cell and explain how the structure of each reflects its function.
- Explain cell differentiation and cell memory.

THE EUKARYOTIC CELL

All organisms consisting of cells with a membrane-bound nucleus are classified in the eukaryote superkingdom. The animalia, plantae, protista, and fungi kingdoms all belong within this group. The typical eukaryotic cell (Fig. 2.1) shows the following features:

- a complex series of inner membranes that separate the cell into distinct compartments that perform specific functions
- cell division by mitosis
- specialized organelles, such as centrioles, mitotic spindles, mitochondria and microtubules.

Unlike prokaryotes, eukaryotic cells are capable of endocytosis (see Ch. 4). This allows patches of the plasma membrane to pinch off to form membrane-bound vesicles, delivering nutrients from the external environment to compartments deep within the cell. Endocytosis liberates eukaryotic cells from the constraints of simple diffusion, allowing them to sustain a relatively small surface area to volume ratio. As a result, eukaryotic cells are generally much larger than prokaryotic cells (Fig. 2.2). In the typical animal cell, the various specialized organelles occupy about half the total cell volume.

STRUCTURE AND FUNCTION OF EUKARYOTIC ORGANELLES

Eukaryotic cells have a complex ultrastructure comprising membranous and non-membranous organelles. These structures serve specific functions within the cell.

Plasma membrane

The plasma membrane is a selectively permeable barrier that surrounds the eukaryotic cell forming a dynamic interface between the cytosol and the environment. Non-polar (lipid soluble) molecules diffuse across the lipid bilayer by passive transport, while polar molecules are transported between the cell and the extracellular fluid by proteins embedded within the bilayer (either by facilitated diffusion or active transport – see Ch. 3).

Membranous organelles

Membranous organelles are enclosed within a phospholipid bilayer. They maintain discrete biochemical environments that contain characteristic sets of enzymes.

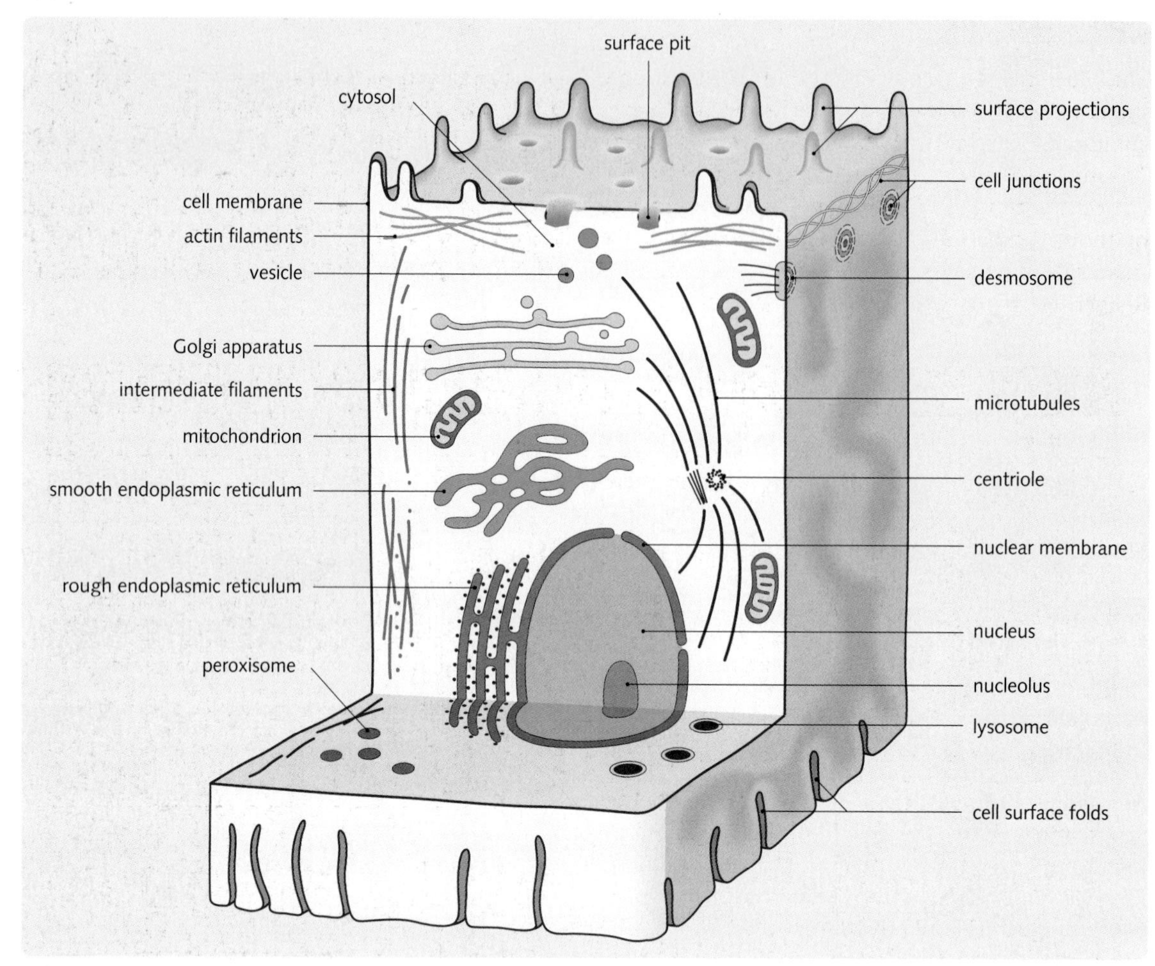

Fig. 2.1 Structure of a typical eukaryotic cell. Genetic material is contained within the nuclear space surrounded by the nuclear membrane. Membrane bound organelles serve as compartments for specific cellular functions, permitting greater cellular specialization and diversity. Cytoskeletal components maintain cell shape and facilitate dynamic functions such as endocytosis. (Adapted from Stevens and Lowe, 1997.)

Fig. 2.2 Basic features of prokaryotic and eukaryotic cells.

Fig. 2.2 Prokaryotic compared with eukaryotic cells

Prokaryotic cells	Eukaryotic cells
Includes bacteria and blue–green algae	Four major groups: Protista, fungi, plants and animals
No true nucleus	True nucleus
DNA circular and free	DNA linear and within nucleus
No membrane-bound organelles	Internal compartmentalization with organelles, hence division of labour (specialization)
Simple binary reproduction	Mitotic reproduction (and meiotic)
No development of tissues	Tissue and organ systems common
Multicellular types rare	Independent unicellular organism or part of multicellular organism
Size: 1–10 μm	Size: 10–100 μm

Nucleus

The nucleus (Fig. 2.3) is bound by a double membrane, with a distinct space between which is continuous at points with the endoplasmic reticulum. The nucleus contains the genetic material of the cell (chromosomes – see Ch. 6). It may also contain one or more dense-staining areas called nucleoli, the main role of which is the biosynthesis of ribosomal RNA (rRNA) and the assembly of ribosomes.

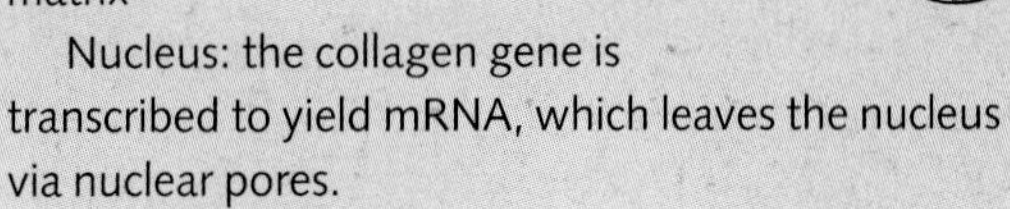

Collagen – from gene to extracellular matrix

Nucleus: the collagen gene is transcribed to yield mRNA, which leaves the nucleus via nuclear pores.

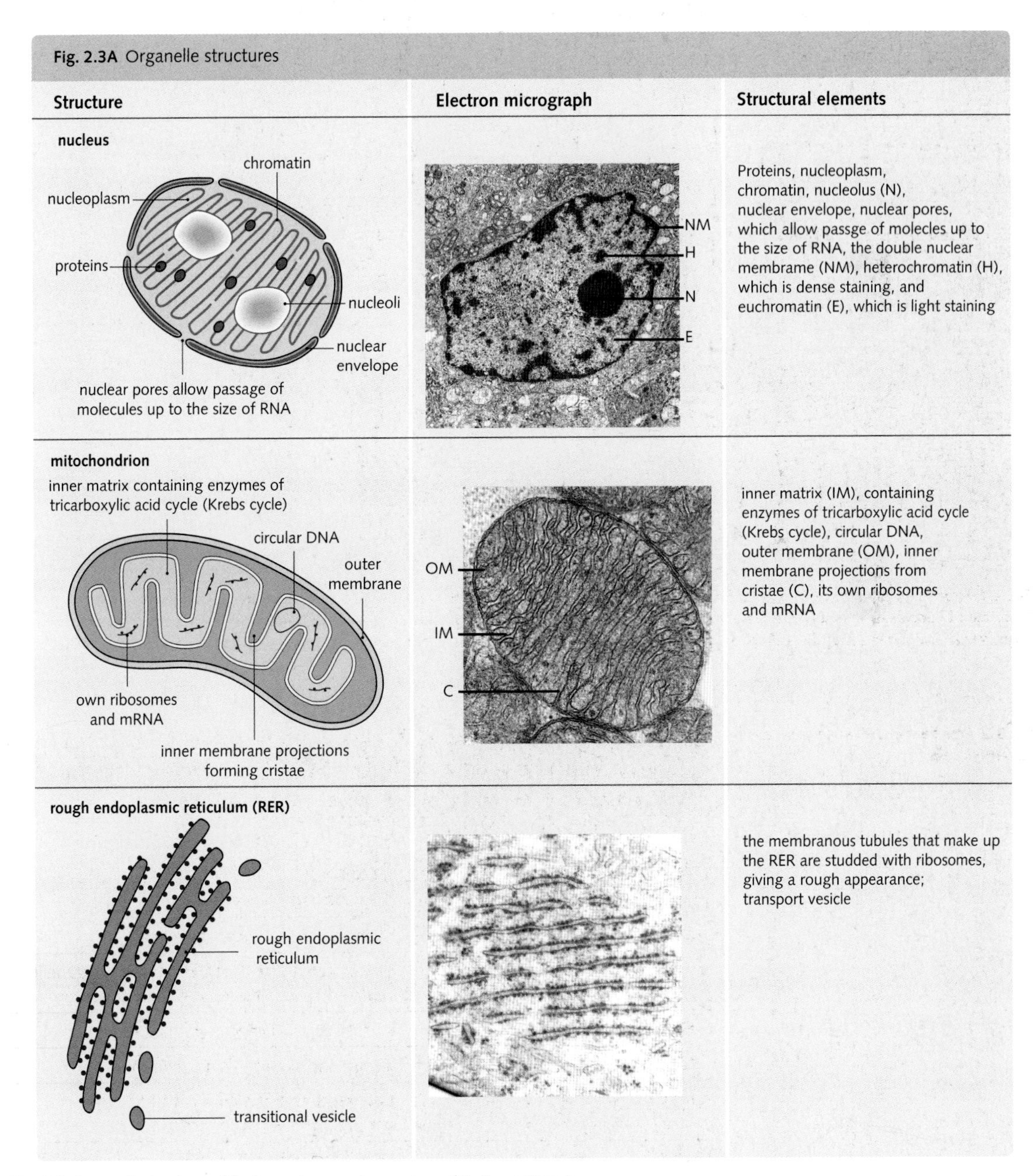

Fig. 2.3 Organelle structures. (Electron micrographs courtesy of Dr Trevor Gray.)

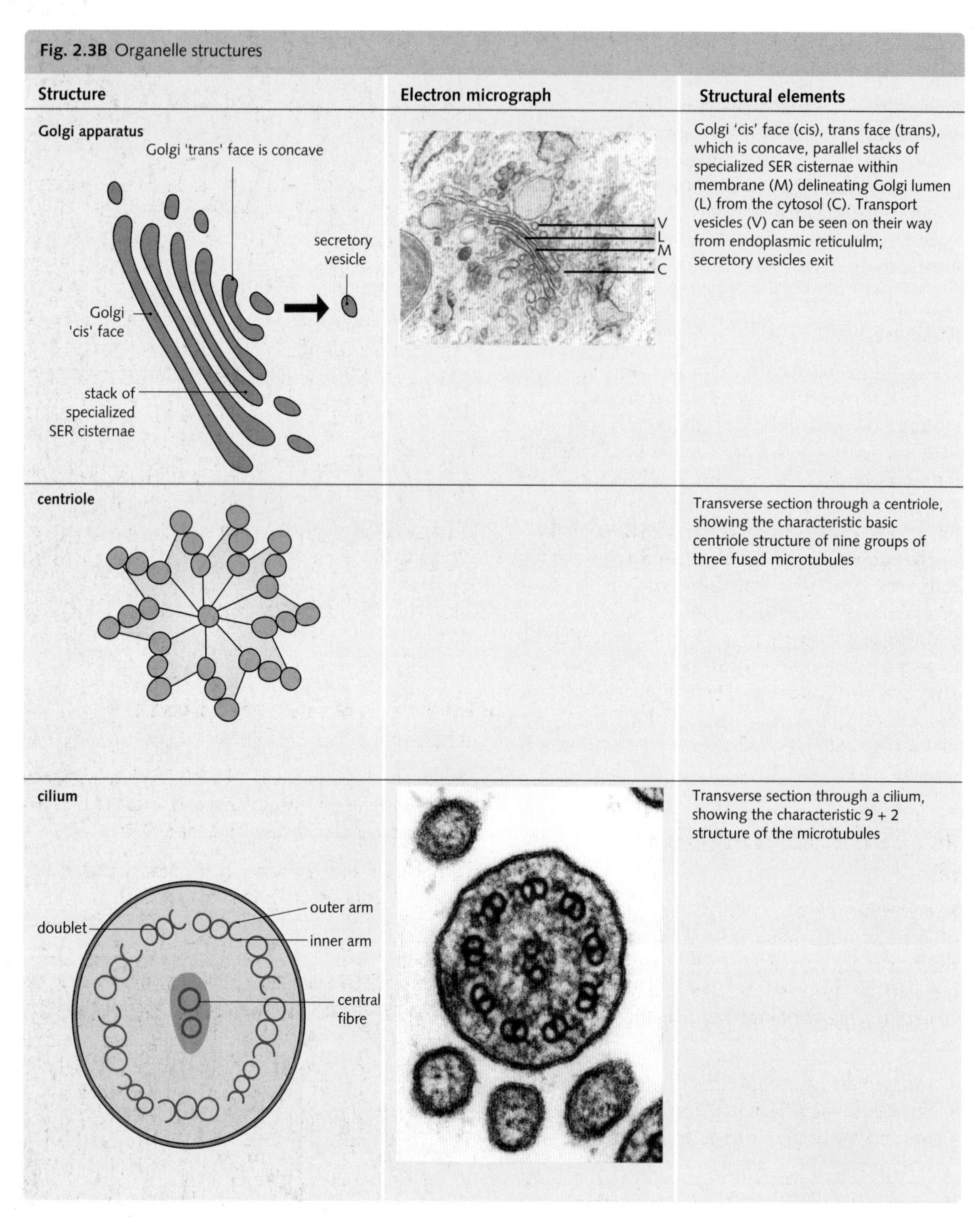
Fig. 2.3B Organelle structures
Structure
Electron micrograph
Structural elements
Golgi apparatus
Golgi 'trans' face is concave
secretory vesicle
Golgi 'cis' face
stack of specialized SER cisternae
V
L
M
C
Golgi 'cis' face (cis), trans face (trans), which is concave, parallel stacks of specialized SER cisternae within membrane (M) delineating Golgi lumen (L) from the cytosol (C). Transport vesicles (V) can be seen on their way from endoplasmic reticululm; secretory vesicles exit
centriole
Transverse section through a centriole, showing the characteristic basic centriole structure of nine groups of three fused microtubules
cilium
doublet
outer arm
inner arm
central fibre
Transverse section through a cilium, showing the characteristic 9 + 2 structure of the microtubules

Fig. 2.3—Cont'd

The nucleus is the location of a number of events, including:

- sequestration and replication of DNA
- transcription and modification of RNA
- facilitated selective exchange of molecules, such as RNA, e.g. transfer RNA (tRNA), with the cytoplasm via nuclear pores
- production of ribosomes within the nucleolus.

Mitochondria

The structure of a mitochondrion is illustrated in Fig. 2.3. Mitochondria are semi-autonomous and self replicating organelles, with their own ribosomes, RNA and several copies of a circular DNA molecule – the mitochondrial genome (see Ch. 6). Mitochondrial DNA shows a maternal inheritance pattern, and mutations within can yield genetic disease.

The main function of mitochondria in aerobic cells is oxidative phosphorylation and the production of energy though synthesis of ATP. In addition to this crucial role, mitochondria are also important in apoptosis and cellular Ca^{2+} handling. They are also potent producers of free-radicals, a process implicated in many pathological states, including Alzheimer's disease and cardiovascular disease.

Rough (granular) endoplasmic reticulum

Rough endoplasmic reticulum (RER) is a labyrinth of membranous sacs, called cisternae, to which ribosomes are attached giving a 'rough' appearance on electron microscopy (Fig. 2.3). The cisternal ribosomes make polypeptides, which are then in turn:

- inserted into the membrane
- released into the lumen of the cisternae
- transported to the Golgi complex or elsewhere.

Proteins made within RER are kept within vesicles or secreted, and cells that make large quantities of secretory protein have large amounts of RER (e.g. pancreatic acinar cells, plasma cells).

Collagen – from gene to extracellular matrix

Rough endoplasmic reticulum (RER): the mRNA is translated to produce the immature 'preprocollagen' polypeptide on the surface of the RER. Preprocollagen is a precursor molecule, consisting of procollagen and an N-terminal 'signal peptide', required to carry the polypeptide across the membrane of the RER. Within the lumen of the RER, the signal peptide is cleaved, yielding procollagen. From here, procollagen undergoes a series of covalent modifications, including hydroxylation, to yield hydroxyproline and hydroxylysine, and glycosylation of hydroxylysine. Groups of three procollagen α chains are arranged into a trimeric molecule and subsequently into a triple helical conformation. Molecules are then packaged in secretory vesicles for their onward journey to the cis face of the Golgi.

Smooth (agranular) endoplasmic reticulum

Smooth endoplasmic reticulum (SER) is a labyrinth of cisternae with many enzymes attached to its surface or found within its cisternae. SER:

- makes steroid hormones (e.g. in the ovary)
- detoxifies body fluids (e.g. in the liver).

Liver cells have very highly developed smooth endoplasmic reticulum (SER), which contains a large proportion of the body's detoxifying enzymes, including the P450 cytochromes. Enzyme induction by certain drugs may be followed by proliferation of the SER, accompanied by the production of more detoxifying enzymes, which can result in enhanced breakdown of a drug. This has an impact upon the dose required to reach therapeutic threshold and leads to a loss of therapeutic effect.

Golgi apparatus

The Golgi apparatus (Fig. 2.3) is a polarized system of membranous flattened sacs, each with a cis and a trans face. They are involved in modifying (e.g. by glycosylation), sorting and packaging macromolecules for secretion or delivery to other organelles. Protein sorting and packaging occurs and the trans face. Cells that produce many secretory products have well-developed Golgi apparatus (e.g. hepatocytes).

Collagen – from gene to extracellular matrix

Golgi apparatus: transport through the Golgi apparatus is accompanied by modification of oligosaccharide groups.

Together with the endoplasmic reticulum, the Golgi makes up the endomembrane system.

Lysosomes

Lysosomes are the primary components of intracellular digestion (see Ch. 4), and are derived from the trans face of the Golgi apparatus. Cells specializing in phagocytosis (e.g. macrophages) have many lysosomes which:

- contain granular amorphous material and about 60 types of hydrolytic enzymes
- vary in size from 50 nm to over 1 μm
- digest material with hydrolases that are active at acid pH.

Peroxisomes

Peroxisomes are vesicular bodies that are smaller than lysosomes, and contain specific enzymes. They are derived from the endoplasmic reticulum. Peroxisomal functions include biogenesis reactions (i.e. cholesterol, bile and plasmalogens), the degradation of fatty acids (by way of β-oxidation) and the breakdown of excess purines to urea. They are also required for the breakdown of toxic compounds, and so are found in high abundance in the liver and kidney. Diseases where peroxisomes have been implicated include X-linked adrenoleukodystrophy and Zellweger syndrome.

Zellweger syndrome is a rare, autosomal recessive disorder characterized by the reduction or absence of peroxisomes, which leads to multiple disturbances of lipid metabolism due to a build up of very long chain fatty acids. It also leads to severe neurological dysfunction, suggesting lysosomes are required for normal central nervous system neuronal migration. Death usually occurs within 6 months of onset, highlighting the physiological importance of these organelles, usually as a result of respiratory distress, gastrointestinal bleeding or liver failure.

Secretory vesicles

Secretory vesicles are organelles that deliver their contents, such as hormones and neurotransmitters, to the outside of the cell by fusing with the plasma membrane. They are derived from the trans face of the Golgi apparatus and their production may be:

- constitutive (e.g. collagen from fibroblasts and albumin production by hepatocytes)
- regulated (e.g. insulin from β-cells of pancreatic islets).

Collagen – from gene to extracellular matrix

Secretory vesicle: modified procollagen is secreted constitutively by fibroblasts. Upon secretion, uncoiled terminal ends of procollagen are cleaved to form tropocollagen, which themselves aggregate with the crosslinking of lysine and hydroxylysine residues are crosslinked to form a collagen fibril.

Non-membranous organelles

Ribosomes

Eukaryotic cells possess '80S' ribosomes, consisting of a small 40S subunit and a large 60S subunit. Ribosomes that synthesize proteins for use within the cytosol of the cell are found suspended within in the cytosol, whereas proteins destined for the plasma membrane or cell vesicles are attached to the cytosolic face of the membranes of the endoplasmic reticulum (see p. 27).

The cytoskeleton

The cytoskeleton is the internal framework of the cell, consisting of filaments and tubules. Cytoskeletal structures maintain and change cell shape by rearrangement of the cytoskeletal elements. They are essential for endocytosis, cell division, amoeboid movements, and contraction of muscle cells. There are several classes of cytoskeletal structural components:

- microfilaments formed from actin
- microtubules formed from tubulin
- intermediate filaments formed from intermediate filament proteins, such as keratin.

These structures may be cross linked by other proteins into networks or specialized organelles, the most common of which are as follows:

- Centrioles – these usually occur in pairs, which in non-dividing cells are aligned at right angles to each other. The basic centriole structure is one of nine groups of three fused microtubules arranged as a cylinder around a central cavity (Fig. 2.3). As cells prepare to divide, the centriole pairs separate and go to opposite ends of the cell, where they act as the site of spindle assembly in cell division.
- Cilia – used by some cells to aid the movement of a cell or substance over the surface of cells (e.g. fallopian cells move ova towards the uterus). They are attached to structures known as basal bodies, identical in structure to centrioles, which are anchored to the cytoplasmic side of the plasma membrane. Microtubules, arising from the basal bodies, are arranged in a '9 + 2' arrangement consisting of nine microtubule doublets surrounding two single microtubules (Fig. 2.3). Dynein side arms extend between adjacent doublets and hydrolyse ATP to generate a sliding force between them. This action underlies ciliary beating. Cilia can also be non-motile (sensory), as seen in the rod cells of the eye and the terminal fibres of olfactory neurons, where they have a '9 + 0' arrangement, with the central pair of microtubles being absent.
- Flagella – very long cilia used for propulsion by spermatozoa. Eukaryotic flagella have a '9 + 2' structure and, therefore, have both a different structure and origin to prokaryotic flagella.
- Microvilli – non-motile extensions of plasma membrane supported by actin, which increase the surface area of the cell (e.g. the small intestine brush-border).
- Pseudopodia – although not true specializations, pseudopodia are extensions of the plasma membrane, formed by actin polarization. They are commonly seen in phagocytes, such as macrophages.
- Junctions – these are points of adhesion between cells and other cells, and between cells and their basement membrane. They are discussed in more detail in Chapter 4.

Kartagener's syndrome is an autosomal recessive syndrome typified by situs inversus, chronic sinusitis and bronchiectasis. Cilia motility is produced by dyneins and other related microtubule proteins, ultrastructural defects in which leading to ciliary diskinesia have been hypothesized as being at the root of Kartagener's syndrome. Impaired ciliary function leads to reduced or absent mucus clearance in the lungs, and susceptibility to chronic, recurrent respiratory infections. Situs inversus (a mirror image arrangement of the organs) is thought to arise as interplay between motile and sensory cilia is required for determination of left–right axis in early vertebrate development.

CELL DIVERSITY IN MULTICELLULAR ORGANISMS

Cell specialization

It is thought that multicellular organisms evolved as a result of specialized cells acting together and combining to form a single organism, able to exploit ecological niches not available to any of its component cells acting alone.

Similar types of specialized cells combine together to form tissues, of which there are four main types each adapted to a specific function (Fig. 2.4). By definition, specialized cells show structural features

Fig. 2.4 Cell specialization in tissues

Tissue structure	Function	Specialized cell types
Epithelial tissue Consists of continuous sheets of cells that are bound together by tight junctions	Epithelial tissue lines inner and outer surfaces of the body to form a selectively permeable barrier Epithelial surfaces may be specialized for: • absorption • substance movement • secretion	Absorptive cells—the luminal plasma membrane is folded into microvilli to increase the surface area (e.g. intestinal villi) Ciliated cells—the luminal plasma membrane is coated with cilia that beat in synchrony (e.g. tracheal mucociliary escalator) Secretory cells—the RER and Golgi apparatus are highly developed (e.g. chief cells in the stomach)
Muscle tissue Consists of groups of cells containing fibrillar proteins arranged in an organized manner in the cytoplasm and linked by intermolecular bonds Skeletal and cardiac muscle appear striated	Muscle functions to produce movement. Contraction results from the rearrangement of internal bonds between fibrillar proteins Muscle tissue is specialized to allow: • voluntary movement of the skeleton • involuntary movement of substances through the viscera • continuous synchronous contraction of the heart	Skeletal muscle cells—each muscle fibre is an enormous multinucleated cell that extends the full length of the muscle (nuclei are located at the cell periphery). Thus, excitation results in simultaneous contraction of the full length of muscle in a longitudinal direction Visceral muscle cells—cells are relatively small with tapered ends and only a single nucleus. The cells are arranged in layers at right angles to one another to facilitate peristalsis Cardiac muscle cells—cells are Y shaped with one nucleus that is centrally located. The longitudinal branches of adjacent cells join at intercalated discs. This structure allows for the rapid spread of contractile stimuli from one cell to another
Connective tissue Consists of cells and extracellular material. The extracellular material is secreted by the cells and determines the physical properties of the tissue. It is composed of ground substance, fibres (collagen and elastin), and structural glycoproteins	Connective tissue provides structural and metabolic support for other tissues and organs Loose connective tissue acts as biological packing material Dense connective tissue provides tough physical support. For example, it forms the skeleton, the dermis and organ capsules	Fibroblasts—synthesize and maintain extracellular material. They are active in wound healing where specialized contractile fibroblasts (myofibroblasts) bring about shrinkage of scar tissue Adipocytes—store and maintain fat. They are found in clumps in loose connective tissue and form the main cell type in adipose tissue. Fat stored in adipocytes forms a large droplet that occupies most of the cytoplasm Chondroblasts and chodrocytes—produce and maintain cartilage Osteoblasts, osteocytes and osteoclasts—specialized cells that produce, maintain and break down bone, respectively Blood cells—leukocytes (white blood cells), platelets (thrombocytes), and erythrocytes (red blood cells). Functions include immune defence, blood clotting and oxygen carriage, respectively
Nervous tissue Peripheral nervous tissue is composed of neurons and Schwann cells Central nervous tissue consists of neurons and neuroglial cells (oligodendrocytes, astrocytes, microglia, and ependymal cells). It is divided macroscopically into grey and white matter. White matter consists of tracts of myelinated nerves. Grey matter contains neuronal cell bodies	Nervous tissue detects changes in the internal and external environments. By transmitting and processing this information it coordinates the activities of the multicellular organism to produce an appropriate response Nervous tissue is specialized for: • sensing environmental change • conducting information • integrating and analysing information	Sensory receptors—there are numerous cell types specialized for detecting environmental change, for example, Pacinian corpuscles (mechanoreceptor that detects skin pressure) Neurons—these are specialized for receiving and transmitting information. Therefore, they synthesize neurotransmitters and neurotransmitter receptors. Multiple dendrites allow communication with many neighbouring cells and function as sites of information input. The axon may be extremely long, and it facilitates transmission of information to distant sites. Terminal boutons arise at the end of the axon and communicate with other nerve cells or the effector organ Schwann cells/oligodendrocytes—specialist cells that wrap around the neuronal axon to form the myelin sheath and provide structural and metabolic support

Fig. 2.4 Cell specialization in epithelial, connective, nervous and muscle tissue.

that enable them to perform their designated function. In order to cooperate and coordinate their activities in the multicellular animal:

- cells are bound together by adhesions between their plasma membranes and the extracellular matrix (see Ch. 4)
- cells interact and communicate with one another (see Ch. 3).

Differentiation

Over 200 types of cell are identifiable in human tissue, differing in terms of their structure, function and chemical metabolism. All cell types are derived from a single cell (the zygote) following conception. The zygote is described as being totipotent since it is ultimately able to differentiate into all the cell types that make up the adult organism. As there is no loss of genetic material from somatic cells during human development (erythrocytes are an exception), different cell types arise as a result of mechanisms such as imprinting and differential gene expression.

During development, cell differentiation is driven by successive cascades of proteins that regulate the DNA in each cell, restricting transcription from specific sections of the genome, and promoting it in others. The developmental signals that initiate differentiation in a human cell come from the cells that surround it. However, even when removed from its normal environment the differentiated cell and its progeny will retain many of its functional characteristics. This concept is termed cell memory, because the DNA modifications restricting transcription ('epigenetic' changes) persist and are transmitted to daughter cells.

Objectives

By the end of this chapter you should be able to:

- Draw a labelled diagram of the fluid mosaic model of the plasma membrane.
- Discuss the chemical properties of phospholipids and their significance in the plasma membrane.
- Describe the modes of movement available to phospholipids and proteins in the plasma membrane.
- Define diffusion, osmosis, osmotic pressure, isotonic, hypotonic and hypertonic.
- Describe the relative concentrations of K^+, Na^+, Cl^- and Ca^{2+} across the resting cell membrane.
- Understand the differences between facilitated, primary active and secondary active transport.
- Outline the structure of the Na^+/K^+ ATPase, define what type of transport it mediates and describe its function with reference to ionic gradients.
- Define the terms endocrine, paracrine and autocrine in relation to cell signalling.
- Understand the process of signal transduction and the classes of molecules involved.
- Understand how steroid hormone receptors differ from cell surface receptors.

STRUCTURE OF THE CELL MEMBRANE

Fluid mosaic model

The fluid mosaic model, first proposed by Singer and Nicholson in 1972, is one of a biological membrane consisting of a phospholipid bilayer with proteins embedded in it (Fig. 3.1). The model has been verified by both freeze fracture and freeze etching electron microscopy. The phospholipid molecules that are the major component of the bilayer are amphipathic, i.e. they have hydrophobic and hydrophilic regions (Fig. 3.2). The phospholipid molecules form a stable bilayer in aqueous solutions as a result of:

- hydrophilic interactions of polar head groups with the extracellular and intracellular aqueous environments
- hydrophobic interactions of the fatty acid molecules in the bilayer interior.

The membrane is a dynamic structure and many proteins are able to move freely through it like 'icebergs floating in a sea of phospholipids'.

Under normal cellular conditions, the turnover of the plasma membrane is dependent upon the delivery of new membrane components to the membrane in late endosomes and lysosomes.

> The fluid mosaic model is a common topic in exams; remember to draw a diagram including peripheral and integral proteins, even for short-answer questions. Mention that the membrane is dynamic, and in essay questions discuss the movement of phospholipids and proteins within it.

Components of the biological membrane

Lipids

A lipid is a molecule that is soluble in an organic solvent (e.g. chloroform), but only sparingly soluble in water. Triglycerides, which are hydrophobic, consist of three fatty acids attached to a glycerol backbone by ester linkages.

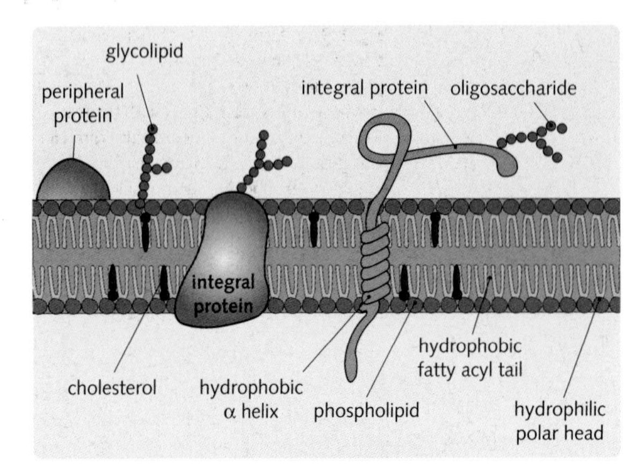

Fig. 3.1 Fluid mosaic model of the cell membrane.

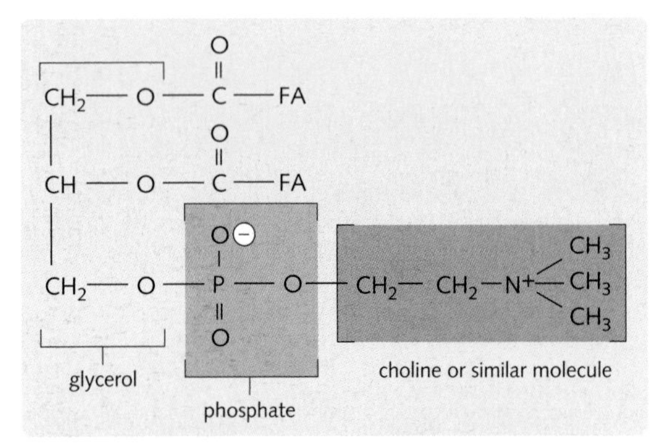

Fig. 3.2 Structure of a phospholipid. The purple shaded areas correspond to the hydrophilic parts of the molecule. FA, fatty acid.

Phospholipids

The hydrophilic moiety in a phospholipid arises as a result of the substitution of one of the fatty acid chains for an amine-containing polar group linked to glycerol (at C3) by a phosphodiester bond (see Fig. 3.2). There are four commonly occurring phospholipids in the plasma membrane (Fig. 3.3):

1. Phosphatidylethanolamine
2. Phosphatidylserine
3. Phosphatidylcholine
4. Sphingomyelin (Note: not a true phospholipid as it possesses an acylated sphingosine back bone [ceramide] rather than a glycerol backbone.)

Phospholipids are arranged asymmetrically, with almost all of the phosphatidylcholine and sphingomyelin occuring in the outer monolayer, while phosphatidylethanolamine and phosphatidylserine occur predominantly in the inner monolayer. Since they aggregate into a continuous sheet impervious to ions, the molecules must pack very closely. Most lipids pack as cylinders or slightly truncated cones, although lysophospholipids, intermediates formed during digestion of dietary and biliary phospholipids and which lack one fatty-acid chain, are shaped like cones (Fig. 3.4). The presence of a double bond in a fatty-acid side chain introduces a kink, which disrupts van der Waals forces and reduces the ability of the molecule to fit tightly with its neighbours, increasing membrane fluidity.

The plasma membrane is a dynamic structure, with a constant turnover of constituents. Niemann-Pick (NP) disease is a collection of lysosomal storage diseases displaying an autosomal recessive inheritance pattern. Types A and B result from a deficiency in lysosomal acid sphingomyelinase required to hydrolyse spent membrane sphingomyelin to yield ceramide and phosphocholine, leading to its accumulation in reticuloendothelial foam cells within the spleen, liver, lungs, bone marrow and brain.

Cholesterol

Cholesterol is a lipid molecule consisting of four hydrophobic rings and a hydrophilic hydroxyl group (Fig. 3.5). It orientates in the membrane such that the rings lie parallel to the hydrophobic fatty-acid groups, with the hydroxyl group forming a hydrogen bond with the carboxyl group on an adjacent phospholipid. The net result is that:

- at physiological temperature, cholesterol restricts the movement of the fatty-acid chains and stabilizes the membrane by reducing fluidity
- at low temperatures it inhibits phospholipid packing, which increases membrane fluidity.

Membrane proteins

Integral proteins

Integral proteins span the membrane and they have intracellular and extracellular domains:

- The membrane-spanning domains are rich in hydrophobic amino acid residues and traverse the membrane as α-helical loops.
- The cytosolic and extracellular domains are rich in polar amino acid residues.
- Extracellular domains may be glycosylated.

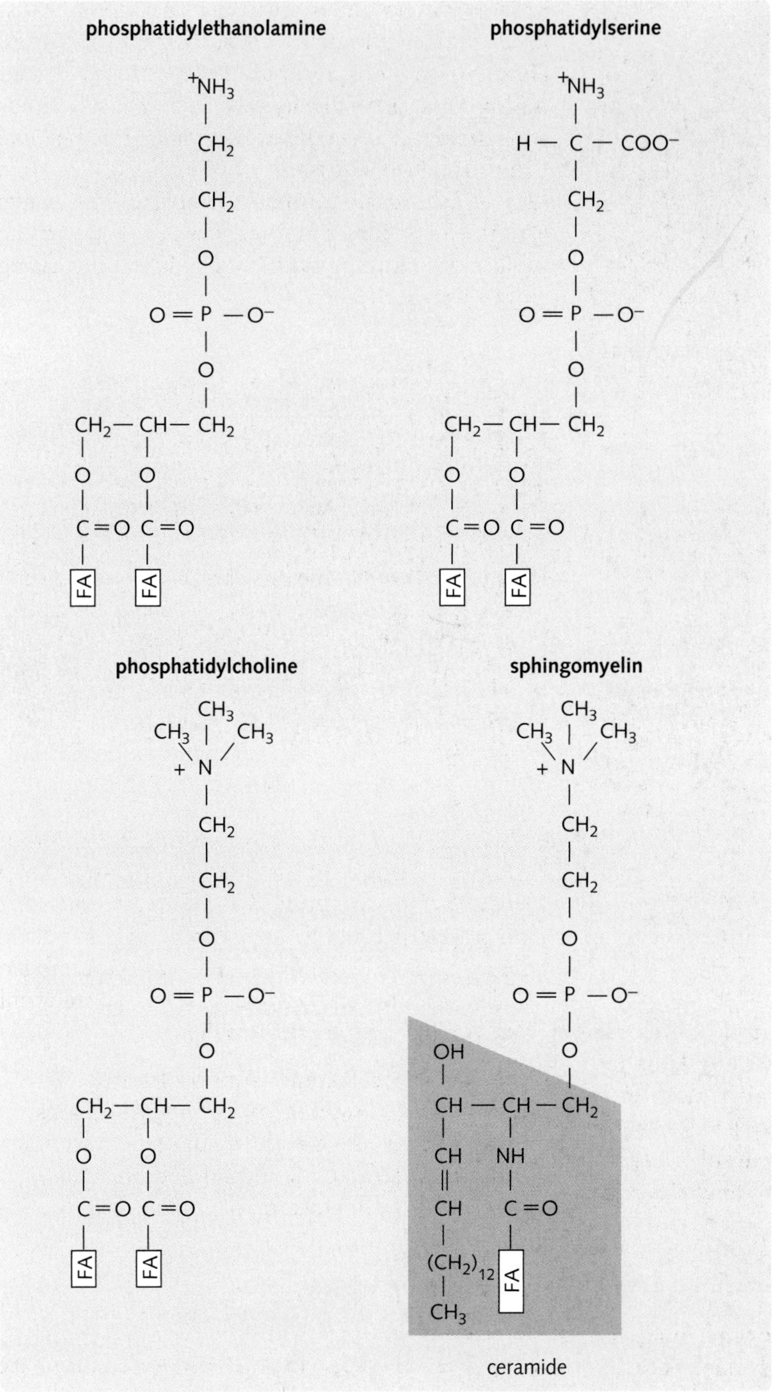

phosphatidylethanolamine

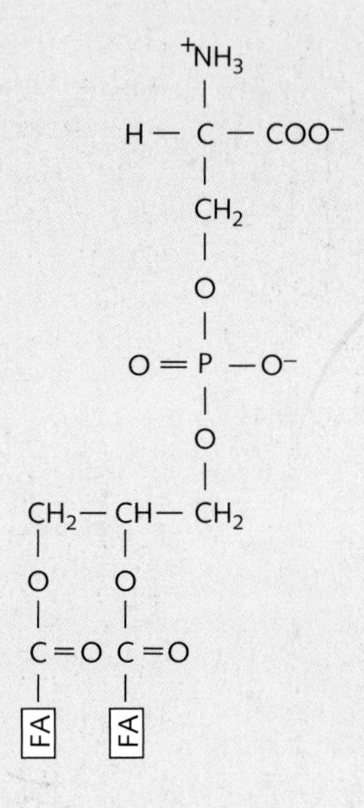

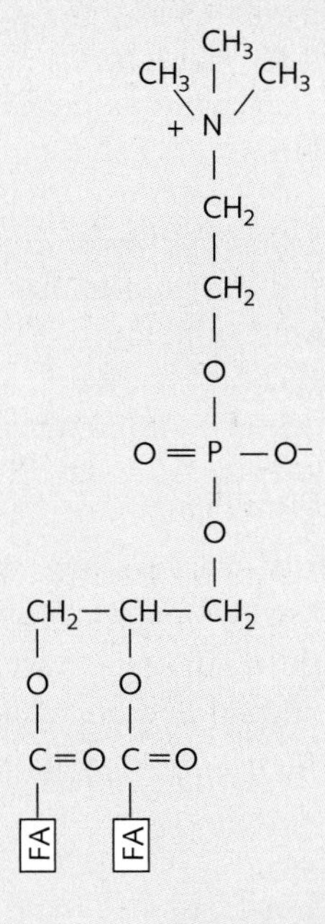

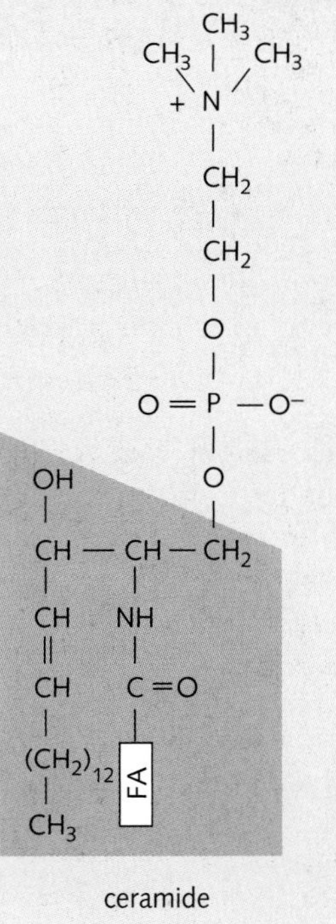

Fig. 3.3 The three major phospholipids, phosphatidylethanolamine, phosphatidylserine and phosphatidylcholine, differ with respect to their polar head groups. Sphingomyelin is a sphingolipid as opposed to a phospholipid and it consists of a choline group attached via a phosphate molecule to a ceramide group. FA, fatty acid.

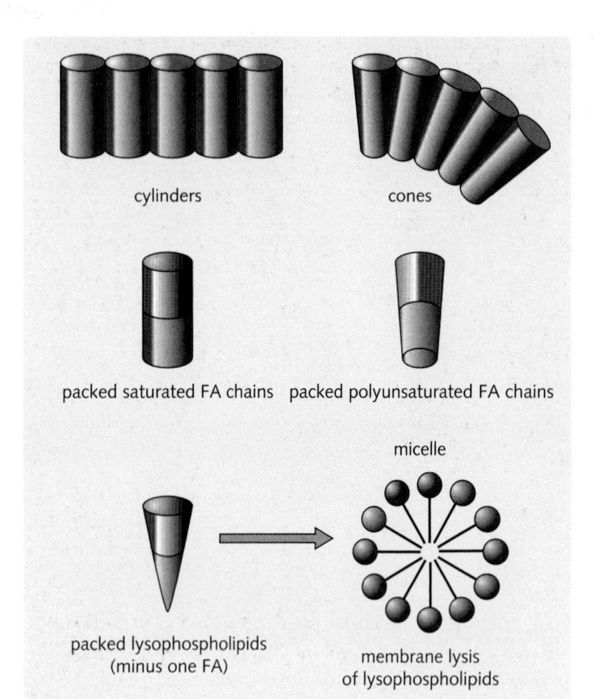

Fig. 3.4 Packing of phospholipids. The nature of the fatty-acid side chains influences their packing. Lysophospholipids, which lack one fatty acid, form micelles preferentially. However, the favoured structure for phospholipids with two fatty-acid chains in aqueous solution is a lipid bilayer because they are too bulky to form micelles. FA, fatty acid.

Monotopic integral proteins traverse the membrane once, while bitopic and polytopic proteins pass through the membrane twice and many times, respectively.

Polycystin 2, an integral membrane protein encoded by the PKD2 gene, is required for normal tubulogenesis in the kidney. It is found to be mutated in approximately 15% of all cases of autosomal dominant polycystic kidney disease (ADPKD). Normally, polycystin 2 interacts with polycystin 1 to act as a calcium permeable cation channel, and mutations in either of these proteins cause virtually indistinguishable clinical presentations. Defective polycystins appear to affect epithelial cell maturation, resulting in the development of cysts of varying sizes throughout the cortex and medulla.

Peripheral proteins

Peripheral proteins are associated with either the cytoplasmic or extracellular leaf of the lipid bilayer. They may be attached to the membrane by:

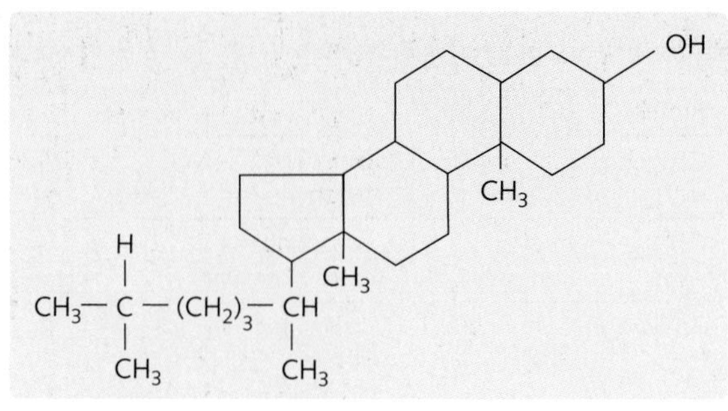

Fig. 3.5 Structure of cholesterol.

- electrostatic attachment to integral proteins
- covalent attachment to non-protein components of the cytoplasmic layer
- covalent attachment to phospholipids via an oligosaccharide linker in the extracellular layer.

Functions of membrane proteins

Membrane proteins have several functions, including:

- Markers – the carbohydrate chains of glycoproteins aid in self-recognition.
- Enzymes – water-soluble enzymes associate with the polar heads of membrane phospholipids (e.g. phospholipases).
- Anchors – membrane proteins act as attachment points to the cytoskeleton and extracellular matrix, aiding strength and adhesion.
- Transport – proteins function as carrier molecules, channels and porins act to transport ions and nutrients in and out of the cell (see pp. 41–42).
- Receptors – receptor proteins serve as binding or attachment sites for molecular messengers, such as hormones (see pp. 46–51).

Examples of membrane proteins, their function, and their means of attachment to the membrane are included in Fig. 3.6.

Properties of biological membranes

Fluidity

The transition temperature is the temperature at which the membrane transforms from a rigid gel-like structure to a relatively disordered, fluid state (Fig. 3.7). In its fluid state, proteins embedded in the membrane are free to interact. The membrane is heterogeneous, and ordered regions alternate with more fluid ones.

Fig. 3.6 Examples of membrane proteins and their functions. CFTR, cystic fibrosis transmembrane regulator.

Fig. 3.6 Examples of membrane proteins

Protein	Type	Bonding with membrane	Function
Cadherin	Monotopic integral	Hydrophobic with phospholipids	Mediates cell–cell adhesion
CFTR	Polytopic integral	Hydrophobic with phospholipids	Gated chloride channel in epithelial tissue
Ankyrin	Peripheral	Electrostatic with the anion exchange protein on the cytoplasmic surface of the lipid bilayer	Maintains erythrocyte structure by forming a link between spectrin and the anion exchange protein (band 3)
Ras	Peripheral	Covalent attachment to the cytoplasmic layer of the lipid bilayer	GTP-binding protein that relays signals from the cell surface to the nucleus

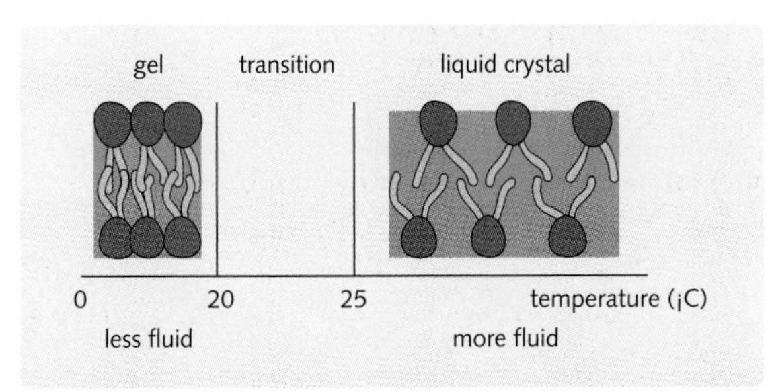

Fig. 3.7 Membrane transition from gel to liquid crystal.

Fluidity is determined by factors that influence the interactions of phospholipids (Fig. 3.8):

- Temperature – membranes are more fluid at high temperatures when phospholipid molecules have more kinetic energy.
- Saturation of fatty-acid chains – membranes are more fluid if they have a high proportion of unsaturated fatty acids.
- Cholesterol – the effects of cholesterol depend on temperature.

Mobility of membrane components

Phospholipids

Phospholipid molecules show four different modes of movement to varying degrees:

1. Intrachain movements, such as flexing of fatty-acid chains
2. Axial rotation of the molecules in the plane of the bilayer
3. Lateral diffusion within the plane of the bilayer
4. Movement from one half of the bilayer to the other ('flip-flop').

Lateral diffusion of phospholipids occurs readily, resulting in a fluid two-dimensional membrane. Flip-flop is rare in the absence of the enzyme 'flipase', enabling asymmetry of phospholipid composition between membrane layers. Membrane asymmetry is functionally important, for example, phosphatidylserine is concentrated on the cytoplasmic side facilitating its ability to interact with protein kinase C.

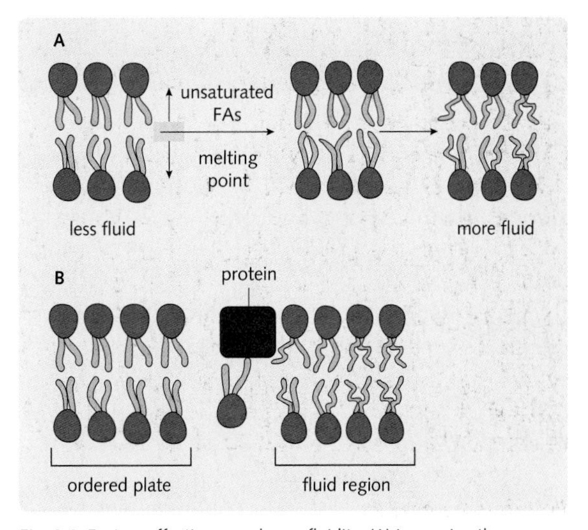

Fig. 3.8 Factors affecting membrane fluidity. (A) Increasing the concentration of unsaturated fatty acids (FAs) decreases the melting point of the membrane and so increases fluidity. (B) Uneven distribution of membrane lipids affects the fluidity.

Proteins

Proteins show three types of movement:

1. Lateral movement, though this may be restricted by interactions with the cytoskeleton or covalent attachment to lipid molecules
2. Axial rotational movements in the plane of the membrane
3. Conformational changes.

Flip-flop is thermodynamically unfavourable, since it would require hydrophilic protein moieties to rotate through the bilayer and it is, therefore, rare. Protein directionality is important for function, and it ensures, for example, that cell surface receptors are correctly orientated.

Permeability

There are three main forms of transport across the membrane:

1. Passive diffusion
2. Facilitated diffusion
3. Active transport.

Though permeable to lipid-soluble compounds, lipid bilayers are impermeable to ionic and polar substances, and these require dedicated channels to cross the membrane. The exception is water, which, although polar, is membrane permeable due to its small size.

TRANSPORT ACROSS THE CELL MEMBRANE

Concepts

Concentrations are measured in moles per litre and the dissociation of ions is not taken into account.

1 mol = 6.02 × 10^{23} molecules (of any kind)

= Avogadro's constant (the number of atoms in exactly 12 g of carbon 12)

For example:

- A molar solution of NaCl contains 1 mol of Na atoms and 1 mol of Cl atoms in 1 L.
- A molar solution of sucrose contains 1 mol of sucrose in 1 L.
- A molar solution of $CaCl_2$ contains 1 mol of Ca^{2+} and 2 mol of Cl in 1 L.

Diffusion is the movement of particles from a region of high concentration to a region of low concentration until they are evenly distributed. Osmosis is the movement of solvent molecules (usually water) across a semi-permeable membrane from a region of high solvent concentration to a region of low solvent concentration. The osmotic pressure is the pressure required to prevent the net movement of pure water into an aqueous solution across a semi-permeable membrane. In osmotic pressure the dissociation of ions is important.

1 osmol = the amount of substance that dissociates in solution to form 1 mol of osmotically active particles

For example (and assuming full dissociation):

- 1 Osmol/L of NaCl contains 0.5 mol of the ion Na^+ and 0.5 mol of the ion of Cl^-.
- 1 Osmol/L of sucrose contains 1 mol of sucrose.
- 3 Osmol/L of $CaCl_2$ contains 1 mol of the ion Ca^{2+} and 2 mol of the ion Cl^-.

The osmolarity of plasma is critical, as changes affect plasma volume, cell volume, and water and ion homeostasis. The range for normal plasma osmolarity is 280–295 mosmol/L. Serum plasma values are often given as osmolality: the concentration of osmotically active particles in solution per kilogram of solvent (osmol/kg). Dissociation is affected by pH, temperature and binding of ions to compounds (e.g. Ca^{2+} binding to myosin during muscle contraction). In living systems, K^+, Na^+, and Cl^- are fully dissociated, whereas Ca^{2+}, Mg^{2+} and H^+ are only partially dissociated.

Dehydration leads to a relative increase in serum solute load and, therefore, an increase in osmolality. Detected in the hypothalamus, anti-diuretic hormone (ADH) release from the anterior pituitary is triggered. ADH increases the permeability of the distal tubules and collecting ducts of the kidney, reducing the amount of water excreted and leading to a relative decrease in serum solute load and a lowering of serum osmolality. If the osmolality of the blood plasma becomes too low, then the output of the kidney is enhanced by the release of atrial natriuretic peptide (ANP), returning the solute load to the normal range.

The term 'tonicity' relates to the behaviour of cells immersed in a solution. It is the effective osmolality and is equal to the sum of the concentrations of the solutes that have the capacity to exert an osmotic force across the membrane:

- Isotonic extracellular solutions have the same osmotic pressure as the inside of the cell, so osmosis does not occur and the cell remains the same size.
- Hypotonic solutions are less concentrated, so water will pass into the cell and it will swell.
- Hypertonic solutions are more concentrated, so water will pass out of the cell and it will shrink.

Distribution of ions across the cell membrane

Life's essential chemical reactions can only occur within narrow physiological parameters, so the cell must regulate the entry and exit of intracellular molecules. Some biological processes, such as muscle contraction, depend upon an electrochemical gradient across the cell membrane. The distribution of ions across the cell membrane is shown in Fig. 3.9. Distribution is influenced by:

- the semi-permeable membrane concept
- electrochemical gradient
- pumps.

Semi-permeable membrane concept

Polar molecules cannot diffuse through the lipid bilayer, and they rely on the proteins embedded in the membrane for transport. These proteins are generally specific for particular molecules. The cell regulates the activity of membrane transport proteins such that only certain molecules can get through at any one time.

Fig. 3.9 Distribution of ions across the cell membrane

Component	Outside	Inside
K^+ (mmol/L)	4.5	140 (varies with cell type)
Na^+ (mmol/L)	140	10
Ca^{2+} total (mmol/L)	3	1
Ca^{2+} free (μmol/L)	1	0.1
Cl^- (mmol/L)	110	3
HCO_3^- (mmol/L)	24	10
pH	7.35	7
Amino acids, proteins	10	120

Fig. 3.9 Distribution of ions across the cell membrane.

Electrochemical gradient

It is thermodynamically favourable for ions to move from areas of high concentration to low concentration and for positively charged ions to move to negatively charged environments.

Pumps

Pumps are used to maintain energetically unfavourable concentration gradients. For example, the cell is able to maintain an energetically unfavourable sodium gradient because it expends energy in the form of ATP to drive sodium out of the cell.

Transport across the membrane

Transport across a biological membrane is summarized in Fig. 3.10.

Passive (simple) diffusion

Passive diffusion is the free movement of molecules across a membrane down a concentration gradient. Small non-polar molecules (e.g. O_2 and CO_2) and uncharged polar molecules (e.g. urea) may diffuse directly through the lipid bilayer by this means. No energy is required and diffusion continues until equilibrium is reached. Saturation does not occur because no binding sites are involved. The diffusion rate is directly proportional to the ion gradient, hydrostatic pressure and electrical potential, and is summarized by Fick's law of diffusion:

$$\text{Rate of diffusion} = D \times \text{Area} \times \Delta\,\text{conc}$$

where D = diffusion constant, A = membrane area and Δ conc = concentration gradient.

Facilitated diffusion

As charged molecules cannot diffuse directly through the lipid bilayer, they depend on specific proteins. The transport of molecules by a protein receptor down a concentration gradient is called facilitated diffusion, which is a form of passive transport that continues until equilibrium is reached. Proteins mediating facilitated diffusion may be channels or carrier proteins. Transport with carrier proteins shows Michaelis–Menten kinetics (Fig. 3.11):

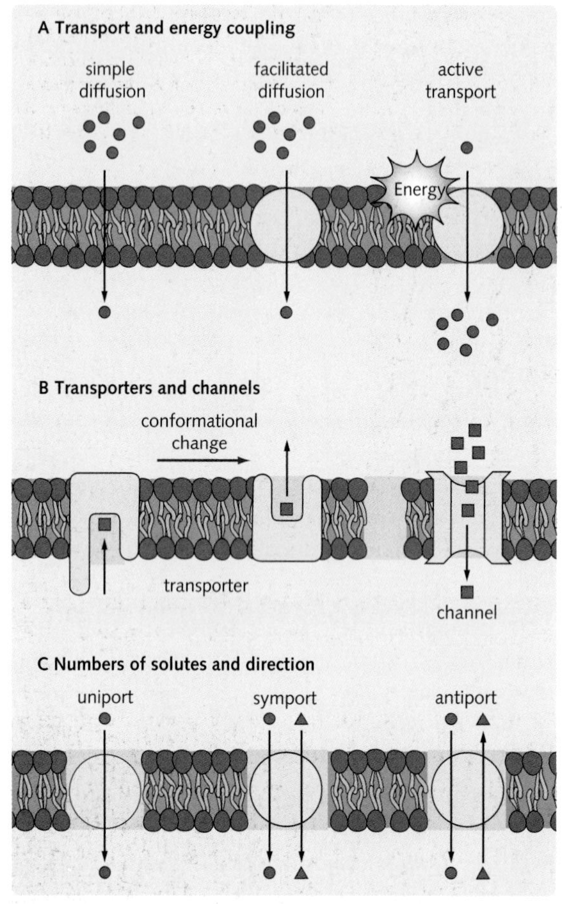

Fig. 3.10 A summary of solute movement across membranes. (Adapted from Baynes and Dominiczak, 1999.)

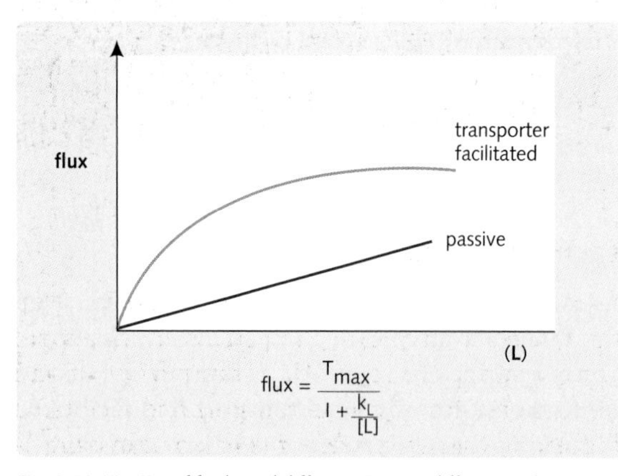

Fig. 3.11 Kinetics of facilitated diffusion. Passive diffusion is slower than transporter facilitated diffusion and its rate is directly proportional to substrate concentration. Transporter facilitated diffusion behaves like an enzyme and becomes saturated. (T_{max}, maximum transport rate; k_L, affinity for ligand; [L], concentration of ligand.)

- substrate specificity or selectivity, affinity for a particular ligand (measured as K_L)
- saturability of ligand binding (B_{max})
- transferability (T_{max}), the maximum rate of molecule transfer across the membrane
- inhibition (e.g. transport of glucose into erythrocytes).

The rate of movement of ions through a membrane via channels depends on the concentration gradient, the speed with which the ion moves through the channel (a constant) and the number of open channels. It is, therefore, analogous to passive diffusion, with the number of open channels being equivalent to the surface area. Channels may be gated such that the cell controls when they are open. Irrespective of the electrochemical gradient, an ion cannot cross the membrane if there are no open channels.

Active transport

Active transport couples the movement of molecules against an unfavourable electrochemical gradient to a thermodynamically favourable reaction:

- Primary active transport is coupled directly to the hydrolysis of ATP.
- Secondary active transport is coupled indirectly to the hydrolysis of ATP.

Primary active transport

Primary active transport directly uses energy to transport molecules across a membrane. Sodium and potassium are examples of ions that are transported across the cell by primary active transport via the Na^+/K^+ dependent ATPase. For every ATP hydrolysed, this transporter pumps three Na^+ ions outward and two K^+ ions inward, against their respective concentration gradients (see pp. 42–43).

Secondary active transport

Secondary active transport does not use ATP directly, but takes advantage of a separate existing concentration gradient. The action of the Na^+/K^+ ATPase establishes K^+ and Na^+ concentration gradients across the membrane. Movement of Na^+ into the cell, down its electrochemical gradient, is thermodynamically favoured, and in secondary transport this is coupled to the movement of a second ion against its gradient. The Na^+ electrochemical gradient may drive the transport of ions in either direction across the membrane (Fig. 3.12):

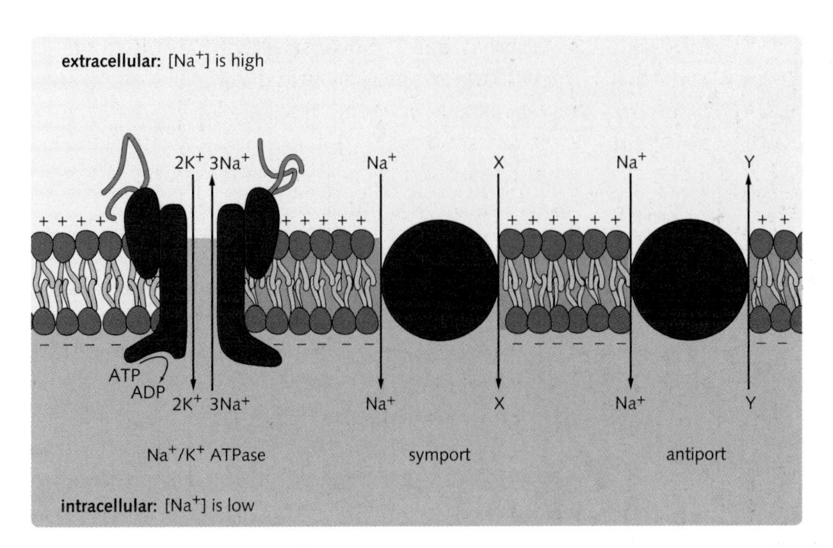

Fig. 3.12 Secondary active transport. The action of the Na^+/K^+ ATPase results in a sodium concentration gradient across the membrane. This potential energy may be used to drive molecules into the cell via symports or out of the cell via antiports.

- Symports transport both ions in the same direction.
- Antiports transport the ions in opposite directions.

Summary of types of transport

In summary:

- Passive diffusion is through the plasma. membrane, and it does not expend energy.
- Facilitated diffusion requires specific proteins, but energy is not expended.
- Active transport requires specific proteins and energy expenditure.

Transport mechanisms

Membrane transport proteins may be channels or carriers (see Fig. 3.10). Most transport proteins are reversible and, depending on the prevailing conditions, may transport ions into or out of the cell.

Ion channels

Ion channels are proteins that span the membrane and have central water-filled pores. The pores are specific, allowing either cations or anions through. Transport speed is greater than 10^6 ions/s, and it is always down a concentration gradient. Potassium channels are the most common type. One type is perpetually open, with the leakage of K^+ through these channels being critical to the membrane potential. Defects or damage can cause muscular dysfunction (e.g. periodic paralysis). Many channels are 'gated', and open and close under specific conditions.

Cystic fibrosis is an autosomal recessive disease characterized by chronic lung disease, exocrine pancreatic insufficiency and male infertility. It results from mutations in the cystic fibrosis transmembrane conductance regulator (CFTR) gene, the product of which encodes a cAMP regulated gated ion channel. The channel is primarily responsible for controlling the movement of chloride from outside the cell into the cell. In the absence of a functional CFTR gene product, transport of chloride ions across apical epithelial cell membranes is impaired, leading to an accumulation of chloride outside the cell and of water and sodium in the cell, causing mucus or watery secretions outside the cells to be too thick.

Carrier proteins

Carrier proteins bind specific ligands (the transported molecule) and undergo conformational change during transport. They transport polar and ionic molecules by active transport and facilitated diffusion. Carrier protein mediated transport is hundreds of times slower than that via ion channels, and they can become saturated, limiting the rate of transport. Uniports transport single molecules across the membrane (Fig. 3.10). Coupled transporters transfer molecules across the membrane with simultaneous transfer of another molecule (symports and antiports). Different cells have different carrier protein populations, and so they have different permeabilities.

Glucose transporter

Most cells transport glucose by facilitated diffusion through uniports, as the concentration of glucose is greater outside the cell. However, in the intestine and kidney some cells absorb glucose from low extracellular concentrations, mediated by secondary active transport via symports co-transporting sodium.

Active transporters

Active transporters are carrier proteins that are linked to a source of energy, such as ATP or an ionic gradient.

The Na^+/K^+ ATPase pump

The sodium pump is an example of an active transporter. It is a heterodimer consisting of an α-subunit and a glycosylated β-subunit (Fig. 3.13). The glycosylated subunit is important for the assembly and localization of the pump. The α-subunit is the catalytic unit, and it has binding sites for sodium and ATP on its intracellular surface, and potassium on its extracellular surface. Binding of sodium causes phosphorylation of the cytoplasmic side and a conformational change, which transfers the sodium outside the cell. Binding of potassium causes dephosphorylation, so the subunit returns to its original state, transferring the potassium inside the cell simultaneously (Fig. 3.14). The pump has several functions:

- maintaining the intracellular sodium concentration at a low level
- maintaining a constant cell volume
- providing a sodium gradient as an energy source for co-transport. The gradient is exploited by many body processes, including the transporters that regulate intracellular pH and those that drive glucose into kidney cells
- generation of membrane potential.

MEMBRANE POTENTIAL

Definition

A membrane potential (E_m) is defined by the difference in electrical charge on each side of a membrane. It is very important in the functioning of excitable cells, especially nerve and muscle cells. These cells use the controlled opening of gated ion channels to cause a change in their membrane potential. There are three major types of gated channels in excitable cells:

1. Voltage-gated channels (e.g. voltage-gated sodium channels used in action potential generation)
2. Chemically-gated channels (e.g. acetylcholine receptor channels in neuromuscular transmission)
3. Mechanical receptors (e.g. touch receptor channels in sensory neurons).

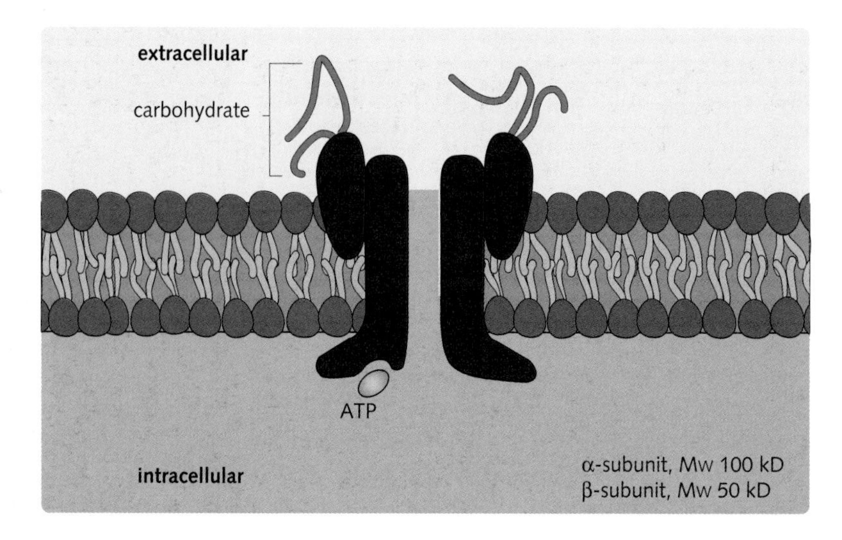

Fig. 3.13 Structure of the sodium pump.

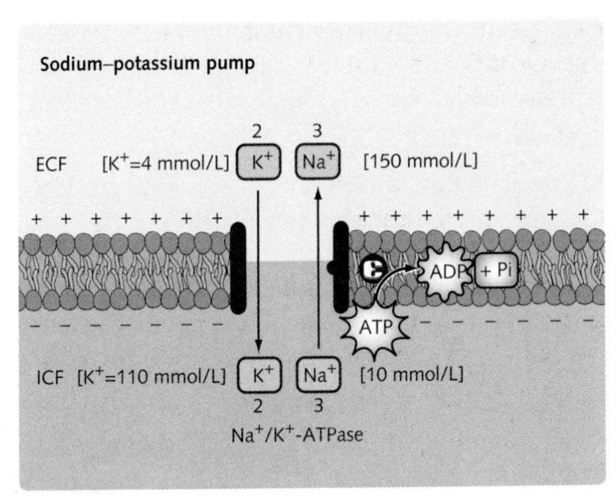

Fig. 3.14 The Na^+/K^+ ATPase. The hydrolysis of one molecule of ATP is associated with the transport of three sodium ions out of the cell and two potassium ions into the cell against their respective concentration gradients. ECF, extracellular fluid; ICF, intracellular fluid. (Adapted from Baynes and Dominiczak, 1999.)

Maintenance of membrane potential

Electrochemical potential difference of ions

Ions have both an electric charge and a chemical concentration. When a solution is not at equilibrium, the movement of its ions is influenced by both these gradients. If these factors operate in different directions across a cell membrane, net ion flow will tend to be down whichever gradient is the steepest. Since the electrochemical potential difference ($\Delta\mu$) of ions is defined as the electrochemical potential of the ion on side A minus that of the ion on side B (Fig. 3.15):

Fig. 3.15

The electrochemical potential difference ($\Delta\mu$) of an ion X^+ across a membrane separating compartments A and B can be calculated as follows:

$$\Delta\mu\,(X^+) = (RT\ln[X^+]_A/[X^+]_B + zF\,(E_A - E_B))$$

(Where $\Delta\mu$ is the electrochemical potential difference between A and B; R is the ideal gas constant; T is absolute temperature; $[X^+]_{A/B}$ is the concentration of X^+ in A and B; z is valency; F is Faraday's number; $E_A - E_B$ is the electric potential difference across the membrane.)

Fig. 3.15 Calculation of the electrochemical potential difference. This equation includes the contributions of both the concentration difference and the electrical potential difference to the tendency for the ion to flow across the membrane.

- If $\Delta\mu$ is positive, ions move from A to B.
- If $\Delta\mu$ is negative, ions move from B to A.
- If $\Delta\mu$ is zero, there is no net movement of ions (the solution system is at equilibrium).

When a potential difference for an ion exists across a membrane there is a tendency for it to move down a chemical or electrical gradient. This potential energy can be harnessed, which is the basis of secondary active transport.

The Nernst equation

When a reaction is at equilibrium there is no net movement of ions across the cell membrane, i.e. the concentration gradient and the electrical gradient are balanced. In this situation, the electrochemical potential difference equation above can be rearranged to give the Nernst equation (Fig. 3.16):

$$E_A - E_B = (60\ \text{mV}/z)\,(\log([X^+]_B/[X^+]_A))$$

When moving down a chemical gradient, an ion is moving from an area of high concentration to low concentration. Ion movement down an electrical gradient is dictated by charge. An ion crossing a membrane down an electrochemical gradient is responding to both chemical and electrical gradients.

When no net movement of X^+ across a membrane occurs it is at equilibrium and the electropotential difference ($\Delta\mu$) for X^+ is zero, therefore:

$$RT\ln\frac{[X^+]_A}{[X^+]_B} + zF(E_A - E_B) = 0$$

Solving for $E_A - E_B$ gives:

$$E_A - E_B = -\frac{RT}{zF}\ln\frac{[X^+]_A}{[X^+]_B} = \frac{RT}{zF}\ln\frac{[X^+]_B}{[X^+]_A}$$

A convenient form of the equation is obtained by converting to a form that involves $\log_{10}$ ($\ln y = 2.303 \log y$). At 29°C the quantity 2.303 RT/F is equal to 60 mV:

$$E_A - E_B = \frac{60\text{mV}}{z}\log\frac{[X^+]_B}{[X^+]_A}$$

Fig. 3.16 Derivation of the Nernst equation. $E_A - E_B$, the electric potential difference across the membrane; R, ideal gas constant; T, absolute temperature; $[X^+]_{A/B}$, concentration of X^+ in A and B; z, valency; F, Faraday's number.

The Nernst equation can be used to calculate:

- the electrical potential difference that must exist between two chambers for an ion to be in equilibrium across the membrane
- the direction an ion will flow when the reaction is not in equilibrium, given an experimentally derived electrical potential difference.

The electrochemical potential difference when an ion is at equilibrium is called the equilibrium potential, e.g. $E_{Cl} = -70mV$.

Gibbs-Donnan equilibrium

The Gibbs-Donnan equilibrium describes the electrochemical equilibrium that develops when two solutions are separated by a membrane that is impermeable to at least one of the ionic species present.

In an experimental system consisting of two compartments containing equimolar solutions of KCl (compartment B) and KY (compartment A), separated by a membrane permeable to K^+ and Cl^-, but impermeable to the Y^- anions, the permeant ions will redistribute as follows (Fig. 3.17A):

- Cl^- moves down its concentration gradient from B to A.
- Electroneutrality is preserved, because K^+ follows, moving from B to A.

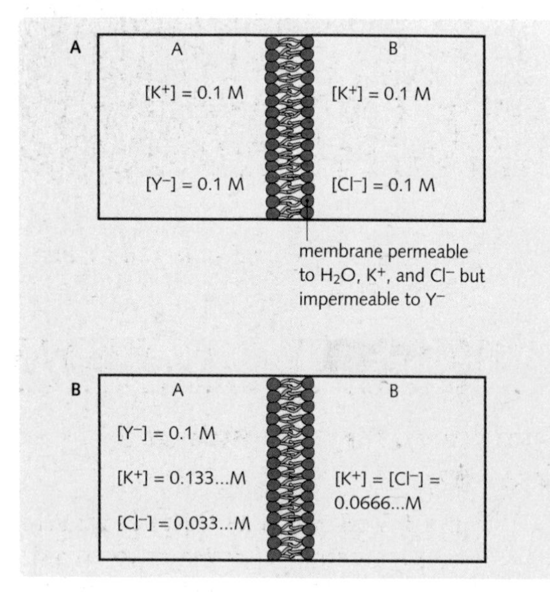

Fig. 3.17 Gibbs-Donnan equilibrium. (A) Ion concentrations at the start of the experiment. (B) Ion concentrations when Gibbs-Donnan equilibrium has been reached. (Adapted from Berne et al., 1998.)

- Y^- cannot diffuse across the membrane, and it is trapped in compartment A.
- Therefore, K^+ remains trapped in compartment A and electroneutrality is preserved.

Electroneutrality is preserved when ions cross the membrane because the movement of Cl^- sets up a local electrical potential that draws K^+ across the membrane. The movement of ions continues until the reaction is at equilibrium (Fig. 3.17B) at which point:

- the tendency for Cl^- to move down the concentration gradient from B to A is offset by its tendency to move down the electrical gradient from A to B
- the tendency for K^+ to move down its concentration gradient from A to B is offset by its tendency to move down the electrical gradient from B to A.

At equilibrium, $\Delta\mu\ K^+$ and $\Delta\mu\ Cl^-$ both equal zero. This is the basis for the derivation of the Gibbs-Donnan equation, which states that the product of the concentrations of both permeant ions is the same in each compartment:

$$[K^+]_A\,[Cl^-]_A = [K^+]_B\,[Cl^-]_B$$

The Gibbs-Donnan equation holds for any univalent anion cation pair in equilibrium between two chambers. A system in Gibbs-Donnan equilibrium has a number of important features at equilibrium:

- The compartment containing the impermeant ion contains more osmotically active ions (Fig. 3.17B).
- The compartment containing the impermeant anion has a negative electropotential. (Note: a compartment containing an impermeant cation would have a positive electropotential.)

Living cells resemble the experimental Gibbs-Donnan equilibrium above in a number of respects:

- They contain impermeant ions in the form of proteins and nucleic acids.
- The membrane is permeable to K^+ and Cl^-, and these ions are abundant.

However, there are significant differences:

- The cell is sensitive to osmotic gradients.
- The cell membrane is not entirely impermeable to positively charged ions, such as Na^+ and Ca^{2+}, and these leak into the cell down an electrochemical gradient.
- If allowed to accumulate, they would exert osmotic pressure and the cell would swell.

- Osmotic effects are avoided by actively transporting such ions out of the cell.

Cell swelling is avoided because the Na^+/ K^+ ATPase pumps three Na^+ ions out of the cell, while only two K^+ ions are pumped back in.

Resting membrane potential

The membrane potential (E_m) of a cell is proportional to the concentration gradient of the dominant ions (Na^+, K^+, and Cl^-) and the membrane permeability to each one. If an ion is freely permeable across the cell membrane, it will tend to force E_m towards its own equilibrium potential. Resting excitable cells are most permeable to K^+, so E_m reflects the balance between K^+ leaking out of the cell down its concentration gradient and being pulled in down the electrical gradient (i.e. E_K). This movement is not accompanied by the extrusion of an anion because the membrane is only permeable to Cl^-, which has an opposing concentration gradient. Thus, E_m is proportional to the concentration of K^+ on each side of the membrane (Fig. 3.18).

Resting E_m (−70mV) is not quite equal to E_K (−90mV) because the membrane is slightly permeable to other ions, such as Na^+. Sodium influx makes the membrane potential slightly more positive because E_{Na} is positive (+60mV), reflecting its tendency to move into the cell down its electrochemical gradient. However, note that Na^+ is not at equilibrium across the resting cell membrane because it is not freely permeable to this ion.

The action of the Na^+/K^+ ATPase, being electrogenic, contributes a small amount to the membrane potential directly. However, the majority of the membrane potential arises from the indirect action of the pump and reflects the movement of K^+ and Na^+ down the concentration gradients that it has established.

The excitable cell may manipulate its resting potential to control activity:

- Depolarization means the E_m becomes less negative, so there is a decrease in the potential difference. A depolarized nerve cell is more likely to fire.
- Hyperpolarization means that the potential difference increases in magnitude, by increasing the relative negative charge inside the cell. A hyperpolarized nerve cell is less likely to fire.

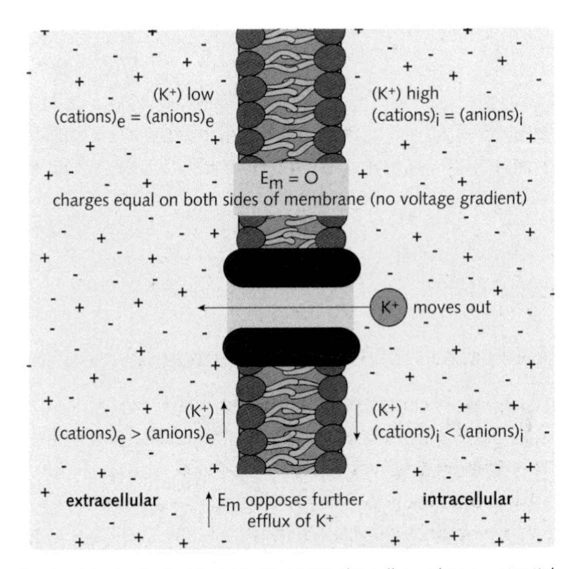

Fig. 3.18 Role of potassium in generating the cell membrane potential. Em, membrane potential; e, extracellular; i, intracellular.

> Membrane potential is highly sensitive to the concentration of K^+. An increase in the extracellular concentration of K^+ (hyperkalaemia) will partially depolarize excitable cells, bringing the resting potential closer to the threshold potential. Moreover, since K^+ efflux after an action potential is inhibited repolarization is impeded. Hypokalaemia (a decrease in the extracellular concentration of K^+) will tend to hyperpolarize cells, making them less excitable. Both conditions affect cardiac cells, causing arrhythmias.

(Action potentials are discussed in detail in *Crash Course, Nervous System.*)

RECEPTORS

Concepts of transmembrane signalling

It is essential that cells in a multicellular organism are able to communicate in order to coordinate their activities. Signal transduction pathways regulate multiple cell activities including division, differentiation, migration and degranulation.

Such processes enable responses to be made to external factors governing cell activity. The pathway begins at cell surface receptors and ends in the nucleus with proteins that regulate gene expression. Since different cell types may respond differently to the same signal at the level of transcription, these processes facilitate the coordination of the whole organism's response to a stimulus.

Only certain lipid-soluble molecules can cross the cell membrane directly (e.g. steroid hormones); other molecules transfer their signal by binding cell surface receptors. The signal transduction pathways are formed by interacting proteins, which can amplify, dampen, or process signals before passing them downstream. Each cell may be confronted with many different signals coming from its cellular neighbours, environment, substratum contact, and the presence of growth factors and hormones. The resulting signal pathways are integrated so that the cellular response is appropriate. When signal pathways malfunction, the cell may multiply uncontrollably, and this may result in malignancy.

Important concepts include:

- Cell surface receptors are specific proteins that selectively bind a signalling molecule and convert this binding into intracellular signals, which alter the cell's behaviour.
- Intracellular receptors are largely ligand-activated transcription factors which, when activated, migrate to the nucleus and bind to DNA, stimulating or suppressing gene transcription.
- The second-messenger system is a set of intracellular molecules that are activated by cell surface receptors and affect cell function, producing a physiological response. Second messenger systems produce a signal cascade that amplifies the initial signal and facilitates a variety of cellular responses, the details of which vary between cell types.

The three mechanisms of cell signalling to surface receptors are (Fig. 3.19):

1. Endocrine
2. Paracrine
3. Autocrine.

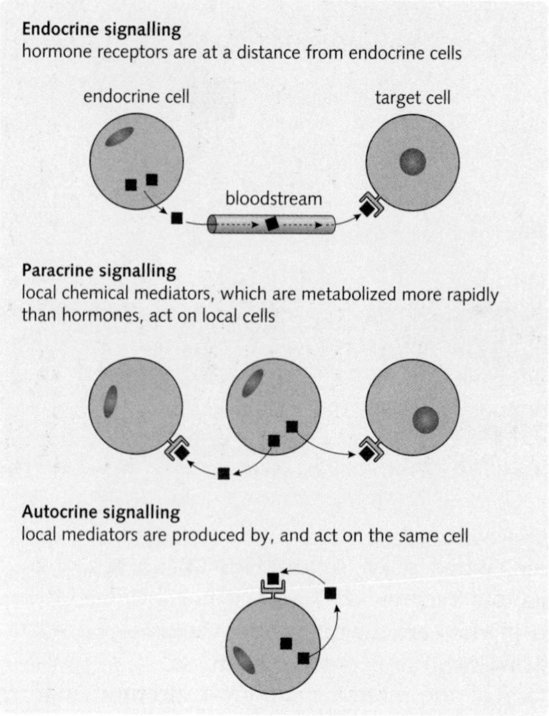

Types of receptor

The presence or absence of a specific receptor on a cell governs the responsiveness of that cell to signalling molecules. The majority of cell-surface receptor proteins belong to one of three main families:

Fig. 3.19 The three mechanisms of cell signalling. A single signalling molecule may fall into more than one of these categories depending on where it is synthesized and released.

1. Ionotropic receptors
2. Metabotropic G protein-coupled receptors
3. Enzyme-linked receptors.

Ionotropic receptors

Ionotropic receptors or ligand-gated ion channels are similar to other ion channels, but contain a ligand-binding receptor site within their structure. They are composed of several subunits that, when activated by ligand binding, directly affect the activity of a cell by opening the ion channel. They are predominantly found in the nervous system and characteristically mediate fast excitatory or inhibitory neurotransmission (e.g. nicotinic acetylcholine receptor; Fig. 3.20).

Metabotropic receptors (G-protein coupled)

Metabotropic receptors are membrane receptors that influence the activity of cells indirectly, with the transduction of an extracellular signal (ligand binding) to an intracellular one and the activation of second messenger molecules. This receptor class is defined by a common structural feature: they consist of a single polypeptide chain that spans the

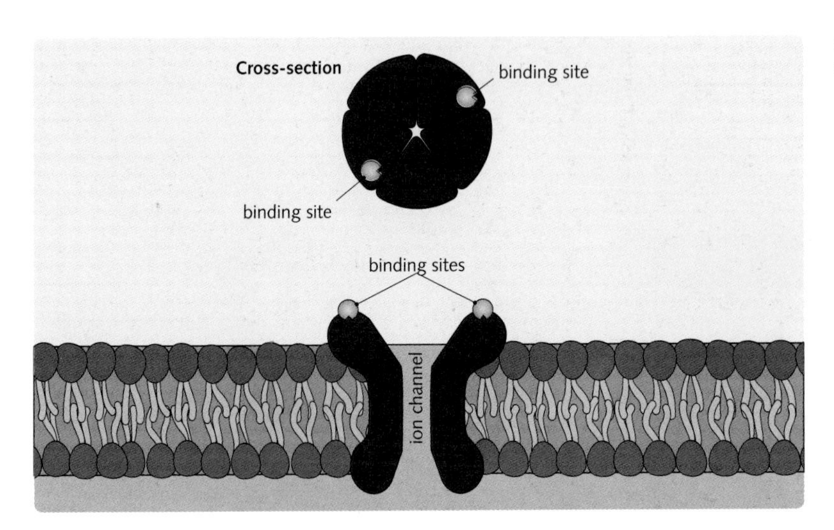

Fig. 3.20 Structure of the nicotinic acetylcholine receptor.

membrane seven times (Fig. 3.21). Metabotropic receptors activate GTP-binding proteins (G-proteins). G-proteins are heterotrimeric, being composed of α, β and γ subunits, and are bound to the cytoplasmic face of the plasma membrane. Receptor binding triggers a conformational change in the G-protein, with the dissociation of the α subunit from the β and γ subunits, ultimately leading to the hydrolysis of GTP. Thus, they function like a binary switch (Fig. 3.22). Activated G-proteins subsequently activate intracellular second-messenger pathways, for example:

- cyclic AMP (cAMP) (Fig. 3.23)
- calcium (directly by opening of calcium channels in the plasma membrane, or indirectly by the inositol lipid pathways)
- inositol lipid (IP_3) pathways (Fig. 3.24).

Different G-proteins activate different pathways, for example:

- G_s increases cAMP
- G_i decreases cAMP
- G_q activates the inositol lipid pathway.

When GTP on the α subunit is hydrolysed to GDP and P_i, it reassociates with the β and γ subunits and the binary switch is turned off. An example of a metabotropic receptor is the β-adrenergic receptor.

The TK receptor is commonly mistaken for a metabotropic receptor as it has a G-protein-linked transduction mechanism. However, it is not a metabotropic receptor as it does not have a seven-pass membrane structure.

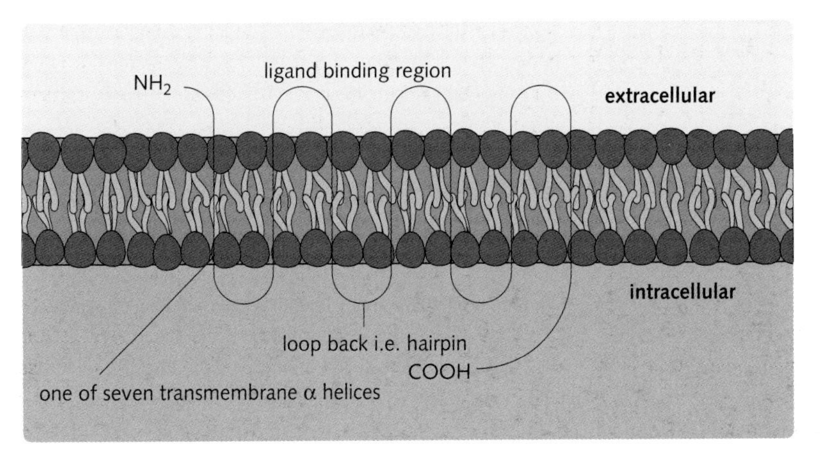

Fig. 3.21 Structure of a metabotropic receptor, e.g. a β-adrenergic receptor.

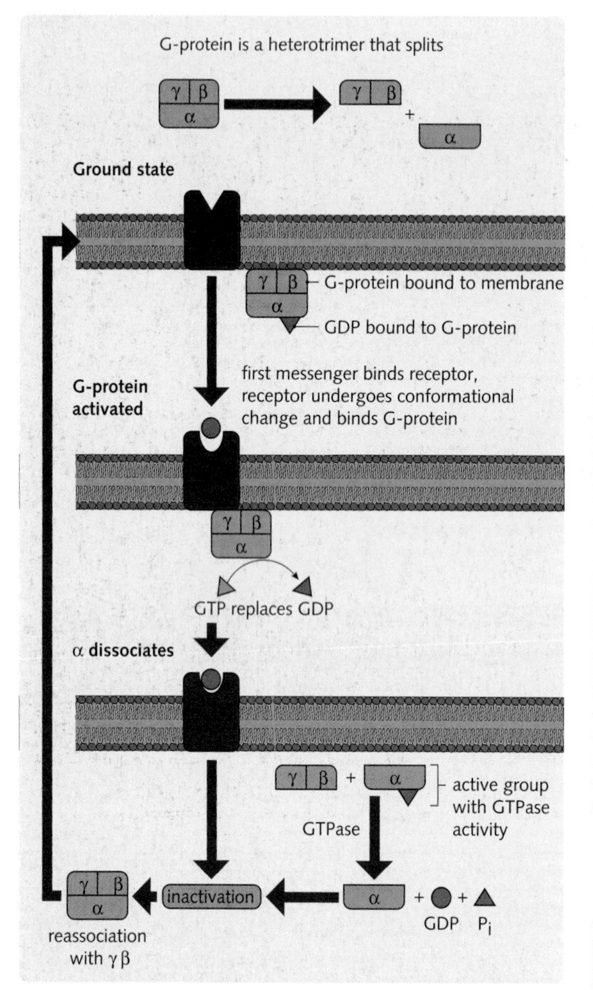

Fig. 3.22 Activation of G-proteins. The G-protein in its inactive state is a heterotrimer that has a GDP bound via its α-subunit. Interaction of the receptor with its ligand drives the exchange of GTP for GDP. This induces a conformational change in the α-subunit that results in its dissociation from both the receptor and the β-γ-subunits. The dissociated subunits are free to interact with effectors that generate secondary messengers. Eventual hydrolysis of GTP by the α-subunit permits the regeneration of the inactive heterotrimer. GDP, guanosine diphosphate; GTP, guanosine triphosphate; P_i, inorganic phosphate.

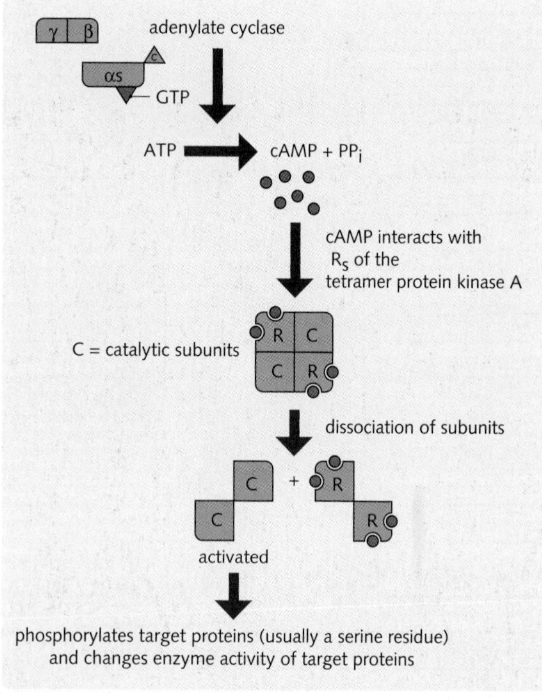

Fig. 3.23 Adenylate cyclase pathway. cAMP, cyclic adenosine monophosphate; ATP, adenosine triphosphate; C, catalytic subunits; GTP, guanosine triphosphate; PP_i, pyrophosphate; R, regulatory subunits.

Laron syndrome is an autosomal recessive disorder characterized by marked short stature. It results from molecular defects of the growth hormone receptor gene (type 1), or post-receptor defects in the signal transduction required to produce insulin-like growth factor-1 (IGF1) (type 2). Ordinarily, the interaction of growth hormone and the growth hormone receptor leads to activation of cytoplasmic tyrosine kinases. As mutations yielding Laron Syndrome involve, or are downstream of, the growth hormone receptor, treatment with growth hormone does not increase the growth rate, and treatment with IGF1 is required.

Enzyme-linked receptors

Enzyme-linked receptors are single-pass transmembrane proteins, with an extracellular and intracellular domain (Fig. 3.25). The intracellular domain either possesses intrinsic enzyme activity or associates directly with an enzyme, and is activated when the appropriate ligand binds to the external portion of the receptor. There are six classes of enzyme-linked receptor, mediating the actions of a large number of ligands, including peptide hormones, growth factors and cytokines (Fig. 3.26):

- receptor tyrosine kinases
- tyrosine-kinase-associated receptors
- receptor-like tyrosine phosphatases
- receptor serine/threonine kinases
- receptor guanylyl cyclases
- histidine-kinase-associated receptors.

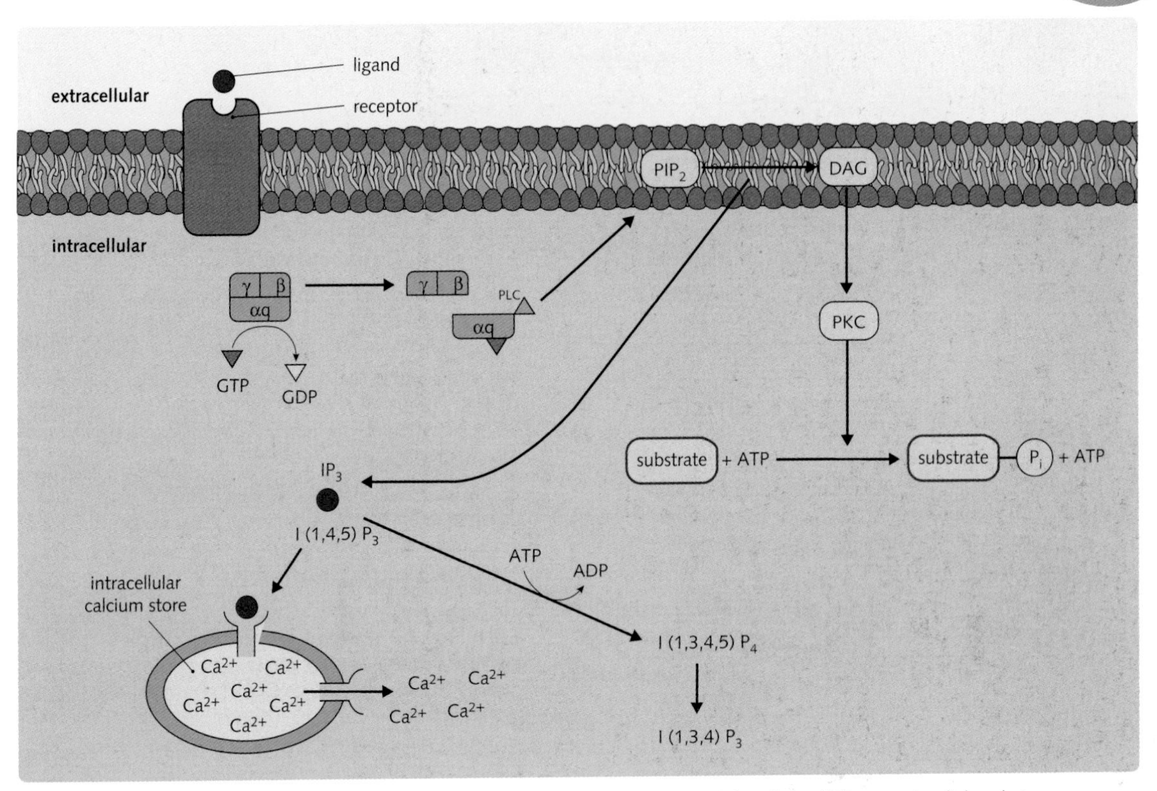

Fig. 3.24 Inositol phospholipid signalling pathway. PLC, phospholipase C; GTP, guanosine triphosphate; ADP, guanosine diphosphate; PIP2, phosphoinositol diphosphate; DAG, diacylglycerol; PKC, phosphokinase C; P_i, inorganic phosphate; IP_3, inositol triphosphate; Ca^{2+}, calcium ions.

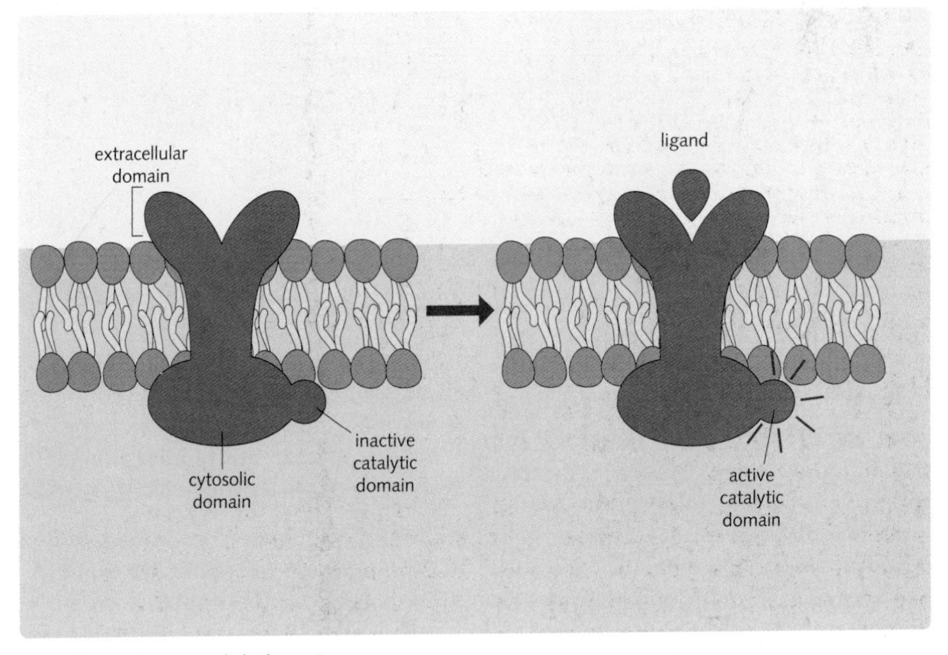

Fig. 3.25 Structure of a generic enzyme-linked receptor.

Fig. 3.26 Enzyme-linked receptors

Class	Mechanism of activation	Examples
Receptor tyrosine kinases.	Activation of the receptor by ligand leads to oligomerization of the receptor and enzymatic activation of an intrinsic kinase resulting in the phosphorylation of tyrosine residues in intracellular signalling proteins.	The binding of epidermal growth factor (EGF) to the epidermal growth factor receptor leads to the proliferation of various cell types. The binding of insulin to the insulin receptor leads to the stimulation of carbohydrate utilization and protein synthesis
Tyrosine-kinase-associated receptors.	These receptors possess no catalytic activity of their own, and associate with cytoplasmic tyrosine kinases. Kinases associated with these receptors either belong to the Src family or the JAK family.	The binding of ligand, such as α interferon, induces the noncovalent association of the two separate cytokine receptor subunits and activation of the associated cytoplasmic kinase. Receptors include cytokine receptors e.g. TNF, antigen receptors e.g. CD4 and CD8, the growth hormone receptor and prolactin receptor.
Receptor-like tyrosine phosphatases.	Activation of the receptor leads to the dephosphorylation of tyrosine residues on cytosolic signalling proteins.	The leucocyte common antigen (CD45), when cross-linked by extracellular antibodies, becomes activated with the removal of phosphotyrosine residues from specific target proteins.
Receptor serine/threonine kinases.	Activation leads to the phosphorylation of serine/threonine residues on intracellular signalling proteins.	Following ligand binding, type I and type II TGFβ receptors form a heterotetrameric complex that activates intracellular second messengers of the Smad family. The Smads then translocate to the nucleus and initiate gene transcription. TGFβ superfamily members have important roles in a wide range of developmental processes including tissue differentiation, morphogenesis, proliferation, and migration. Loss of TGFβ activity has been implicated in tumorigenesis.
Receptor guanylyl cyclases.	Activation catalyses the cytosolic production of cGMP, which activates cGMP dependent kinase (PKG), which phosphorylates specific serine/threonine residues in target proteins.	Atrial natriuretic peptide (ANP) binding to the ANP receptor leads to transcriptional regulation of key genes and to counter the blood pressure-raising effects of the renin–angiotensin system.
Histidine-kinase-associated receptors.	A two stage reaction in which autophosphorylation of a histidine residue yields Pi, which is then transferred to a asp residue on a cytosolic signaling protein. Also known as two-component activation. Such systems are utilized for bacterial chemotaxis and are also seen in yeast and plants, but have not been not identified in man.	The Hog 1 osmoregulation pathway of *S. cerevisiae.*

Fig. 3.26 Features and examples of enzyme-linked receptors.

Intracellular (steroid) receptors

These are not membrane bound receptors, but soluble intracellular proteins (Fig. 3.27). There are two classes based on cellular localization. Class I (classical) steroid receptors are found cytoplasmically and are held in complexes with other proteins, such as heat shock proteins. They include receptors for the sex hormones and the steroid hormones of the adrenal cortex. Class II steroid receptors are located in the cell nucleus, and bind retinoid and thyroid hormones, and vitamin D.

When steroid hormones interact with their receptor, a characteristic series of events occurs:

- The receptor undergoes conformational changes and becomes competent to bind DNA.
- Activated receptors bind to hormone response elements (HREs) located in promoters of hormone responsive genes.

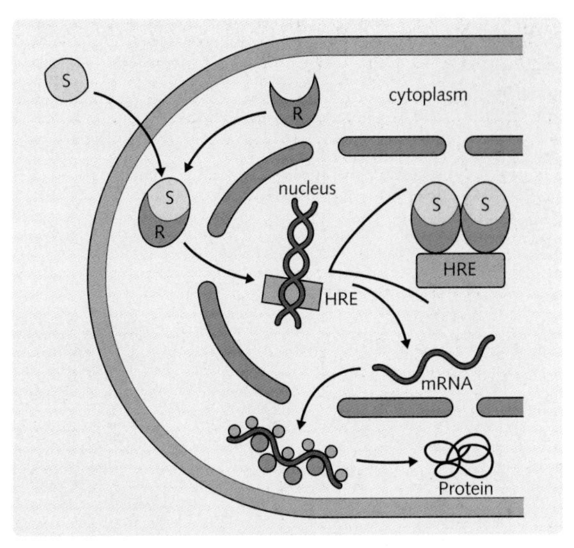

Fig. 3.27 Steroid receptor activations and cellular response. HRE, hormone-responsive element; R, receptor; S, steroid.

- The hormone-receptor complex functions as a transcription factor, either activating or repressing transcription of the associated gene.
- The product of the gene may in turn regulate the transcription of further genes (the secondary response).

As such, members of the steroid hormone receptor family have three distinct domains (Fig 3.28):

1. A ligand-binding domain
2. A DNA-binding domain
3. A transcriptional regulatory domain.

Receptors and drugs

Many drugs produce their pharmacological effect by acting on cell surface receptors. The effect produced depends upon whether the drug acts as an agonist or antagonist:

- Agonists are molecules that activate receptors. They may be pharmacological or physiological agents.

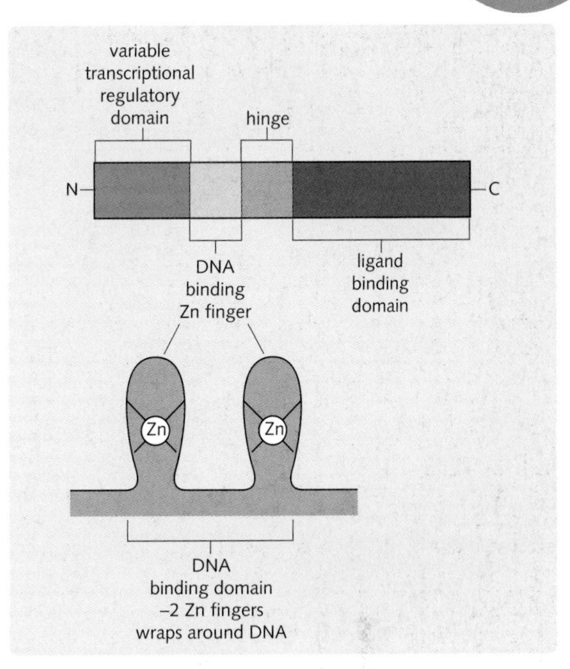

Fig. 3.28 Structure of a steroid receptor. Zn, zinc.

- Antagonists are molecules that bind receptors, but do not activate them. They block the receptor's ligand from binding, so preventing its action.

Reversible antagonists, also called competitive antagonists, compete with the ligand for the receptor. Competitive antagonism can be reduced by increasing the concentration of agonist. Irreversible antagonists cannot be removed from the receptor; thus, they reduce the effective number of receptors, and increasing the agonist concentration has no effect.

The specificity of a drug reflects its ability to combine with one receptor type. The desired action of a drug is to combine with a specific receptor in the targeted tissue. Adverse effects may be caused by non-specific binding to other receptor types, or by binding with the desired receptor, but in a different tissue.

Objectives

By the end of this chapter you should be able to:

- Describe the basic structure of actin, microtubules and intermediate filaments.
- Understand the role of the cytoskeleton in the structure and function of cells and their specializations.
- Understand the structure and functons of lysosomes.
- Describe the basis of endocytosis, pinocytosis and phagocytosis.
- Understand the function of the lysosome and appreciate the different mechanisms leading to lysosomal storage diseases.
- Appreciate the function of the different types of cell–cell and cell–matrix junction.
- Understand the structure and function of adhesion molecules.
- Identify the components of the extracellular matrix and summarize their functions.

CYTOSKELETON AND CELL MOTILITY

Concepts

The cytoskeleton is a dynamic system of structural proteins that support the topography of the cell membrane, organizing the cytoplasmic components into defined areas. Its major functions are:

- determining cell shape
- providing mechanical strength
- organelle anchoring and polarity determination
- motility (and migration)
- anchoring of the cell to external structures
- metabolic functions
- separating duplicated chromatids and homologous chromosomes into separate cells at mitosis and meiosis respectively.

The major components of the cytoskeleton are:

- microfilaments – polymers of globular actin (G-actin)
- microtubules – polymers of α and β tubulin dimers
- intermediate filaments – polymers of a family of fibrous proteins that include the lamins and keratins
- in general, microfilaments have a structural role or are associated with cell movement; microtubules appear to be important in organelle organization and intracellular transport; and intermediate filaments provide the cell with mechanical strength.

Both actin and tubulin subunits have a pair of appropriately orientated, complementary binding sites that allow each subunit to bind to other monomers. In this way, long, helical structures are formed. As each subunit is asymmetrical, the resulting filament is polarized. The filaments are dynamic with one end (the plus end) being capable of rapid growth, and the other (the minus end) tending to lose subunits if not stabilized. Actin and tubulin are highly conserved throughout evolution, and they are found in all eukaryotic cells. However, by interacting with a range of accessory proteins, microfilaments and microtubules are able to perform a variety of distinct functions.

In contrast to actin and tubulin, the subunits of intermediate filaments are symmetrical, with a globular domain at each end. These fibrous monomers wind together to form the rope-like intermediate filaments.

The functions of the cytoskeleton are a common topic for exam questions. For top marks in long answer questions, give some examples and consider the role of actin-associated and microtubule-associated proteins.

Components of the cytoskeleton

Microfilaments

Microfilaments, with a diameter of 6.5 nm, form a layer just beneath the plasma membrane called the cortex. The individual globular G-actin subunits polymerize in a reversible process to form helical F-actin filaments (Fig. 4.1). This polymerization/depolymerization reaction is closely regulated by the cell; for example, extracellular signals may influence polymerization via G-protein coupled cell surface receptors, facilitating cell processes, such as chemotaxis in neutrophils.

The actin protein is encoded by a family of related genes. Mammals have at least 6 actin isoforms, which fit into three classes: α actins, which are generally found in muscle, and β and γ isoforms, which are prominent in non-muscle cells. These isoforms may allow distinct protein interactions in accordance with the differing functions of the muscle and non-muscle forms. Actin has a contractile function in muscle cells. In non-muscle cells, actin:

- maintains structure of microvilli
- is a component of a specialized region of the cell cortex, the terminal web, which lies beneath microvilli and desmosomes
- facilitates movement of macrophages by gel–sol transitions of the actin network (gel phase – where the actin in the cytoskeleton is polymerized, to a sol phase – where it is soluble), mediated by actin-binding proteins
- facilitates movement of fibroblasts and nerve growth cones by controlled polymerization and rearrangement of actin filaments.

Various actin-binding proteins cause changes in the molecular forms of actin, and they can be classified into groups according to their function:

- Severing proteins such as gelsolin will cleave actin filaments in the presence of calcium ions. This property, when required, allows the cell to break up the cell cortex to facilitate processes such as phagocytosis.
- Linking proteins that bind actin strands together. Actin may be bound into tight arrays of parallel strands by 'bundling proteins', such as fimbrin and α-actinin. Alternatively, it may be arranged into a loose gel by 'gel-forming proteins', such as filamin, that bind crosswise intersections between strands.
- Myosin proteins are members of a protein family that move groups of oppositely orientated actin filaments past each other. This is the basis for contraction in muscle cells, but it is also important in non-muscle cells, where a transient assembly of actin and myosin produces the contractile ring that separates the cells in cell division. Other accessory proteins, such as troponin, affect actin and myosin interactions (see p. 56).
- Attachment proteins mediate linking of actin filaments to the plasma membrane – this group includes fodrin, talin and vinculin.

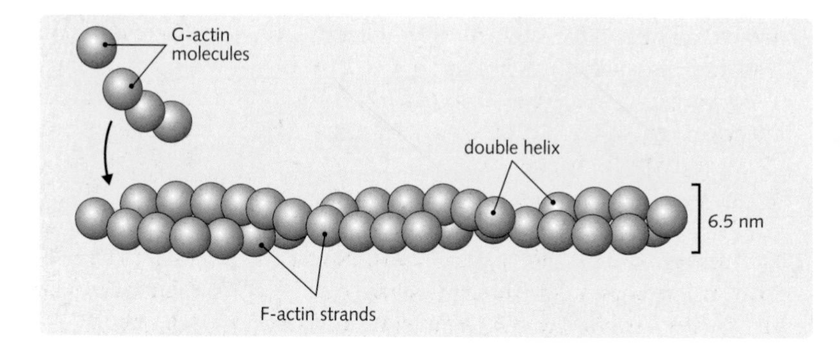

Fig. 4.1 Structure of a microfilament (actin). Actin filaments consist of a tight helix of uniformly orientated actin molecules. The filament is extended as globular actin polymerizes at the plus end. Because of its appearance when complexed with myosin, the minus end is also referred to as the 'pointed end' and the plus end as the 'barbed end'.

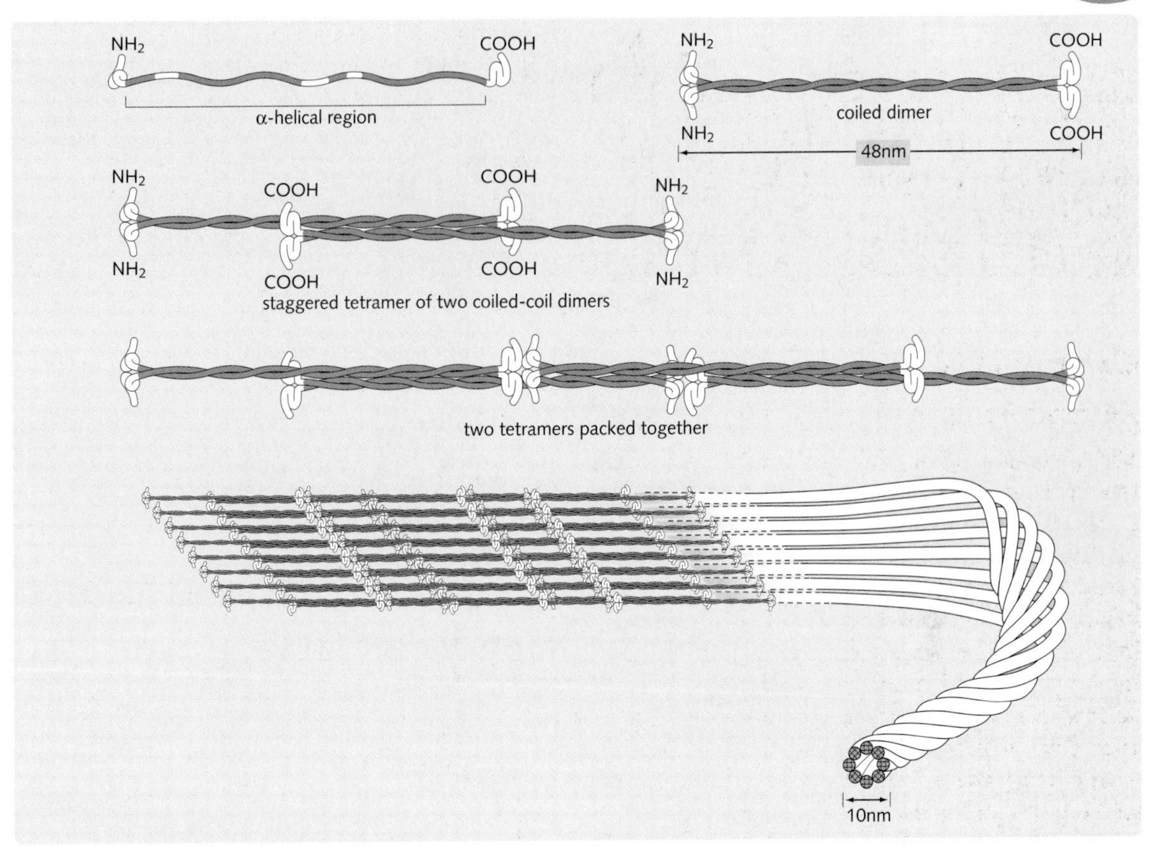

Fig. 4.2 Structure of an intermediate filament. Identical monomers bind in a parallel fashion to form dimers. Two dimers associate in antiparallel arrays to form tetramers, which wind together in groups of eight to produce the final rope-like intermediate filament. Since the association of the dimers in tetramers is antiparallel, intermediate filaments are not polarized. (Adapted from Norman and Lodwick, 1999.)

Intermediate filaments

Intermediate filaments (IFs) are 8–11 nm wide (Fig. 4.2). They are generally more stable than microfilaments and microtubules, and do not dissociate into monomers under physiological conditions. IFs are thought to be the major structural determinants in cells. There are various types of IFs, distinguishable by the protein from which they are made. These include:

- Keratins – there are many isoforms of keratin, which can be divided into soft 'cytokeratins' and hard 'hair keratins'. There are about 10 hard keratins, giving rise to nails and hair, and about 20 cytokeratins found more generally in epithelia lining internal body cavities.
- Lamins – these are found exclusively in the nucleus, which localize to two nuclear areas: the nuclear lamina and the nucleoplasmic veil. The nuclear lamina lines the inner surface of the nuclear envelope and is responsible for the disarrangement and reassembly of the nuclear envelope into vesicles during mitosis or meiosis.
- Neurofilaments – these are found in neuron axons. They are responsible for the radial growth of an axon and thus determine axonal diameter. They may account for the strength and rigidity of the axon.
- Glial fibrillary acidic protein (GFAP) – this is found in glial cells surrounding neurons.
- Vimentin – this is expressed in mesenchymal cells, such as fibroblasts, and in endothelial cells. These fibres often end at the nuclear membrane and desmosomes. They are closely associated with microtubules, and they form cages around lipid droplets in adipose tissue.
- Desmin – this is found predominantly in muscle cells. It forms an interconnecting network perpendicular to the long axis of the cell. Desmin fibres anchor and orientate the Z bands in myofibrils, thus generating the striated pattern.

Cells usually contain only one type of intermediate filament. Rapidly growing cells and myelin-producing glial cells do not have intermediate filaments.

Microtubules

Microtubules are hollow tubules and are 25 nm wide (Fig. 4.3). They are polymers of tubulin dimers (α-β dimer), and they extend from microtubule organizing centres, such as centrosomes, which stabilize the negative pole of the extending polymer. With the exception of mature erythrocytes, all cells have microtubules. They are particularly abundant in neurons, where they direct axon elongation.

There are many microtubule-associated proteins (MAPs), which have specific interactions with tubulin, and different microtubules associate with different MAPs, e.g. Tau in the nerve axon:

- Some MAPs function as ATP-dependent molecular motors (e.g. dynein and kinesin). Such motors may carry a cargo, such as an organelle or transport vesicle, along the microtubule to its designated location in the cell.
- Some MAPs influence the polymerization of tubulin, e.g. centrioles.

Microtubules form cilia in the respiratory tract and the flagella of spermatozoa, both of which move via cycles of ATP-powered dynein arm linkage.

The formation of microtubule spindles is essential for cell division. Nocodazole, taxol and vinblastine are antimitotic cancer chemotherapy drugs that interfere with the exchange of tubulin subunits between the microtubules and the free tubulin pool.

Myosin

Myosin is an actin accessory protein that functions as a molecular motor. It is composed of two heavy chains and four light chains (Fig. 4.4). The two essential light chains have ATPase action, while the two regulatory light chains determine the binding of calmodulin to myosin. Actin and myosin interact to produce contraction, which is regulated by:

- troponin in skeletal muscle
- calmodulin in non-muscle cells.

There are several isoforms of myosin, and muscle and non-muscle forms have slightly divergent amino acid sequences. (see *Crash Course, Musculoskeletal System* for further details.)

Examples of cytoskeletal function

Erythrocyte cytoskeleton

Erythrocytes have a very rigid, but malleable shape. The erythrocytic cytoskeleton is atypical, being present in only a thin strip below the cell membrane. The cell shape is indirectly maintained by spectrin, which directly links actin to ankyrin and band 4.1, which are, in turn, bound to integral transmembrane proteins (Fig. 4.5).

Hereditary spherocytosis is a disorder of the red blood cell membrane, ultimately resulting in haemolytic anaemia. The key defects are cytoskeletal, most commonly spectrin deficiency. This causes membrane instability, the loss of erythrocyte surface area and abnormal cellular permeability to sodium, resulting in the production of rigid, spherical cells. These cells are fragile and susceptible to spontaneous haemolysis. They have a reduced lifespan in the circulation, as they are generally unable to pass through the splenic microcirculation.

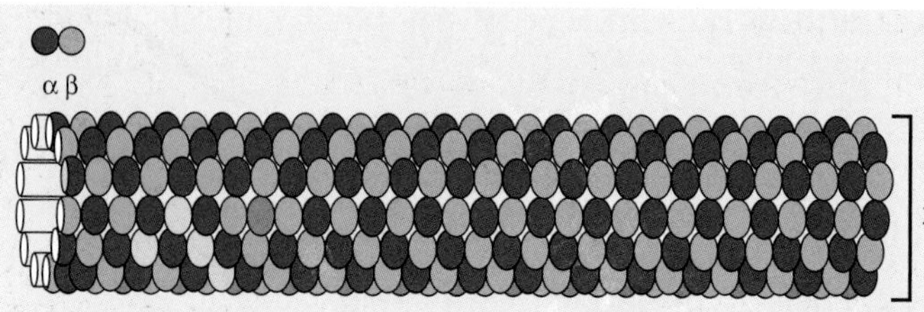

Fig. 4.3 Structure of a microtubule. There are normally 12–13 tubulin units per turn in the assembled microtubule. (Adapted with permission from *Molecular Biology of the Cell*, 3rd edn, by B Alberts et al, Garland Publishing, 1994. Reproduced by permission of Routledge, Inc., part of The Taylor & Francis Group.)

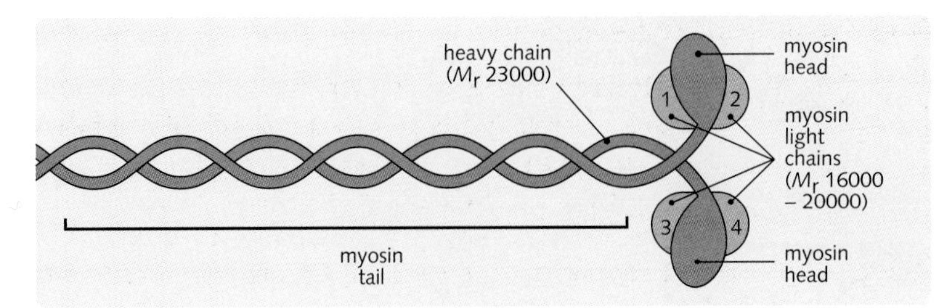

Fig. 4.4 Structure of myosin. Myosin II, which is the form found in muscle cells, is composed of two heavy and four light chains. The α-helices of the two heavy chains wrap around one another to produce a dimer. M_r, relative molecular mass. (Adapted from Stevens and Lowe, 1997.)

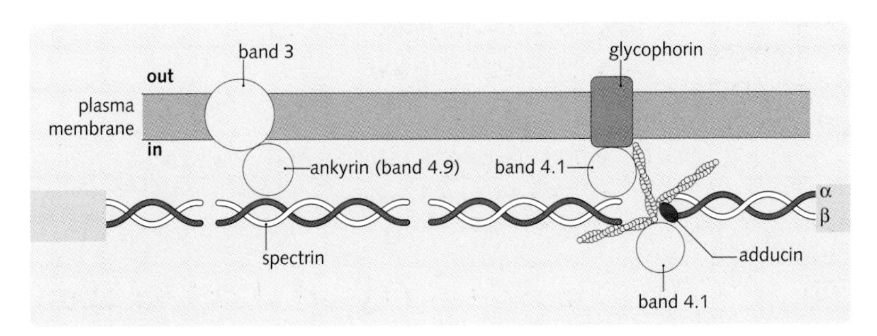

Fig. 4.5 The erythrocyte spectrin-based cytoskeleton. Spectrin is a dimer consisting of antiparallel α- and β-subunits. It is linked to the anion exchange protein (band 3) by ankyrin and to glycophorin by band 4.1, which also binds to actin and adducin (protein band numbers relate to migration in SDS-PAGE electrophoresis). (Adapted from Norman and Lodwick, 1999.)

Cilia

Cilia are formed of a 9+2 arrangement of microtubules with a basal body. Dynein arms connect microtubule pairs, and the sliding mechanism enables the cilia to bend (see p. 29).

Intestinal epithelium

Absorption is increased by microvillous projections, which increase intestinal surface area (Fig. 4.6).

Axonal transport

Kinesin and dynein transport materials along axons, each moving in a different direction. Organelle movement away from the cell body is driven by kinesin, which moves towards the plus end of the microtubule. Conversely, movement towards the cell body is driven by cytoplasmic dynein, which moves towards the minus end of the microtubule. Transport is normally at a rate of 25 mm/day (Fig. 4.7). Vesicles containing newly synthesized neurotransmitters are transmitted to the cell terminal by this means.

Muscle contraction

In skeletal muscle, the arrangement of parallel actin and myosin into sarcomeres allows maximum efficiency of contraction. In smooth muscle, the contractile subunits resemble sarcomeres, but they are not as organized. (See *Crash Course: Muscles, Bone and Skin* for further details.)

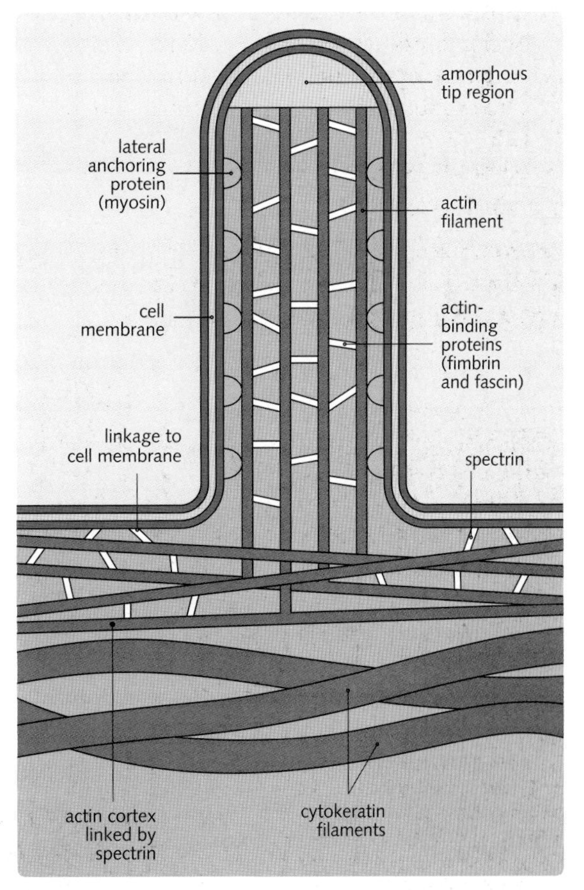

Fig. 4.6 Structure of a microvillus. A helical arrangement of myosin molecules binds the actin bundle to the inner surface of the cell membrane. (Adapted from Stevens and Lowe, 1997.)

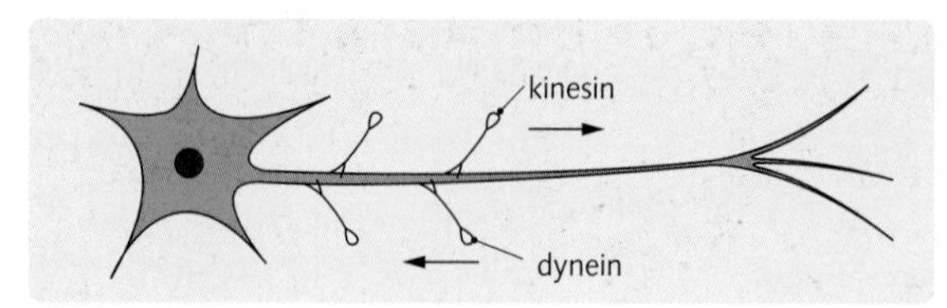

Fig. 4.7 Axoplasmic flow. Kinesin and dynein transport materials along axons.

Motility of phagocytes

Phagocyte motility is achieved by the projection of foot-like pseudopodia, which are associated with actin gel–sol transition at the tip, allowing the pseudopodia to advance.

Mitotic spindle

The spindle is a polar arrangement of microtubules across the equator of the cell. Chromosomes attach to the spindle via a kinetochore protein, at their centromeres. Separation of chromatids occurs as the microtubules contract, pulling them to separate poles (see Ch. 6).

LYSOSOMES

Definition

A lysosome is a membrane-bound organelle that contains acid hydrolases capable of breaking down macromolecules. Confinement of such enzymes in this organelle protects the rest of the cell from their potentially damaging effects.

Lysosomes have:

- diameters ranging from 50 nm to 1 μm
- a single membrane consisting of a phospholipid bilayer that undergoes selective fusion with other membranous organelles
- an ATP driven H^+ pump in the membrane, which acidifies the lysosomal matrix to pH 4.5–5.5, thus activating the hydrolases
- hydrolases in the inner matrix that are active at acid pH and break down carbohydrates, lipids, and proteins.

New lysosomes are derived from the Golgi complex and are called primary lysosomes. Secondary lysosomes are formed from the fusion of a lysosome with a vesicle containing substrate (Fig. 4.8). Most cells have hundreds of lysosomes, with phagocytic cells containing thousands. However, erythrocytes do not contain any lysosomes.

Functions of lysosomes

Lysosome functions are (Fig. 4.8):

- autophagy – digestion of material of intracellular origin (i.e. fuses with vacuoles from inside the cell)
- heterophagy – digestion of material of extracellular origin (i.e. fuses with vacuoles from outside the cell – pinocytic, endocytic, or phagocytic)
- biosynthesis – recycling the products of receptor-mediated endocytosis, which includes the receptor, its ligand and associated membrane.

Endocytosis is uptake of material into the cell, and it can be specific (receptor-mediated endocytosis) or non-specific (pinocytosis). Pinocytosis results in uptake of extracellular molecules at their extracellular concentrations. Phagocytosis is the internalization of membrane-bound particulate molecules by engulfment. It only occurs in specialized cells. The endocytic vesicles that result from endocytosis fuse with primary lysosomes.

The products of enzymatic digestion are transported across the lysosomal membrane by specific receptors. Digestion is facilitated by lysosomal enzymes, of which over 60 exist, including:

- nucleases (e.g. acid RNase, acid DNase)
- glycosidases (e.g. β-glucuronidase, hyaluronidase)
- carbohydrate degradation enzymes (e.g. β-galactosidase, α-glucosidase)
- proteases (e.g. cathepsins, collagenase)
- phosphatases (e.g. acid phosphatase)
- sulphatases (e.g. aryl sulphatase)
- lipases.

Following synthesis in the rough endoplasmic reticulum (RER), lysosomal enzymes are modified by glycosylation in the RER lumen followed by covalent modification in the Golgi apparatus. Covalent modification includes phosphorylation of mannose groups to produce mannose-6-phosphate groups, which act as recognition markers and direct the enzymes specifically to primary lysosomes.

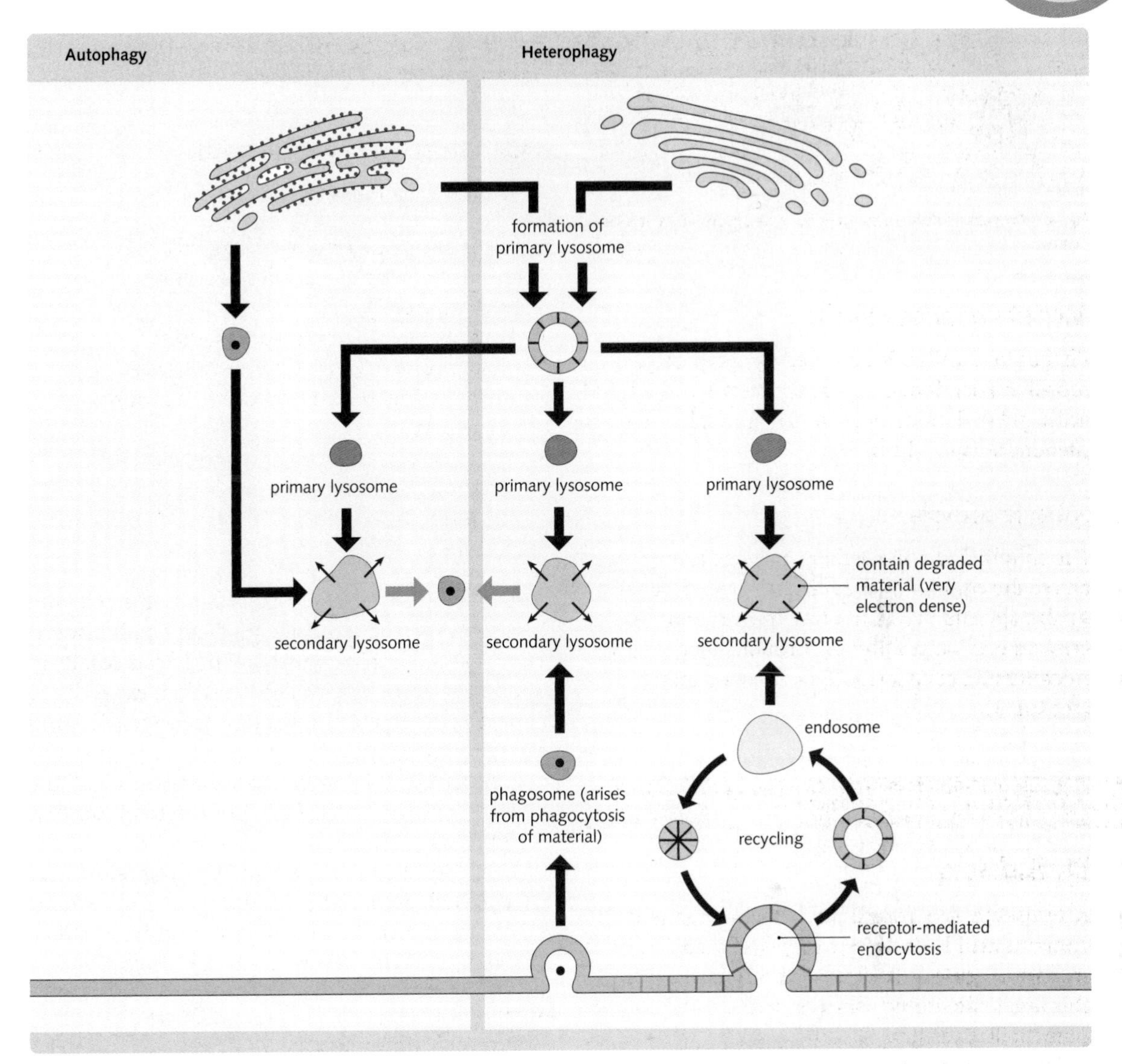

Fig. 4.8 Multiple pathways of exocytosis, endocytosis and membrane recycling. The lysosome is common to all these pathways.

Receptor-mediated endocytosis

Receptor-mediated endocytosis occurs when ligands that bind specific surface receptors are internalized in clathrin-coated pits (Fig. 4.9). In general, the ligand is degraded in the lysosome and its receptor is recycled to the cell surface. A variety of receptors and their ligands undergo receptor-mediated endocytosis (Fig. 4.10).

Lysosomal storage diseases

Lysosomal storage diseases (LSDs) are disorders of lysosomal function that result in macromolecules becoming trapped inside the lysosome. As such macromolecules build up, the lysosomes expand and the tissue enlarges, resulting in cellular dysfunction and pathological features. Such disorders may manifest as a result of:

- enzymatic processes – resulting from deficient or defective acid hydrolases, or absence of a crucial activator
- non-enzymatic processes – caused by transporter defects.

There are some 40 LSDs, each of which is rare, but together they affect 1 in 4800 live births. They are commonly fatal, but can be diagnosed prenatally. All are single gene disorders and, with three exceptions, show autosomal recessive inheritance.

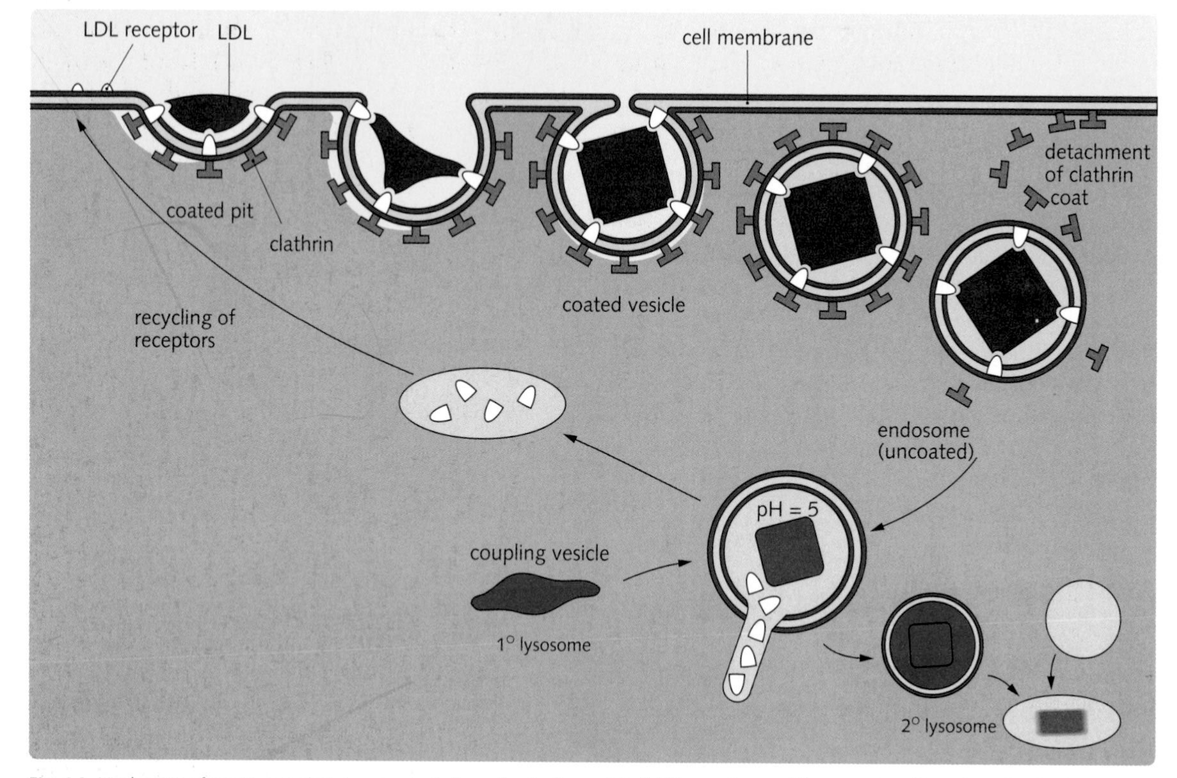

Fig. 4.9 Mechanism of receptor-mediated endocytosis. Low-density lipoprotein (LDL) receptors and their associated ligands are localized in clathrin coated pits and are subsequently internalized in clathrin coated vesicles. The coats are rapidly shed and uncoated vesicles fuse with endosomes. The LDL ligands dissociate from their receptors in the acid environment of the endosome and eventually end up in lysosomes. Meanwhile, the receptors are sequestered in a part of the endosome that is recycled back to the plasma membrane for reuse. (Adapted from Stevens and Lowe, 1997.)

Fabry's and Hunter's diseases are X-linked recessive disorders, while Danon's disease shows X-linked dominant inheritance. Features leading to suspicion of a lysosomal storage disorder are:

- progressive neurological degeneration
- hepato(spleno)megaly
- skeletal dysplasia with or without short stature
- coarse facies

Fig. 4.10 Functions of receptor-mediated endocytosis

Molecules taken up	Function
Low-density lipoprotein (LDL)	Transports TAGs and cholesterol
Transferrin	Transports iron
Insulin	Affects cell metabolism
Fibrin	Removes injurious agents

Fig. 4.10 Functions of receptor-mediated endocytosis.

- eye changes (e.g. cherry red spot, corneal clouding)
- angiokeratoma.

Gaucher's disease

Gaucher's disease is the most common lysosomal storage disorder (incidence 1 in 25 000 live births), with a high incidence seen in Ashkenazi Jews, who have a carrier frequency of 1 in 60. A deficiency of β-glucosidase, which is encoded on chromosome 1q21, results in the accumulation of its substrate – glucocerebroside, principally in the phagocytic cells of the body, but also sometimes in the central nervous system. There are three types of Gaucher's disease, defined by age of onset and brain involvement. Type 1 is the most common:

- Type I – adult type, non-neuronopathic. Lifespan is shortened, but not markedly.
- Type II – severe infantile, rare, neurological signs seen at 3 months, death usually by 2 years of age.

- Type III – 'juvenile' subacute, neuronopathic, variable presentation from childhood to 70 years of age.

Although the exact course of the disease cannot be predicted based on the genotype, key mutations in the β-glucosidase gene are associated with specific types of Gaucher's, and can be useful in making clinical decisions. The substitution mutation N370S has a strong concordance with type I Gaucher's disease, while the homoallelic L444P mutation correlates with severe, neuropathic forms, and the heteroallelic L444P mutation correlates with milder forms of the disease, typically type III, although can also be found in type II.

Tay-Sachs disease

Tay-Sachs results from a deficiency in the hexosaminidase A α-chain, encoded on chromosome 15q22–25, leading to an accumulation of ganglioside GM2. Again, there is an increased incidence in Ashkenazi Jews (at 1 in 25, the carrier frequency is ten times higher than the general population). Affected children appear normal at birth. Symptoms, including a decreased muscle tone, paralysis and blindness, typically manifest between 4 and 8 months of age, and are invariably fatal by 3–4 years of age. Carrier screening is available, using an enzyme assay of the hexosaminidase system. At time of writing there is no effective treatment, and supportive care is all that can be offered.

CELLULAR INTERACTION AND ADHESION

If groups of cells are to combine together to form organ structures, each cell has to be able not only to be held in its proper place, but also to communicate with its neighbours. Interactions between cells, and with the extracellular matrix (ECM), not only carry out a structural role, but may also facilitate cell–cell communication in several biological processes including migration, growth, immunological functioning, permeability, cell recognition, tissue repair, differentiation and embryogenesis.

Generally, cells bind via specific adhesion molecules to the ECM, providing elasticity and resistance to mechanical forces (i.e. as seen in connective tissue). Epithelium has little ECM (only the basement membrane), so cell–cell interactions are adapted to bear tensile and compressive stresses, and show several types of cell–cell junctions.

Cell–cell junctions

Junctions are found between cells, and between cells and the ECM. There are three groups of cell junction (Fig. 4.11), which comprise six types (Fig. 4.12). A junctional complex consists of a tight junction, an adhering junction and a desmosome.

Tight (occluding) junctions

All epithelia act as selectively permeable barriers, with tight junctions blocking diffusion of membrane proteins between apical (top) and basolateral (sides at the bottom) domains of the plasma membrane and sealing neighbouring cells together so that water-soluble molecules cannot leak between cells (Fig. 4.13). Cell–cell contact at these junctions is mediated by the proteins occludin and claudins. The ability to restrict ion passage increases logarithmically with the number of occludin strands (e.g. small intestine tight junctions are 10 000 times more leaky than bladder tight junctions are). The degree of permeability offered by tight junctions is under physiological control and it is influenced by intracellular signals.

Fig. 4.11 Types of cell junction

Group	Members
Occluding junctions	Tight junctions
Anchoring junctions	Actin filament attachment sites: (adherens junctions) cell–cell (e.g. adhesion belts) cell–matrix (e.g. focal contacts) Intermediate filament attachment sites: cell–cell (e.g. desmosomes) cell–matrix (e.g. hemidesmosomes)
Communicating junctions	Gap junctions Chemical synapses

Fig. 4.11 Types of cell junction.

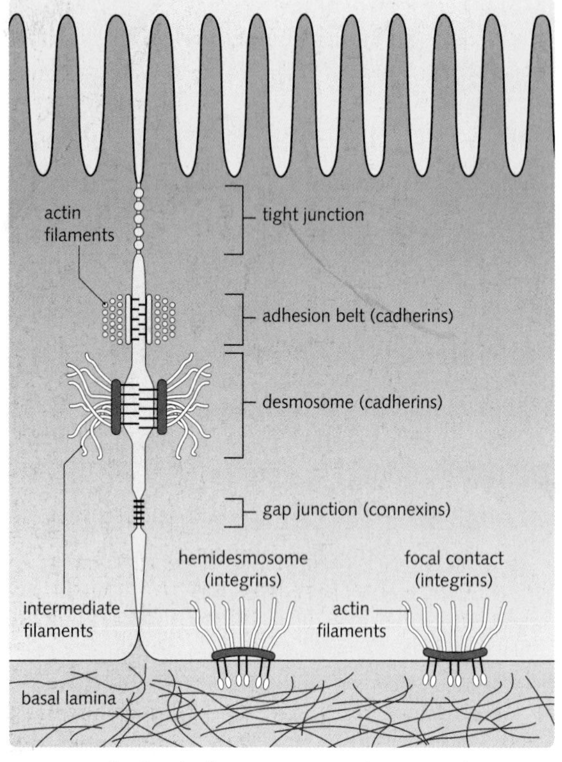

Fig. 4.12 Cell–cell and cell–matrix junctions. There are six distinct types of junctions in epithelial tissue. (Adapted from Norman and Lodwick, 1999.)

Anchoring junctions

Anchoring junctions are responsible for maintaining tissue integrity, and are most abundant in cells under stress (e.g. cardiac muscle). They link the cytoskeletons of adjoining cells to each other or to the ECM, and are made up of:

- intracellular attachment proteins
- transmembrane linker glycoproteins
- ECM or transmembrane linker glycoproteins on another cell (Fig. 4.14).

Anchoring junctions containing actin filament connections are called adherens junctions. They occur as streak-like attachments in non-epithelial cells and as continuous belts just below tight junctions in epithelial cells. These junctions attach the cytoskeletons (actin cell cortex) of adjacent cells together.

Desmosomes act as anchoring sites for intermediate filaments and, thus, provide tensile strength (Fig. 4.15). Cell–cell contact is mediated by desmogleins, which are a type of cadherin (see pp. 65–66). Hemidesmosomes have a similar structure to desmosomes, but they link cellular intermediate filaments to the ECM (basement membrane) via integrin protein attachments (see Fig. 4.12).

Communicating (gap) junctions

In contrast to occluding and anchoring junctions, communicating junctions do not seal membranes

Fig. 4.13 Structure of a tight junction. The tight junction forms a continuous band around the cell and it is, therefore, also called zonula occludens. The integral membrane protein occludin mediates cell–cell interaction. Each junction is made up of multiple pairs of this protein, one of each pair coming from each cell. (Adapted from Stevens and Lowe, 1997.)

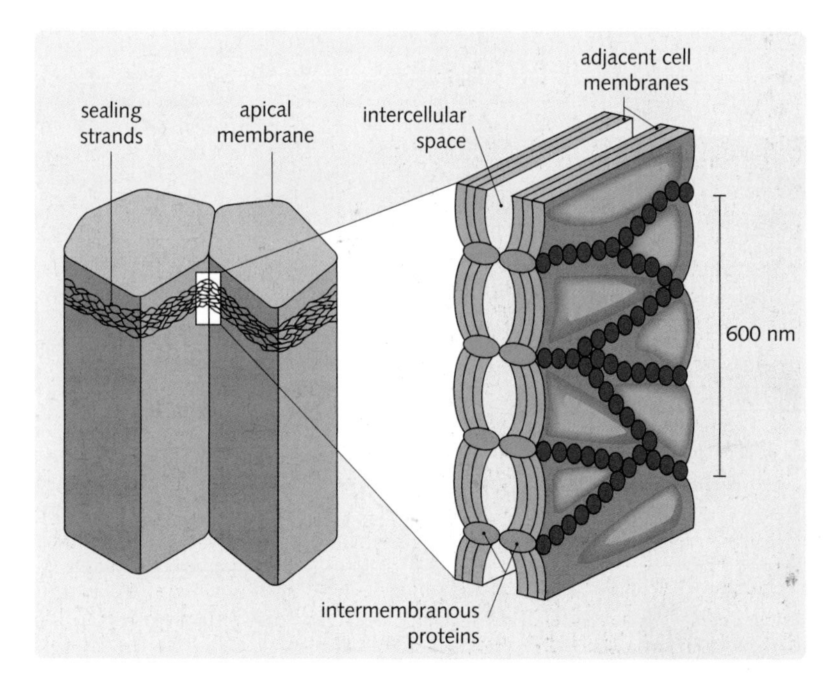

The development of autoantibodies directed against the desmosomal proteins desmoglein-1 and desmoglein-3 leads to the pemphigus family of 'immunobullous' diseases. Autoimmune attack of these proteins leads to the separation of keratinocytes from each other (acantholysis), which float freely in the resultant blister. Autoantibodies to desmoglein-3 lead to pemphigus vulgaris which, if poorly controlled, leads to secondary infections, disturbance of fluid and electrolyte balance, and can be fatal. Autoantibodies to desmoglein-1 results in the more superficial, and usually benign, pemphigus foliaceus.

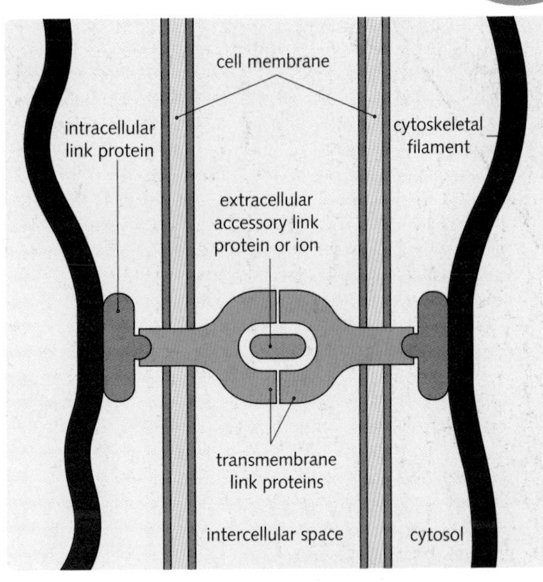

Fig. 4.14 General structure of an anchoring junction. Different (or multiple) link proteins and transmembrane proteins operate for the different classes of junction. (Adapted from Stevens and Lowe, 1997.)

together, nor do they restrict the passage of material between membranes. Gap junctions allow cells in a tissue to respond as an integrated unit. Inorganic ions carrying current and water-soluble molecules are able to pass directly from one cell to another through these structures, permitting electrical and metabolic cell coupling.

One gap junction is composed of two connexons (or hemi-channels), which connect across the intercellular space (Fig 4.16). Each connexon is formed from six connexins, and each connexin consists of four α helices, various combinations of which combine to form gap junctions with different properties. Molecules of up to 1000 Da can pass through the pore, which

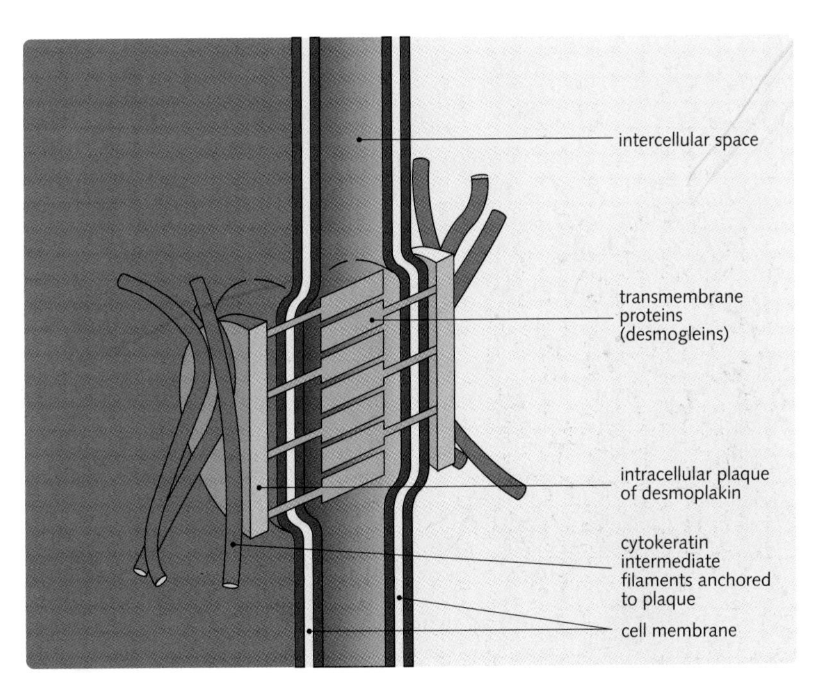

Fig. 4.15 Structure of a desmosome. On the cytoplasmic surface of each interacting cell is a dense plaque composed of desmoplakin that is associated with attached intermediate filaments on one side and desmoglein (a type of cadherin) on the other. Cell–cell interaction is mediated by homophilic binding between adjacent desmoglein proteins. (Adapted from Stevens and Lowe, 1997.)

is typically 1.5–2 nm in diameter. Some pores are gated, with opening related to a three-dimensional change, which is often mediated via extracellular signals. Several thousand connexons form a gap junction. Electrical coupling via gap junctions is important in:

- peristalsis
- synchrony of heart contractions
- coordination of ciliated epithelium.

Gap junctions also play a role in embryogenesis by allowing gradients of morphogens to form across blocks of cells.

Adhesion molecules

Adhesion molecules are cell surface ligands, usually glycoproteins, which mediate cell–cell adhesion. They have demonstrated roles in embryonic development, homeostasis, immune responses and malignant transformation.

There are four major cell adhesion molecule families (Fig. 4.17):

- cadherins
- immunoglobulin superfamily
- selectins
- integrins.

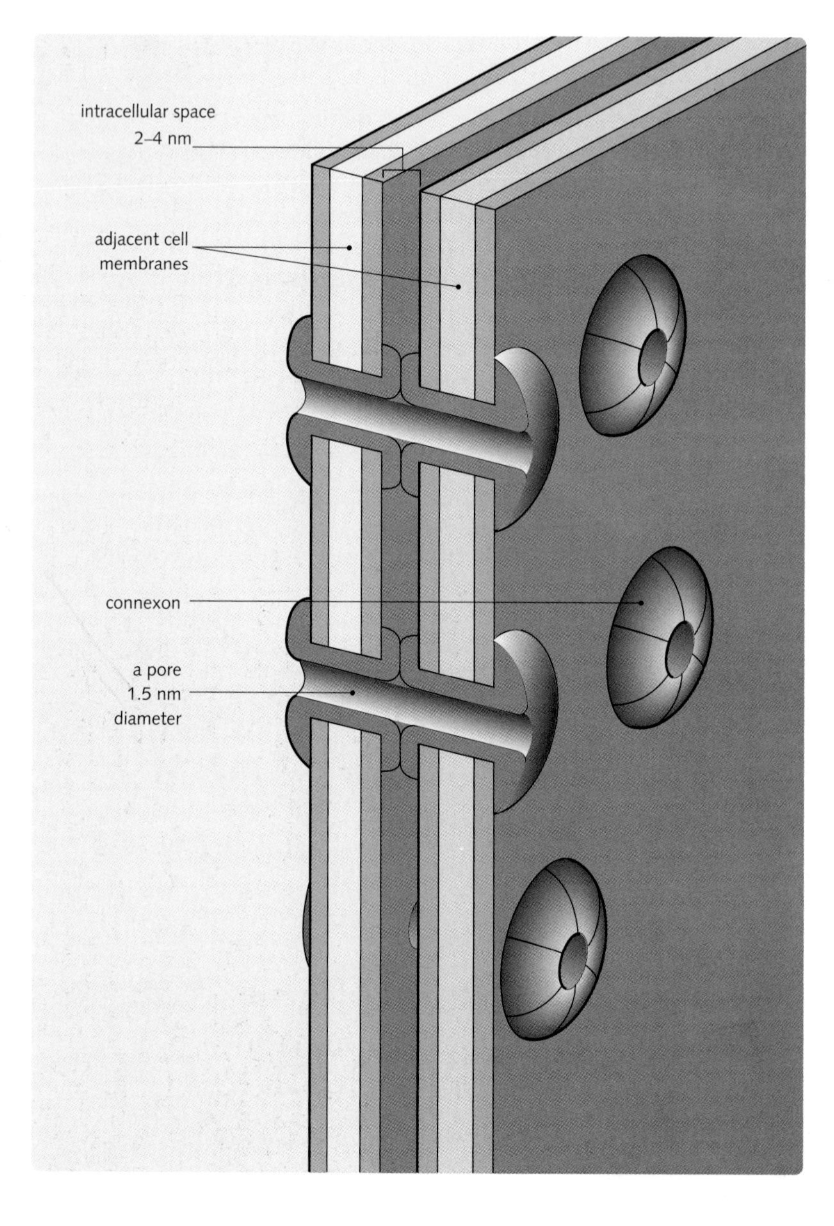

Fig. 4.16 Structure of part of a gap junction. The junction consists of several hundred pores, which are aligned on adjacent cells. Each pore is composed of two connexons, one from each cell, which join across the intercellular gap to form a continuous aqueous channel. This channel facilitates electrical and chemical cellular coupling, since electrical currents and second messengers can pass freely through it. The cell can regulate permeability of gap junctions. (Adapted from Stevens and Lowe, 1997.)

Fig. 4.17 Families of adhesion molecules

Family	Members	Ca^{2+}/Mg^{2+} dependent	Cytoskeletal association	Associated cell function
Cadherins	E-CAD, N-CAD, P-CAD, desmosomal CAD	Yes	Actin filaments	Adhesion belt, desmosomes
Immunoglobulin (Ig) family	N-CAM, V-CAM, L1	Yes	Intermediate filaments (some members)	–
Selectins (blood and endothelial cells only)	P-selectin, E-selectin	No	–	Cell homing
Integrins	LFA-1 (β_2), MAC-1 (β_2)	Yes	Actin filaments, intermediate filaments	Focal contacts, hemidesmosome

Fig. 4.17 Families of adhesion molecules. The characteristics of the four main families of adhesion molecules.

Cadherins

The cadherins are single-pass glycoproteins that mediate communication and adhesion (Fig. 4.18). Cadherins are generally involved in calcium-dependent homophilic interactions (i.e. the protein is both the ligand and the receptor). The N-terminal sequences, which contain a conserved HAV (histidine, alanine and valine) motif, have been shown to be important in ligand binding and specificity. Cadherins are attached to the actin cytoskeleton by a class of linker proteins called the catenins.

Since cadherins are calcium dependent, changing the extracellular Ca^{2+} concentration alters their interactions. The three most widely expressed cadherins are:

- E-CAD (CDH1) – found in the epithelium and early nervous tissue

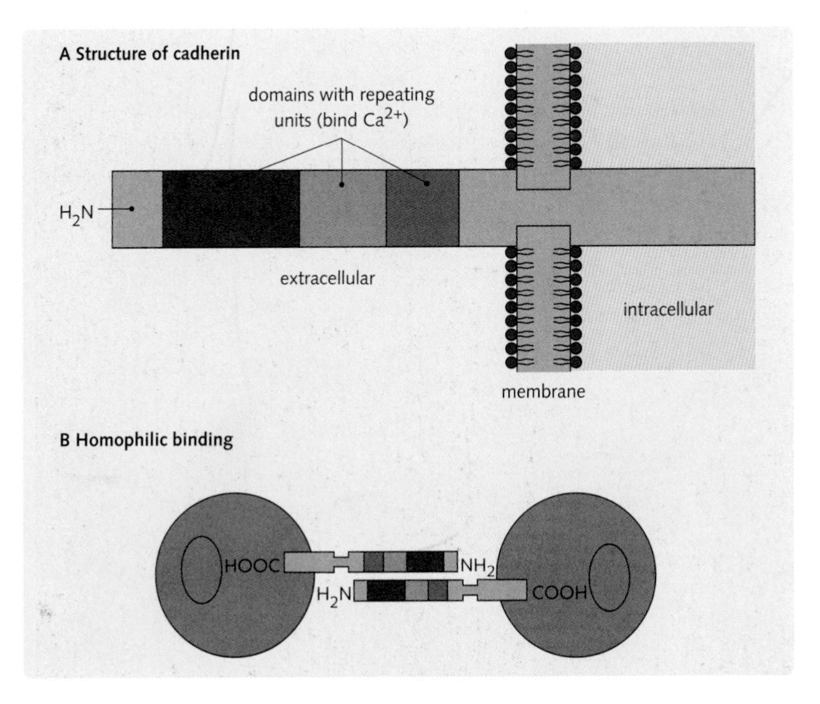

Fig. 4.18 (A) Structure of cadherin. It is composed of five extracellular domains, each 700–750 amino acid residues in length, and one intracellular domain. The intracellular portion is not present in T-CAD (T, truncated). (B) Cadherin exhibits homophilic binding, in which the molecule acts as both ligand and receptor.

- P-CAD (CDH3) – found in the trophoblast (placenta) and epithelium
- N-CAD (CDH2) – found in nervous tissue and skeletal muscle.

The ability of tumour cells for uncontrolled growth, migration, invasion and metastasis is often associated with disruption of cell–cell and cell–extracellular matrix junctions. The loss of cadherins can drive tumour invasion and malignancy by allowing easy disaggregation of cells. Reduced cell-surface expression of E-cadherin has been noted in many types of cancers, and arises either from somatic mutation in the E-cadherin gene, or as a secondary effect of mutations in other genes. Germline E-cadherin mutations yield a predisposition to gastric cancer.

Immunoglobulin (Ig) superfamily

These adhesion molecules are characterized by:

- antibody fold in each domain
- β-barrel structure
- two β-pleated sheets joined by cysteine–cysteine disulphide bonds, which are 60–80 residues apart
- loop regions without β-structure (variable expressed regions).

There is Ca^{2+} independent adhesion, and the Ig proteins have homophilic and heterophilic (protein is either a ligand or a receptor) binding sites. Ig members are involved in:

- adhesion
- signal transduction
- axonal growth and fasciculation (fasciculation means that axons grow along other axons by homophilic binding).

Important Ig family members include: neural cell adhesion molecule (NCAM), vascular adhesion molecule (VCAM), platelet endothelial (PECAM) and intercellular adhesion molecule (ICAM).

Selectins

Selectins are Ca^{2+}-dependent cells and undergo heterophilic binding to carbohydrate ligands (lectins) (Fig. 4.19), initiating leukocyte–endothelial interactions. There are three members of the selectin familiy:

1. L-selectin, which is expressed on leukocytes
2. E-selectin, which is expressed on activated endothelial cells
3. P-selectin, which is expressed on activated platelets and endothelial cells.

Expression of selectins is induced by local chemical mediators:

- E-selectin is activated by tumour necrosis factor (TNF), interleukin-1 (IL-1) and endotoxin.
- P-selectin is activated rapidly by histamine, thrombin, platelet activating factor and phorbol esters, and more slowly by TNF-α and IL-1.

The lectin domain recognizes specific oligosaccharides on the surface of neutrophils: the oligosaccharides Lewis X and sialyated Lewis X are recognized by P-selectin and E-selectin, respectively. These weak affinity interactions allow leukocytes to stick to the endothelial lining of blood vessels until integrins are activated. L-selectin is constitutive on the surface of polymorphonuclear neutrophils, monocytes and lymphocytes, facilitating the homing of these cells to lymph nodes and subendothelial capillaries.

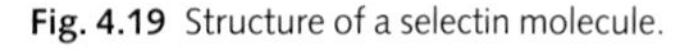

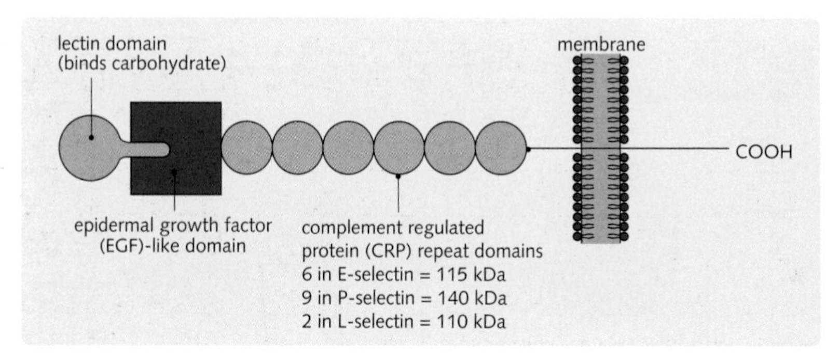

Fig. 4.19 Structure of a selectin molecule.

Do not be confused by the use of old nomenclature by books or tutors referring to the selectins. P-selectin is also called GMP-140, E-selectin is also called E-LAM.

Integrins

Integrins are integral plasma membrane proteins. They are the major receptors for binding to the ECM, and also have a role in signal transduction from the ECM to the cell. They consist of non-covalently associated heteroduplexes of α- and β-glycoproteins (Fig. 4.20). 18 α- and 8 β-glycoproteins have been identified, giving rise to at least 24 α–β pairs. Mammalian integrins form several subfamilies sharing common β subunits that associate with different α subunits. The integrins differ from other receptors in that they bind their ligand with low affinity, and they are present at high concentration. Interactions are heterophilic and Ca^{2+} or Mg^{2+} dependent, depending on the integrin (Fig. 4.21). β-2 and β-1 are important integrins.

β-1 integrins

These are found on most cells, forming dimers with at least 12 different α subunits. The β-1 subunit is denoted as CD29, and β-1 integrins are often referred to as VLA (very late acting) molecules, as early *in vitro* experiments showed that they were expressed on T lymphocytes 2 to 4 weeks after stimulation. The most important is VLA4 (α-4-β-1), which binds VCAM-1 (a vascular adhesion molecule) and is important in homing lymphocytes to the endothelium at sites of inflammation.

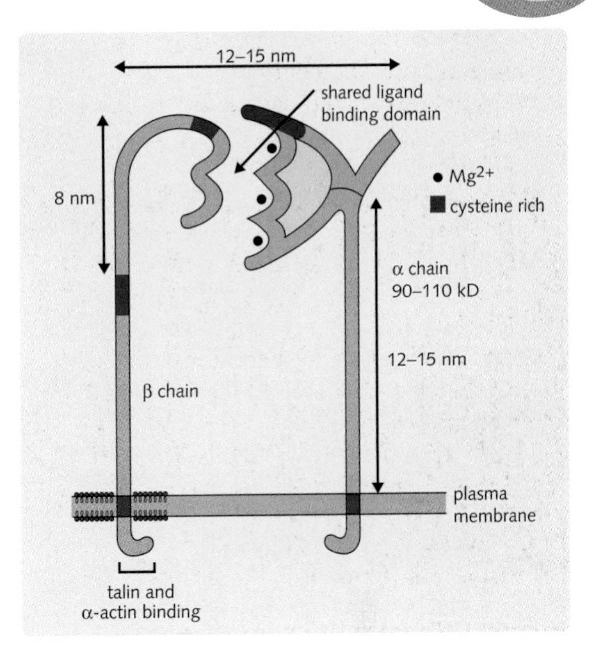

Fig. 4.20 Structure of integrin.

β-2 integrins

These are exclusively expressed on leukocytes and, thus, are also referred to as the leukocyte function-associated antigen-1 (LFA1) family. They form dimers with at least four different types of α subunit. LFA1 is also known as α-L-β-2. It mediates direct cell–cell interactions by binding intercellular adhesion molecules 1, 2 and 3 (ICAM-1, ICAM-2 and ICAM-3). MAC1 (macrophage antigen 1), exclusive to granulocytes and monocytes, is also a β-2 integrin and binds ICAM-1. Surface antigen ('cluster of differentiation') nomenclature is also used (e.g. CD18 is β-2).

Fig. 4.21 Integrins and transmembrane proteoglycans

Family	Members	Ca^{2+}/Mg^{2+} dependent	Cytoskeletal association	Associated cell function
Integrins	Many	Yes	Actin filaments	Focal
Transmembrane proteoglycans	$\alpha_6\beta_4$	Yes	Intermediate	Hemidesmosomes
	Syndecans	No	Actin	None

Fig. 4.21 Integrins and transmembrane proteoglycans.

Failure to express the β-2 integrin (CD18) on leukocytes leads to the condition type 1 leukocyte adhesion deficiency (LAD1). As well as recognizing the ICAM molecules, the β-2 integrin is the receptor for complement (iC3b), and without it not only can leukocytes not migrate from the blood vessels to sites of infection, but they can not recognize or bind complement either. This leads to an immunosuppressive phenotype and recurrent infection. Severe forms of the disease are associated with high mortality; patients typically succumb to bacterial infection within the first year of life.

Integrin binding

Integrins usually bind actin-based cytoskeleton inside the cell. Outside the cell, integrins can bind:

- ECM (e.g. fibronectin)
- cell-surface molecules (e.g. ICAM-1)
- soluble molecules (e.g. fibrinogen).

Integrins have recognition sites for ligands, e.g. Ig-like domains bind ICAMs and the tripeptide RGD (arginine, glycine, aspartic acid), which is a sequence commonly expressed in fibronectin and other ECM proteins. Cells can vary their binding properties by varying integrin affinities and specificities.

Integrins take part in signal transduction in cells (e.g. clustering of β-1-α-5 by fibronectin causes cytoplasmic alkalization).

Basement membrane

The basement membrane, comprising the basal lamina and the reticular lamina, is a sheet of ECM underlying epithelial and endothelial cells and surrounding adipocytes, Schwann cells and muscle cells. It acts to isolate these cells from the mesenchyme or connective tissue. It is composed of type IV collagen, heparan sulphate, proteoglycans, entactin and laminin. Functions of basement membrane are:

- cell adhesion
- to act as a porous filter in the kidney's glomeruli
- to inhibit the spread of neoplasia
- to regulate cell migration
- growth and wound healing
- differentiation.

Extracellular matrix

ECM is a hydrated polysaccharide gel containing a meshwork of glycoproteins. It is composed of:

- proteoglycans
- structural proteins – collagen, elastin
- fibrous adhesive proteins – laminin, fibronectin, tenascin.

ECM components are secreted by local fibroblasts in most tissues, but chondroblasts and osteoblasts are also involved in cartilage. ECM influences cell division, development, differentiation, migration, metabolism and shape. Connective tissues rely on the properties of the local ECM, which is:

- calcified in bone and teeth
- rope-like in tendon
- transparent in the cornea.

Proteoglycans

These form the hydrated polysaccharide gel that acts as a ground substance and allows diffusion of substances, such as nutrients and hormones, from the blood to the tissue and vice versa. Glycosaminoglycans (GAGs) are unbranched polysaccharide chains of repeating disaccharide units (Fig. 4.22). Proteoglycans are formed in the Golgi apparatus where:

- the core protein is linked via a serine to a tetrasaccharide
- glycosyl transferases add sugar residues
- ordered sulphation and epimerization reactions (conversion of some amino acids from their natural L-isomer to the D-Isomer) occur.

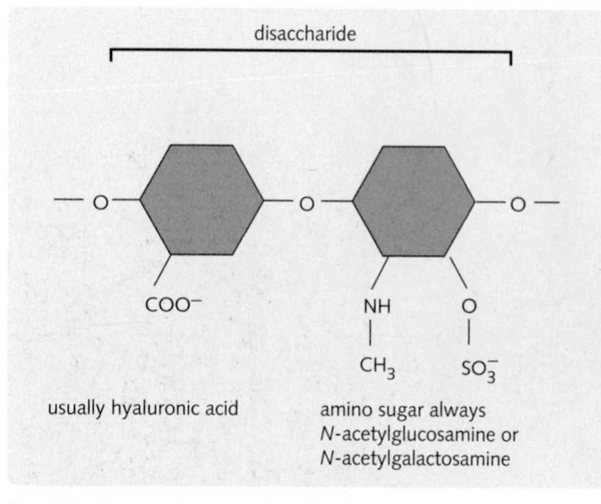

Fig. 4.22 Glycosaminoglycan disaccharide subunit.

The main types of GAGs are:

- hyaluronic acid – found as a lubricant in synovial fluid
- chondroitin sulphate – in cartilage
- dermatan sulphate
- heparin sulphate – an anticoagulant
- heparin
- keratin sulphate – in skin.

Chondroitin sulphate, dermatan sulphate and heparin sulphate are all between 500 and 50 000 Da.

GAGs are very hydrophilic and they have an extended coil structure, which takes up extensive space. GAGs have a negative charge and attract cations, such as the osmotically active Na^+, bringing water into the matrix giving turgor pressure able to withstand forces of many hundreds of times atmospheric pressure.

Proteoglycans function to:

- provide hydrated space
- bind secreted signalling molecules
- act as sieves to regulate molecular trafficking.

Proteoglycans and glycoproteins are compared in Fig. 4.23.

Fig. 4.23 Comparison of proteoglycans and glycoproteins

Proteoglycans	Glycoproteins
Up to 95% carbohydrate	1–60% carbohydrate by weight
Unbranched carbohydrate	Branched carbohydrate
80 sugar residues	13 sugar residues
Larger than 3×10^5 Da	No larger than 3×10^5 Da

Fig. 4.23 Comparison of proteoglycans and glycoproteins.

Collagen

This fibrous protein has great tensile strength, and it is resistant to stretching (Fig. 4.24). It comprises 25% of the protein in mammals, and it is rich in proline (ring structure) and glycine (the smallest amino acid and occurring at almost every third residue so allowing the strands to fit together). Collagen synthesis is carried out in the ER and Golgi, as described in Chapter 2.

To date, over 20 different types of collagen have been described, encoded by a multigene family and divided into two main types: fibrillar and nonfibrillar. Types I, II, and III are fibrillar collagen and they are found in connective tissue (Fig. 4.25).

Fibrils are collagen aggregations of 10–300 nm in diameter and aggregate to form collagen fibres of a few millimetres in diameter. Organization is tissue specific; for example:

- 'wickerwork' pattern in skin resists multidirectional stress
- parallel layers in bone and cornea.

Type IV collagen is nonfibrillar, forming a sheet-like meshwork, and it is only found in the basal lamina (Fig. 4.26).

Osteogenesis imperfecta (brittle bone disease) is a heterogeneous group of conditions characterized by spontaneous fractures, bone deformity and defective dentition. The substitution of a larger amino acid for glycine disrupts collagen triple helix formation and causes a severe, dominantly inherited form of the disease.

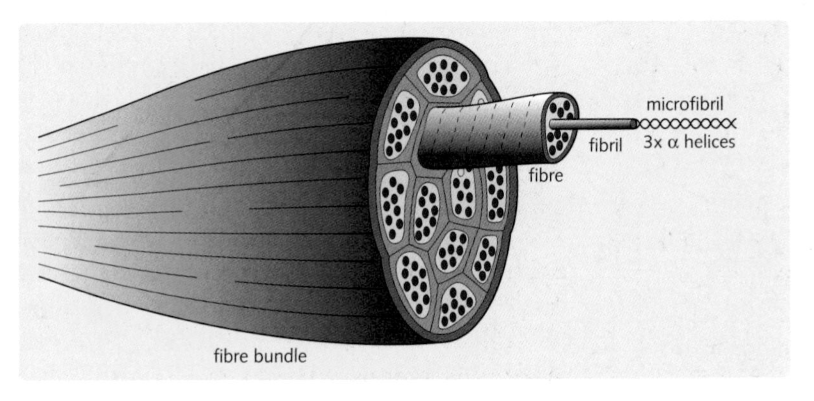

Fig. 4.24 Structure of collagen. Each individual collagen molecule is a left-handed helix. Three collagen molecules twist together to yield a microfibril. Microfibrils bundle together to give fibrils, which, in turn, bundle to give fibres.

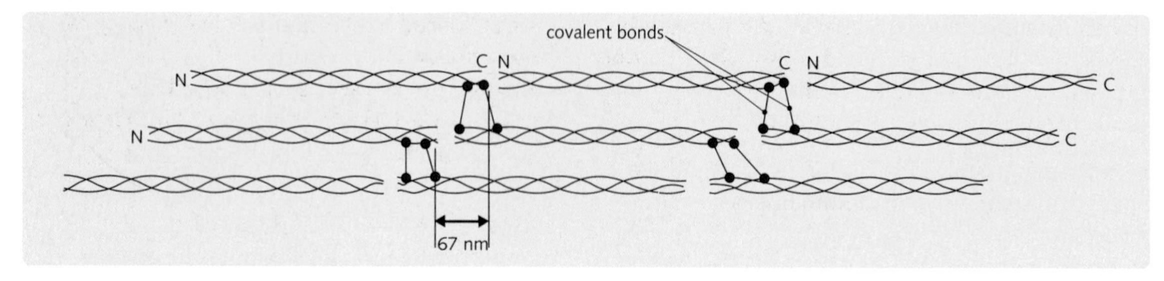

Fig. 4.25 Structure of fibrillar collagen. Collagen molecules are positioned side by side, staggered from adjacent molecules by one-quarter of their length. (Adapted with permission from *Molecular Cell Biology*, 2nd edn, by Darnell, Lodish and Baltimore, Scientific American Books, 1990.)

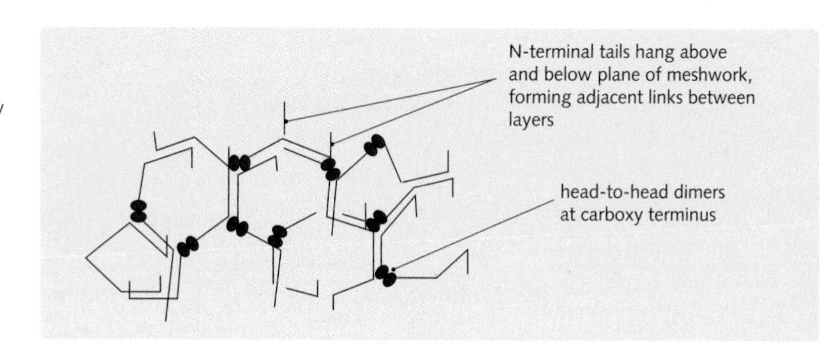

Fig. 4.26 Structure of type IV collagen, which assembles into multilayered sheets. (Adapted with permission from *Molecular Biology of the Cell*, 3rd edn, by B Alberts et al, Garland Publishing, 1994. Reproduced by permission of Routledge, Inc., part of The Taylor & Francis Group.)

Elastin

Elastin is found in places that need flexibility (e.g. skin, blood vessels and lungs). It is a highly glycosylated, hydrophobic protein (Fig. 4.27) that is rich in the non-hydroxylated forms of proline and glycine. One in seven amino acids is valine. The sheets are organized with the help of a microfibrillar glycoprotein, fibrillin, which is secreted before elastin. Fibrillin deficiency results in Marfan syndrome.

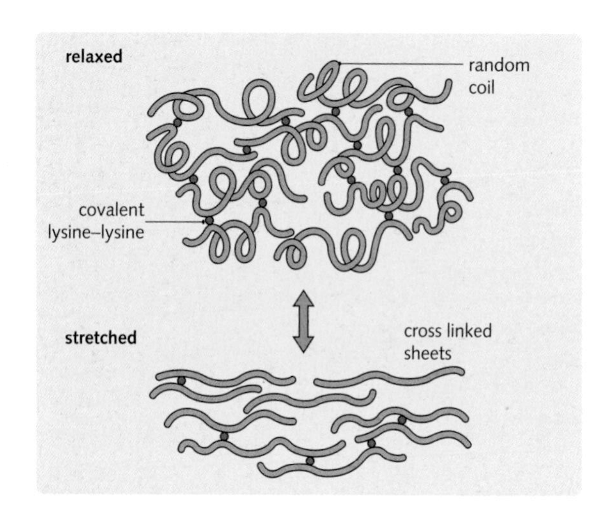

Fig. 4.27 Structure of stretched and relaxed elastin. (Adapted from Stevens and Lowe, 1997.)

Laminin

Laminin anchors cell surfaces to the basement membrane. It is a large heterotrimeric molecule, formed from an α-, β- and γ-chain, held together in a 'cross shape' by disulphide bonds.

Fibronectin

This is an adhesive glycoprotein with binding sites for cells and matrix. It is a dimer of two subunits held together by disulfide bonds, which are folded into globular domains. Forms of fibronectin involved in wound healing and embryogenesis appear to promote cell proliferation and migration.

Tenascin

This protein, which can promote or inhibit cell adhesion and migration, is only produced by embryonic tissue and glial cells.

Role of the fibroblast

Fibroblasts are members of the connective tissue cell family. Members of the connective tissue family are all of common origin and interchangeable under appropriate conditions, other members being chondrocytes, osteocytes, adipocytes and smooth muscle cells. Connective tissue differentiation is controlled by cytokines, especially hormones and growth factors. Interchangeability allows them to support and repair most tissue types. Fibroblasts:

- secrete the fibrous proteins of the ECM in most tissues (except in cartilage and bone where they are produced by chondrocytes and osteocytes, respectively)
- are involved in the organization of ECM, enabling the configuration of ECM into tendons and other structures.

Macromolecules

5

Objectives

By the end of this chapter you should be able to:

- Explain the difference between an essential, semi-essential and a non-essential amino acid, and give named examples.
- Describe characteristics of amino-acid side chains and understand how they influence protein structure.
- Define what is meant by primary, secondary, tertiary and quaternary protein structure.
- Define the following: catalyst, enzyme, substrate, coenzyme and isoenzyme.
- Describe the interaction between the enzyme active site and its substrate.
- Define K_m and V_{max}.
- Describe a typical Michaelis–Menten graph and Lineweaver–Burk graph.
- Define the terms monosaccharide, disaccharide, polysaccharide, homopolysaccharide and heteropolysaccharide, and be able to give named examples of each.

AMINO ACIDS

Amino acids are the subunits of proteins, and they all have the same basic structure (Fig. 5.1):

- a central carbon atom (the α carbon)
- an amino (NH_2) group at the α carbon
- a carboxyl group (COOH)
- a side group (R).

There are 20 naturally occurring amino acids, which differ in their side group, the simplest being hydrogen (H) in the amino acid glycine. All amino acids, except glycine, have an asymmetrical α-carbon atom, giving rise to D or L stereoisomer forms; however, only the L form is found in humans (Fig. 5.2). Amino acids form proteins by joining together through peptide bonds that result from the condensation of the amino group of one amino acid with the carboxyl group of the next (Fig. 5.3). There may be a few to several thousand amino acid residues in a polypeptide, the sequence being determined by the base sequence in DNA. By convention, the free amino group is considered the start of a polypeptide sequence and the free carboxyl group its end.

Essential and non-essential amino acids

There are eight essential, two semi-essential and 10 non-essential amino acids in human biochemistry:

- Essential amino acids cannot be synthesized by the body, so they must be supplied by dietary protein. These are tryptophan, lysine, methionine, phenylalanine, threonine, valine, leucine and isoleucine.
- Non-essential amino acids can be synthesized in the human body, so are not required in the diet. These are alanine, asparagine, aspartate, cysteine, glutamate, glutamine, glycine, proline, serine and tyrosine.
- Semi-essential amino acids are not considered essential because humans can synthesize them *de novo*. However, the rate of biosynthesis does not increase to compensate for depletion or inadequate dietary supply. They become essential during times of growth, as the body cannot produce them in adequate amounts. These are histidine and arginine.

Mnemonic for remembering the essential and semi-essential amino acids: Phe, Val, Thr, Trp, Ile, Met, His, Arg, Leu and Lys (PVT TIM HALL).

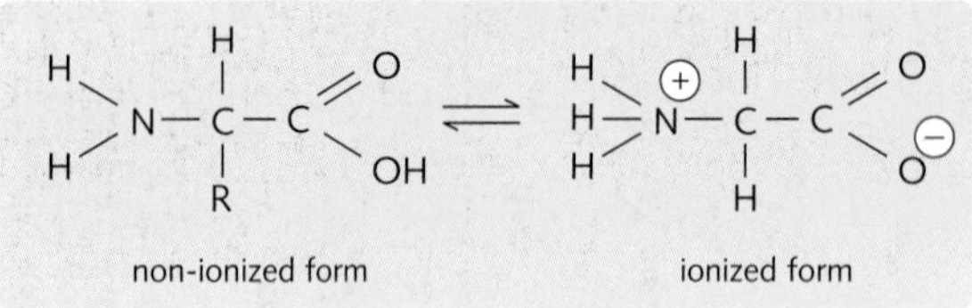

non-ionized form ionized form

Fig. 5.1 Structure of an amino acid. R is the side group. In solution at pH 7.0 the amino acid exists in its ionized form. The charges on the amino and carboxyl groups disappear when they form peptide bonds.

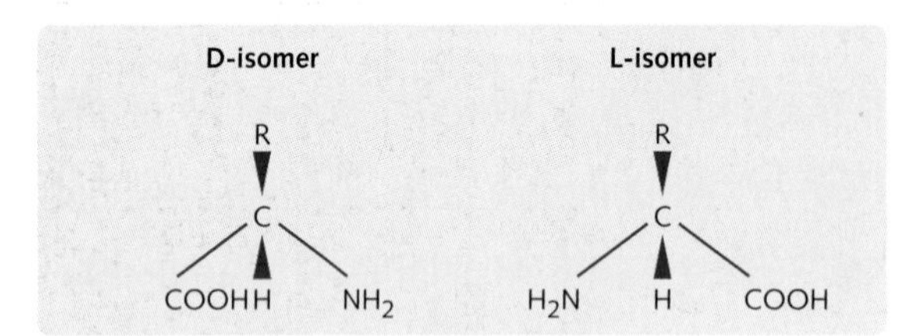

Fig. 5.2 D and L isomers of an amino acid. The stereoisomers are mirror images of one another.

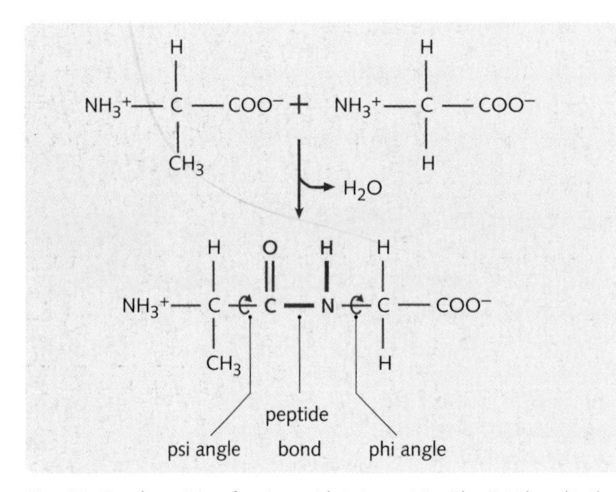

Fig. 5.3 Condensation of amino acids into protein. The C-N bond is the peptide bond. This has a partial double bond character, so the atoms highlighted in bold remain in the same plane. Psi and phi angles are rotational angles about the indicated bonds. (Adapted from Stevens and Lowe, 1997.)

Structure of amino acids

The amino acids can be classified by the nature of their R group or side chain (Fig. 5.4).

Properties of amino acids

The three-dimensional structure of a protein and, therefore, its behaviour, is determined by the characteristics and interactions of the side chains in its amino-acid sequence.

Size and structure

Larger, bulky side chains impede bending of the polypeptide chain, while ring structures prevent it forming the turns required to make α-helices ('steric hindrance'). Small or hydrophobic amino acids favour α-helical secondary structure.

Cross-linkages

Links between amino-acid residues occur through hydrogen bonds, disulphide bridges, hydrophobic bonds and ionic bonds, all of which act to stabilize the protein in its characteristic conformation:

- Hydrogen bonds occur between carbonyl (C=O) and imino (N-H) groups.
- Disulphide bridges are covalent bonds between thiol (-SH) groups of cysteine residues.
- Non-covalent hydrophobic bonds form between two hydrophobic residues.
- Electrovalent (ionic) bonds occur between a negative group of one amino-acid residue and a positive group of another amino-acid residue.

Solubility

In aqueous solution, colloidal proteins, such as cytoplasmic enzymes, are globular with polar R groups, which attract water, arranged on the outside and non-polar groups arranged on the inside.

Covalent modification

Specific side groups can be modified by the addition of a chemical group in post-translational modification reactions (see pp. 113–115), which alters the conformation of the protein, influencing its activity. For example, tyrosine residues are phosphorylated by tyrosine kinases, a property exploited in cell–cell signalling and in the regulation of enzyme activity (see Ch. 4).

Ionization properties of amino acids

Amino acids with non-polar R groups form amphions at neutral pH because the carboxyl group donates a hydrogen to the amino group, producing a molecule that has both a positively and a negatively charged group (Fig. 5.1). This molecule is called a zwitterion, and its overall electrical charge is neutral. Amphions have the ability to act as both donors and acceptors of protons:

Fig. 5.4A Classification of amino acids by side-group type			
Name	**Symbol**	**Stereochemical formula**	**Side-group type**
Aliphatic side chains			
Glycine	Gly (G)	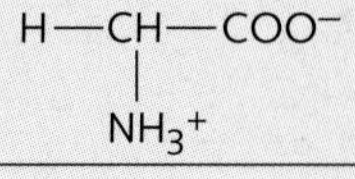 $H-CH(NH_3^+)-COO^-$	Small
Alanine	Ala (A)	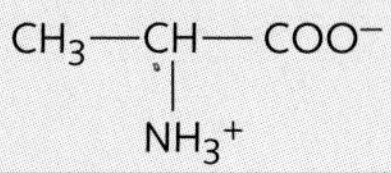 $CH_3-CH(NH_3^+)-COO^-$	Hydrophobic +
Valine	Val (V)	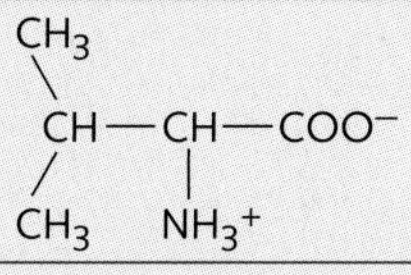 $(CH_3)_2CH-CH(NH_3^+)-COO^-$	Hydrophobic ++
Leucine	Leu (L)	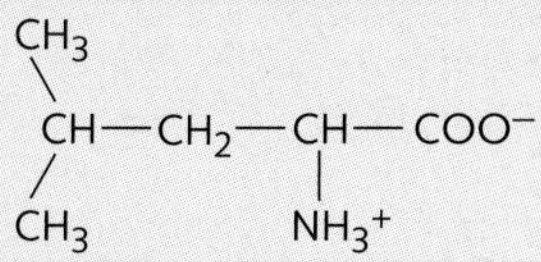 $(CH_3)_2CH-CH_2-CH(NH_3^+)-COO^-$	Hydrophobic ++
Isoleucine	Ile (I)	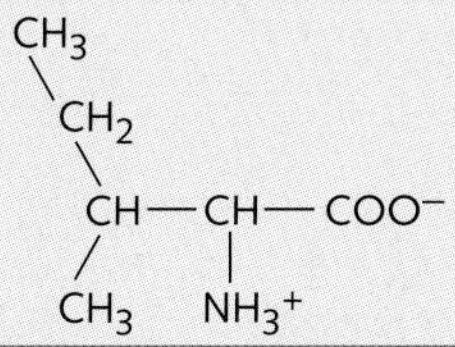 $CH_3-CH_2-CH(CH_3)-CH(NH_3^+)-COO^-$	Hydrophobic +++
Aromatic rings			
Phenylalanine	Phe (F)	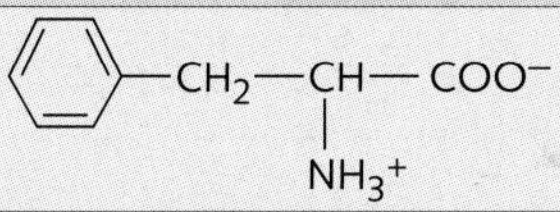 $C_6H_5-CH_2-CH(NH_3^+)-COO^-$	Hydrophobic ++++
Tyrosine	Tyr (Y)	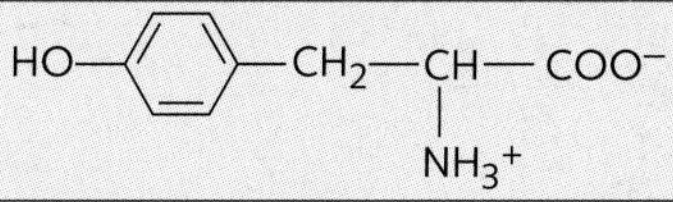 $HO-C_6H_4-CH_2-CH(NH_3^+)-COO^-$	Hydrophobic (polar)
Tryptophan	Trp (W)	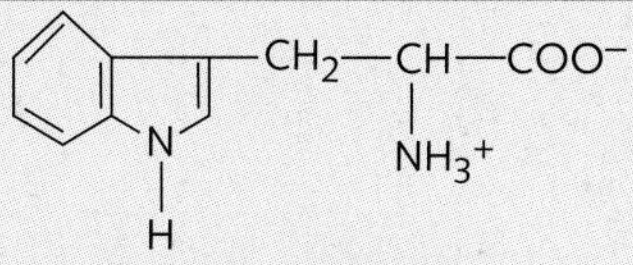 indole (N–H) $-CH_2-CH(NH_3^+)-COO^-$	Hydrophobic
Imino acids			
Proline	Pro (P)	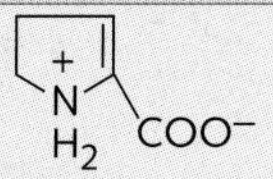 pyrrolidine ring ($\overset{+}{N}H_2$) $-COO^-$	Closed ring
Acidic groups or amides			
Aspartic acid	Asp (D)	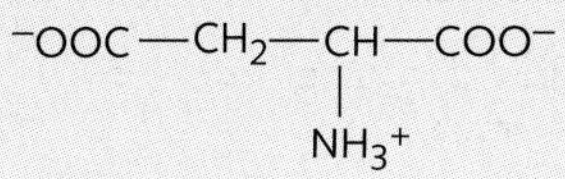 $^-OOC-CH_2-CH(NH_3^+)-COO^-$	Weak acid, pK 4 negative charge

Fig. 5.4 Classification of amino acids by side-group type. The individual side groups are highlighted by boxes.

Fig. 5.4B Classification of amino acids by side-group type

Name	Symbol	Stereochemical formula	Side-group type
Acidic groups or amides (*cont.*)			
Asparagine	Asn (N)	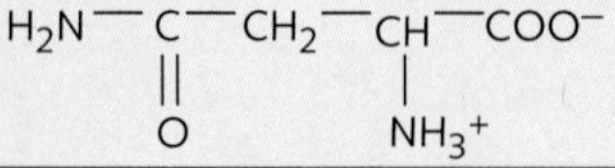 $H_2N-C(=O)-CH_2-CH(NH_3^+)-COO^-$	Polar
Glutamic acid	Glu (E)	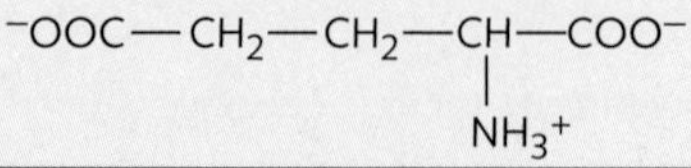 $^-OOC-CH_2-CH_2-CH(NH_3^+)-COO^-$	pK 4 negative charge
Glutamine	Gln (Q)	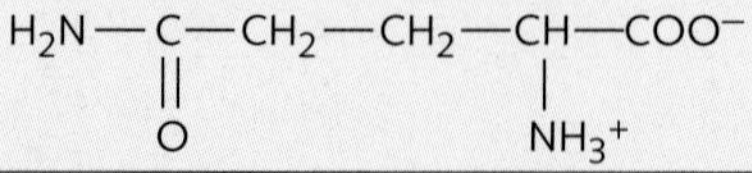 $H_2N-C(=O)-CH_2-CH_2-CH(NH_3^+)-COO^-$	Polar
Basic groups			
Arginine	Arg (R)	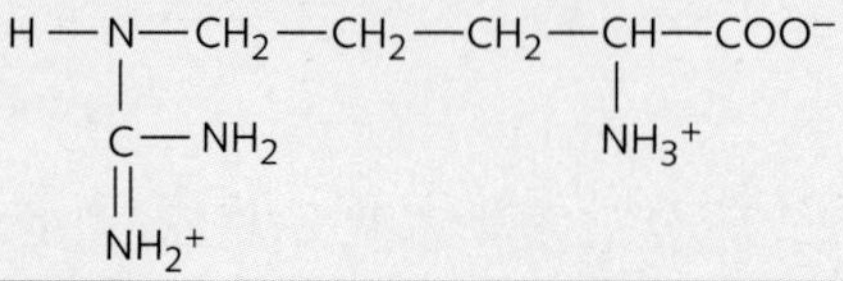 $H-N(C(=NH_2^+)-NH_2)-CH_2-CH_2-CH_2-CH(NH_3^+)-COO^-$	Weak base, pK 12 positive charge
Lysine	Lys (K)	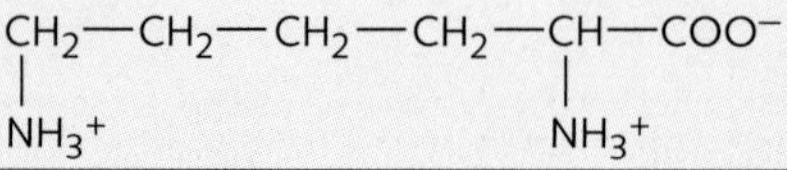 $CH_2(NH_3^+)-CH_2-CH_2-CH_2-CH(NH_3^+)-COO^-$	pK 10 positive charge
Histidine	His (H)	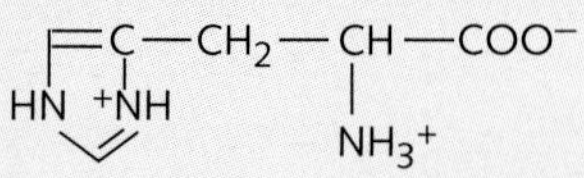 (imidazole ring: HN, $^+$NH) $C-CH_2-CH(NH_3^+)-COO^-$	pK 6 positive charge
Hydroxylic groups			
Serine	Ser (S)	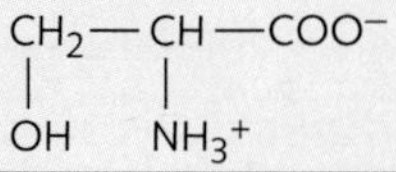 $CH_2(OH)-CH(NH_3^+)-COO^-$	Polar
Threonine	Thr (T)	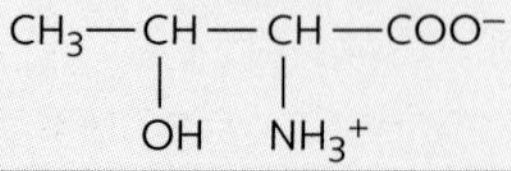 $CH_3-CH(OH)-CH(NH_3^+)-COO^-$	Polar
Sulphur groups			
Cysteine	Cys (C)	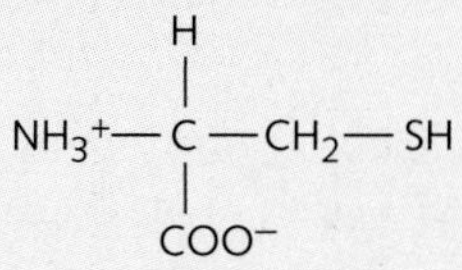 $NH_3^+-C(H)(COO^-)-CH_2-SH$	
Methionine	Met (M)	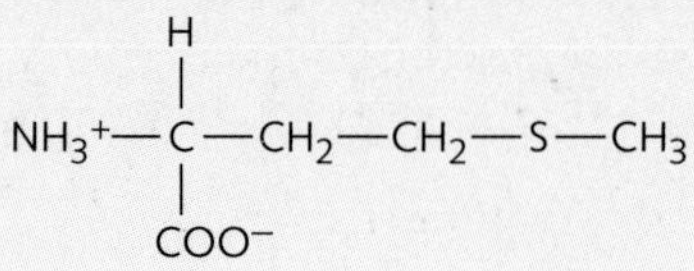 $NH_3^+-C(H)(COO^-)-CH_2-CH_2-S-CH_3$	

Fig. 5.4—Cont'd

- At very low (acidic) pH both the amino group and the carboxyl group are protonated and the molecule (cation) has an overall positive charge.
- At very high (alkaline) pH both amino group and the carboxyl group are deprotonated and the molecule (anion) has an overall negative charge.
- Between these extremes the protonation of the amino and carboxyl groups and the overall charge of the molecule vary with pH.
- The pH at which the net charge on the molecule is neutral, such that the molecule would not move in an electric field, is called the 'isoelectric point'.

The Henderson–Hasselbalch equation

Depending on the pH, both the amino and the carboxyl groups of an amino acid may act as weak acids. A weak acid (HA) is one that only partially dissociates to its anion (A^-) and a proton (H^+). The value of the dissociation constant (K_a) for a weak acid indicates its tendency to dissociate. Rearranging the formula for the dissociation constant gives the Henderson–Hasselbalch equation (Fig. 5.5):

$$pH = pK_a + \log[A^-]/[HA]$$

This equation permits the calculation of:

- the pH of a conjugate acid–base pair, given the pK_a and the molar ratio of the pair
- the value of pK_a for a weak acid given the pH of a solution of known molar ratio.

By the Henderson–Hasselbalch equation, when $[A^-]$ equals [HA], pH equals pK_a, i.e. the pK_a of a weak acid is the pH at which it is half dissociated.

Amino acids as buffers

A buffer consists of a weak acid and its conjugate base. Buffers cause a solution to resist changes in pH when acid or base is added. In amino acids, the amino and the carboxyl groups may both act as buffers. At a pH of 9.8 the amino group of glycine functions as a buffer:

- At pH 9.8 glycine is in equilibrium between the anion and the zwitterions form.
- If protons are added they are removed from solution as they combine with the anion to produce the zwitterion.
- If alkali is added protons dissociate from the zwitterion to produce the anion and a water molecule.

$$\underset{\text{weak acid}}{HA} \rightleftharpoons \underset{\text{proton}}{H^+} + \underset{\text{conjugate base}}{A^-}$$

The dissociation constant for a weak acid is:

$$K_a = \frac{[H^+][A^-]}{[HA]} = [H^+] - \frac{[A^-]}{[HA]}$$

Take the log of each of the terms in this equation.

$$\log K_a = \log [H^+] + \log \frac{[A^-]}{[HA]}$$

Rearrange thus:

$$-\log[H^+] = -\log K_a + \log \frac{[A^-]}{[HA]}$$

Substitute pH for – log [H+] and pK_a for – log K_a.

$$pH = pK_a + \log \frac{[A^-]}{[HA]}$$

Fig. 5.5 Derivation of the Henderson–Hasselbalch equation.

Since at the pK_a of the amino group (9.8) there is an equal concentration of the weak acid (the zwitterion) and its conjugate base (the anion) it functions best as a buffer at this pH. A similar situation arises at pH 2.4 for the carbonyl group, though in this case the cation is the weak acid and the zwitterion its conjugate base. When glycine is titrated with an alkali:

- two pK_a values are observed at pH 2.4 (carboxyl group) and at pH 9.8 (amino group)
- the addition of alkali produces very little change in pH within one pH unit of each pK_a
- the addition of alkali produces a large change in pH at the isoelectric point
- the isoelectric point lies midway between pK_{a1} and pK_{a2}.

Amino acids with ionizable side chains have a third pK_a corresponding to the pH range in which the proton on the side chain dissociates.

Non-protein amino acids

Several other amino acids are found in the body free or in combined states. These non-protein associated amino acids perform specialized functions. Biologically important non-protein amino acids include creatine, homocysteine, γ-aminobutyric acid and thyroxine.

PROTEINS

Functions of proteins

Proteins serve a variety of diverse functions in the human body (Fig. 5.6). Protein isoforms, related proteins encoded by different genes or generated by alternative RNA splicing of transcripts encoded by the same gene, have almost identical biological activity, but they differ in amino acid sequence, for example:

- myosin expressed in heart tissue
- myosin expressed in fast muscle fibres.

Protein conformation is defined by the sequence of its amino-acid residues, which is critical to its function (Fig. 5.7).

Organization of proteins

Primary structure

Primary structure is specified by the linear sequence of amino-acid residues (determined by the base sequence in DNA) linked via peptide bonds.

Secondary structure

Most proteins contain local regions of the polypeptide chain folding (α-helices and β-pleated sheets – see p. 81) resulting from hydrogen bonding interactions between peptide bonds (Fig. 5.8). They are favoured by:

- hydrogen bonding
- repulsion of side groups
- limited flexibility of the polypeptide chain.

Fig. 5.6 Functions of proteins

Protein function	Examples
Construction	Collagen in skin and bone
Contraction	Actin and myosin in muscle
Catalysis	Enzymes
Combat	Antibodies
Carriage	Haemoglobin in blood, transferrin carries iron
Communication	Peptide hormones, receptors, cytoplasmic kinases, major histocompatibility complex

Fig. 5.6 Functions of proteins.

Fig. 5.7 Proteins and ligands

Protein	Ligand
Enzymes	Substrate
Myosin	Actin and other proteins
Antibodies	Antigen
Receptors	Hormones, neurotransmitters, counter-receptors

Fig. 5.7 Proteins and ligands.

Tertiary structure

Folding occurs to form the unique three-dimensional shape of a polypeptide chain (Fig. 5.9). It is determined by interactions between side groups of the amino-acid residues, including disulphide bridges and electrostatic/hydrophobic interactions.

Quaternary structure

Two or more polypeptide chains (subunits) associate to form dimers, tetramers or oligomers (Fig. 5.10). The subunits are held together by the same types of bond that stabilize tertiary structure.

Forces that shape proteins

Peptide bond

The peptide bond is formed between two amino acids by condensation (see Fig. 5.3). This is a strong covalent bond, which is resistant to heat, pH extremes and detergent, with a bond energy of 380 kJ/mol and length of 0.132 nm (1.32Å). The peptide group is planar as it has a partial double-bond character. However, free rotation occurs around the other bonds, giving different phi and psi angles (Fig. 5.3) that allow considerable flexibility in the polypeptide chain. Consequences of the peptide bond are as follows:

- The polypeptide chain has considerable, though restricted, flexibility.
- The partial charge present at the oxygen and nitrogen of the bond enables attraction between two peptide bonds, forming a weak hydrogen bond with a bond energy of 5 kJ/mol.

Many biologically active proteins will spontaneously fold into one conformation that is stabilized by a variety of intramolecular bonds.

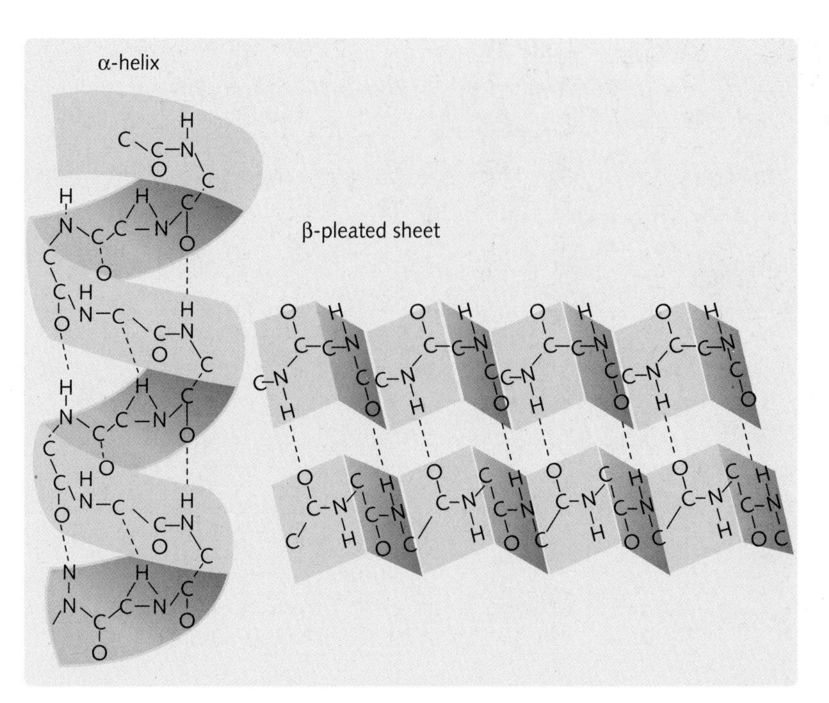

Fig. 5.8 Secondary structure of a protein. In both the α-helix and β-pleated sheet, regions of secondary structure are stabilized by hydrogen bonds (H bonds) between the C=O and N-H groups of the peptide bonds in the protein.

Fig. 5.9 Tertiary structure of a protein. One complete protein chain (β-chain of haemoglobin) is illustrated here.

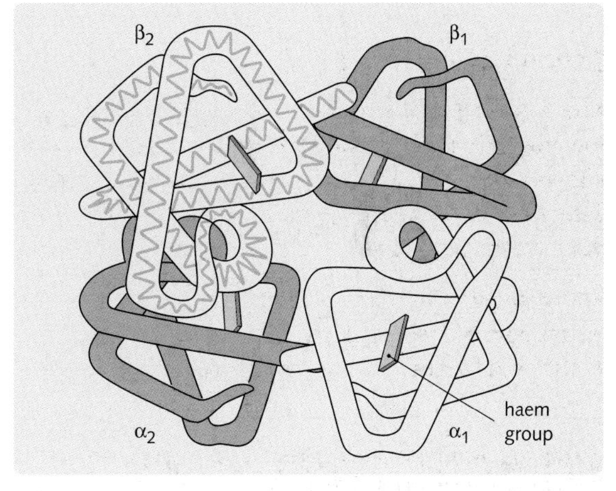

Fig. 5.10 Quaternary structure of a protein. Four separate chains of haemoglobin are assembled into an oligomeric protein.

Hydrogen bonds

Hydrogen bonds occur between peptide bond atoms and polar side groups where a hydrogen atom is shared between two electronegative atoms; they are important in forming secondary and tertiary structures. They have a bond energy of 20 kJ/mol and length of 0.3 nm (Fig. 5.11).

Hydrophobic interactions

Hydrophobic residues (e.g. valine, alanine, leucine and phenylalanine) form interactions, rather than true bonds, where they cluster close together. Bond energy comes from the displacement of water.

Ionic interactions

Ionic interactions result from strong attractions between positive and negative atoms. These bonds are important in tertiary structure and have a bond energy of 335 kJ/mol and length of 0.25 nm (see Fig. 5.11).

van der Waals forces

van der Waals forces (dipole-induced dipole) are weak attractions between two atoms as the electron

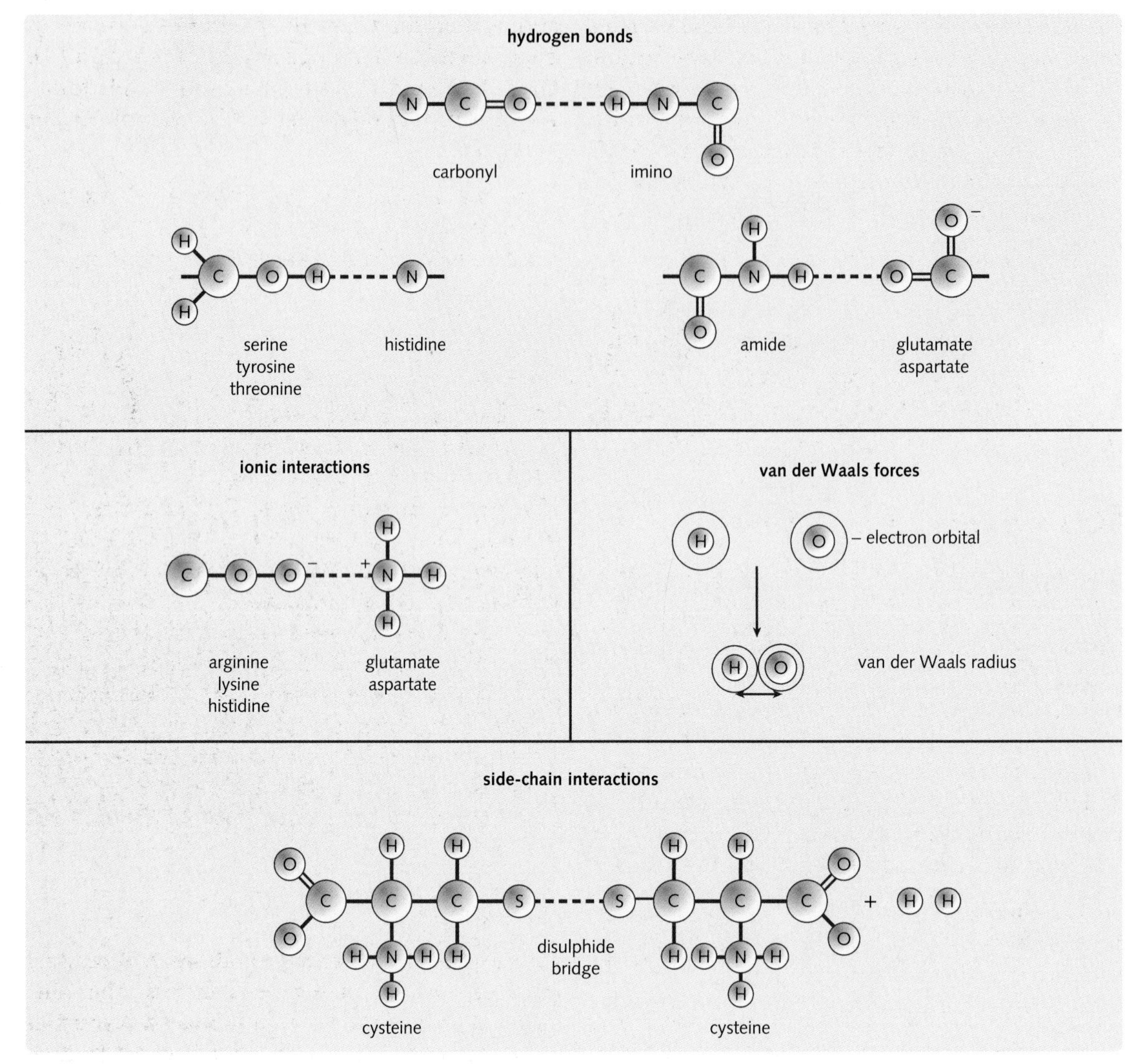

Fig. 5.11 Forces shaping proteins. Proteins are stabilized by chemical bonds that may depend on interactions between the carbonyl and imino groups in the polypeptide chain (e.g. some hydrogen bonds) or on specific side-chain interactions (e.g. disulphide bridges).

orbitals approach each other (see Fig. 5.11). The bond energy is very weak, being 0.8 kJ/mol, and the length is 0.35 nm. Collectively, these bonds 'add' to significant energy in the tertiary structure of large polypeptides.

Side-chain interactions

Side-chain interactions form bonds, the most important being between the thiol groups of two cysteine residues, forming a covalent bond of 210 kJ/mol called a disulphide bridge (see Fig. 5.11). Disulphide bridges are important in tertiary structures and in the secondary structure of elastin.

This reaction is not favoured intracellularly, so disulphide bridges are normally found in exported proteins, for example:

- the digestive enzyme ribonuclease has four disulphide bonds
- the peptide hormone insulin has three disulphide bonds.

Protein folding

The correct 'folded shape' of a protein is determined by the amino acid sequence. Groups far away in the primary sequence may be brought close together in the final three-dimensional structure. Large proteins are composed of several domains linked by flexible regions of polypeptide. Domains have specific tertiary structure associated with a particular function, which may be conserved between proteins. For example, all known nicotinamide adenine dinucleotide (NAD) dependent dehydrogenase enzymes share a conserved NAD binding domain.

Protein structure and folding are often asked about in essay questions.

Structures within proteins

α-helix

The α-helix is a right-handed helix (D form) with a backbone of peptide linkages, from which side chains radiate outwards. Small or hydrophobic amino acid residues favour α-helix formation, so glycine or proline are usually found at the α-helix bends. The structure is stabilized by hydrogen bonds between every first and fourth amino acid, each hydrogen bond being relatively weak, but as all are parallel and 'intrachain' they provide reinforcement. The helix has:

- 0.15 nm rise
- 0.54 nm pitch
- 3.6 residues per turn.

The structure is a rigid rod-like cylinder that is very stable, with side groups pointing out and, therefore, free to interact with other α-helices (see Fig. 5.8). α-helices make up 90–100% of fibrous proteins and 10–60% of globular proteins. For example:

- α-keratin, a fibrous protein that is a component of skin, consists of long rod-like coils made from two identical α-helices wound around one another
- myoglobin, the globular protein that binds oxygen in muscle, has eight α-helical segments, and it is 75% α-helix in total.

β-pleated sheet

β-pleated sheets are extended chains formed from two or more pleated polypeptides joined by hydrogen bonds. In a parallel sheet the terminal amino acids are at the same end whereas, in the more common antiparallel sheet, terminal amino acids are at opposite ends, so forming a more stable structure (see Fig. 5.8). The sheet is a rigid non-elastic 'platform', which is commonly found in fibrous proteins, for example:

- β-keratins in claws, scales, feathers and beaks are made of antiparallel strands
- silk is made of regular β-sheets.

Prion diseases (transmissible spongiform encephalopathies) can be inherited, occur sporadically or be infectious. It is thought that disease-causing prion proteins interact with a normal cellular version of the same protein to produce a conformational change. While the cellular prion protein (PrP^C) has a secondary structure rich in α-helices, the disease producing version (PrP^{Sc}) has a secondary structure that is dominated by β-sheets. PrP^{Sc} then converts PrP^C to more of itself, and this high β-sheet content correlates with infectivity and resistance to enzymatic digestion leading to neuronal loss, spongiform changes and astrogliosis.

Zinc fingers

Zinc fingers are a common motif in DNA-binding proteins such as transcription factors. The 'zinc finger domain' is a folded amino-acid projection surrounding a central zinc atom (Fig. 5.12). Zinc finger proteins recognize and bind to specific DNA regulatory sequences, influencing transcription.

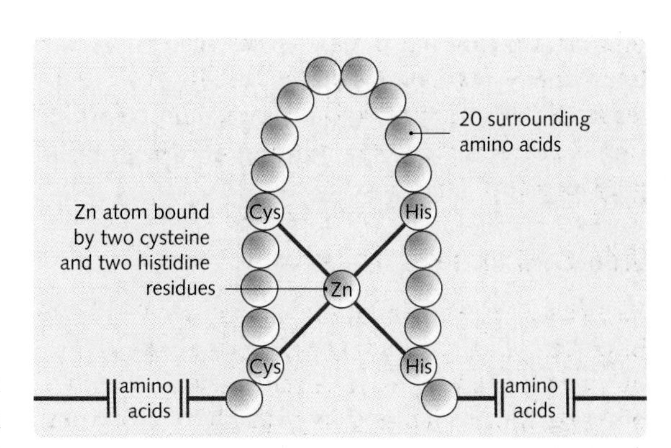

Fig. 5.12 Zinc finger domain. The bonds between the four amino acids and zinc stabilize a loop of polypeptide into a finger-like structure.

Collagen helix

Collagen is a fibrous protein that is a major component of connective tissue. It is composed of three polypeptide chains wound around one another and linked by hydrogen bonds (see p. 79).

> Tropocollagen is constructed from three left-handed helices twisted together into a right-handed helix. It is able to take on this unique shape as the side chain of every third amino acid is very close to the triple helix central axis. Glycine, having the smallest side chain of any amino acid, fulfills this requirement, and substitutions of glycine for bulkier amino acids, disrupting the architecture and collagen quality, have been implicated in osteogenesis imperfecta.

Stability of proteins

Denaturation is the loss of the three-dimensional structure of a protein due to breaking of structural bonds. It is usually associated with loss of biological activity (Fig. 5.13). Any treatment that disrupts chemical bonding (e.g. heat, pH extremes, detergents, oxidation, and physical effects such as shaking) may cause denaturation, thus most proteins only function within narrow environmental limits. If the denaturing conditions are not extreme (e.g. removal of the treatment), most proteins return to their active states when returned to optimal conditions; an example of this is lysozyme. Unlike insulin (Fig. 5.13), all the information required for lysozyme folding (tertiary structure) is present in its primary structure: thus, following denaturation and renaturation, fully functional lysozyme is recoverable.

Complex structures

Individual polypeptides may not be biologically active themselves, but they may serve as subunits in the formation of larger, active complexes. In addition, some proteins are dependent on their interaction with other, non-protein cofactors, such as prosthetic groups and coenzymes.

> Haemoglobin A is a tetramer composed of two α-subunits and two β-subunits each associated with a prosthetic haem group. Each haem group contains an iron atom, responsible for the binding of oxygen. Haemoglobin binds oxygen when pO_2 is high and releases it when it is low. Mutations in the haemoglobin genes result in an altered ability to carry and surrender oxygen to the tissues. The most common of these haemoglobinopathies are sickle-cell disease and β thalassaemia.

ENZYMES AND BIOLOGICAL ENERGY

Properties of enzymes

Enzymes are biological catalysts. Without them, metabolic reactions would proceed too slowly for life. Enzymes have the following properties:

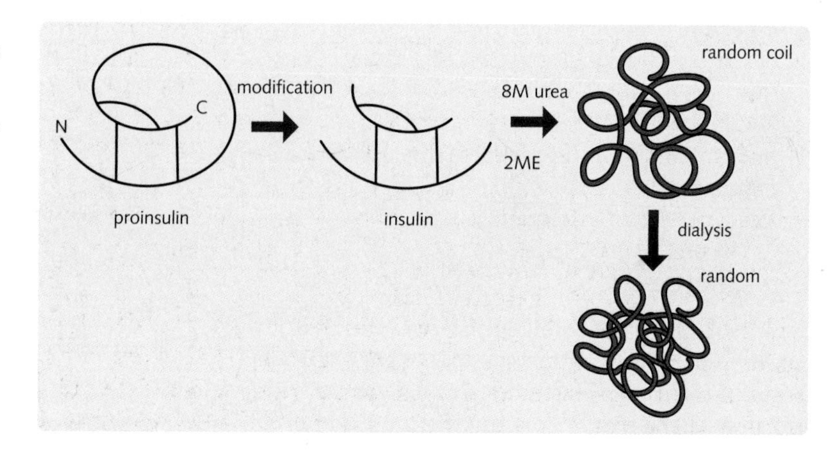

Fig. 5.13 Following denaturation and renaturation of insulin, only a random coil is recoverable, with no functional activity. As insulin is derived from proinsulin, the information required for tertiary structure must be held within the C-peptide.

- They bind specific ligands (substrates) at active sites and catalyse their conversion to products.
- They greatly alter the speed of a reaction.
- They remain in the same chemical state at the end of the reaction as at the beginning, so can be reused.
- They catalyse the forward and the reverse reaction.
- They show great specificity. Some enzymes may only recognize one stereoisomer.

Mnemonic for remembering the six classes of enzyme: oxidoreductases, transferases, hydrolases, isomerases, ligases and lyases (Over The HILL). Enzymes get reactants over the activation energy 'hill'.

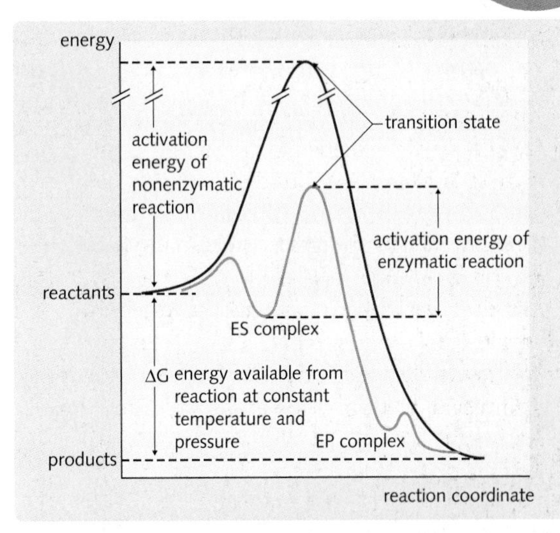

Fig. 5.14 Reaction profile for catalysed and non-catalysed reactions. Activation energy is less for the catalysed reaction. ES complex, enzyme–substrate complex; EP complex, enzyme–product complex. (Adapted from Baynes and Dominiczak, 1999.)

An enzyme's name is normally the name of the substrate with the suffix -ase added, the substrate being the substance on which the enzyme acts. Isoenzymes catalyse the same reaction, but they have different primary structures and may work under different optimum conditions, for example:

- lactate dehydrogenase has heart and muscle isoforms
- creatine kinase has brain and muscle isoforms.

Mechanism of enzyme action

Catalysts and biological energy

In any chemical reaction, when a reactant is converted to a product, highly unstable intermediates are produced as chemical bonds are broken and reformed. These intermediates have higher free energy than the reactant, and so they are not favoured energetically (Fig. 5.14):

- The difference in free energy between the reactant and the unstable intermediate is the activation energy.
- Reactants must have in excess of the required activation energy.

The energy level of individual reactants in a population follows a normal distribution. If activation energy is high, only a few molecules will have sufficient energy to react at any one time and the reaction will be slow.

A catalyst speeds up chemical reaction, but is itself unchanged by it. Catalysts bind transition molecules and stabilize them, so reducing the activation energy for the reaction. Therefore, in a catalysed reaction, more molecules will have the required activation energy and the reaction rate will increase. However, since they accelerate the reaction in both directions the position of the equilibrium is unchanged. A catalysed reaction has the following features relative to a non-catalysed reaction:

- The rate of the reaction is increased.
- The overall change in free energy between reactants and products is the same.
- The position of the equilibrium is the same.

Increasing the temperature of reactants increases their mean energy level. For a reaction catalysed by an inorganic catalyst, rate increases directly proportionally to temperature. However, enzyme-catalysed reactions show maximum activity at 37°C, but much reduced reaction rates at higher temperatures. This is because enzyme activity depends on the active site, and outside physiological parameters the bonds that maintain its structure are disrupted.

Active sites

The active site is the region to which substrate molecules bind. Enzyme specificity occurs because the shape of the active site is such that only substrates with a complementary structure can bind. This has been likened to the fitting of a key into its lock. However, rather than viewing the active site as a rigid structure the 'induced fit' model (Fig. 5.15) proposes that:

- the substrate binds to a substrate binding domain, which induces a conformational change at the active site
- the conformational change in the active site reveals functional groups
- as the product dissociates the enzyme returns to its original conformation and so can bind more substrate.

Non-polypeptide components of proteins are termed prosthetic groups, for example, the haem group in haemoglobin. They frequently form an integral part of the active site. Coenzymes are organic molecules that must be associated with a given enzyme for it to function. They bind to the enzyme, undergo chemical change, and are ultimately released as the reaction is completed. Alcohol dehydrogenase requires the presence of the coenzyme NAD.

Regulation of enzyme activity

The coordinated regulation of enzyme activity allows the organism to adapt to environmental change. In multistep metabolic pathways the slowest enzyme determines the rate at which the final product is produced. The activity of such 'rate determining' enzymes is regulated so that metabolism is coordinated and energy is not wasted. Five mechanisms are involved:

1. Transcription from the enzyme coding gene may be regulated (see p. 109)
2. Enzymes may be irreversibly activated or deactivated by proteolytic cleavage
3. Enzymes may be reversibly activated and deactivated by covalent modifications, such as phosphorylation
4. Enzymes may be subject to allosteric modification. This occurs when the binding of a small molecule to a site distant from the active site alters the conformation of the enzyme and, therefore, its activity
5. The rate of degradation of the enzyme may be regulated.

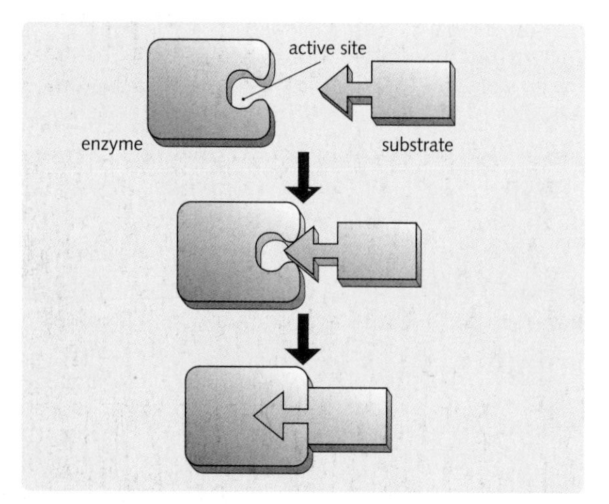

Fig. 5.15 The 'induced fit' hypothesis explains how the binding of the substrate to the enzyme changes the conformation of the active site.

Enzyme kinetics

Enzyme kinetics is the study of the rate of change of reactants and products. Enzyme assays use biosensors, oxygen electrodes, chromogenic substances and other methods to measure the progress of reactions.

Reaction rates

By plotting the amount of product formed against time, the initial velocity of the reaction can be estimated (Fig. 5.16A). The initial velocity (V) equals the reaction rate. Reaction rate varies according to enzyme and substrate concentration. Increasing enzyme concentration increases the reaction linearly, so follows first-order kinetics (Fig. 5.16B).

Increasing substrate concentration increases the reaction rate in an asymptotic, non-linear, fashion; as the enzyme becomes saturated the reaction rate reaches a limit – this is a Michaelis–Menten graph (Fig. 5.17). For simple enzymes the curve is a rectangular hyperbola. At low substrate concentrations the graph is linear (rate is proportional to [substrate]), so follows first-order kinetics. At high substrate concentrations a plateau is reached, so follows zero-order kinetics. The Michaelis–Menten equation relates the reaction rate to the substrate concentration:

$$E + S \underset{K_2}{\overset{K_1}{\rightleftharpoons}} ES \underset{K_4}{\overset{K_3}{\rightleftharpoons}} E + P$$

K_1, K_2, etc., are individual rate constants. K_4 is insignificant so is ignored.

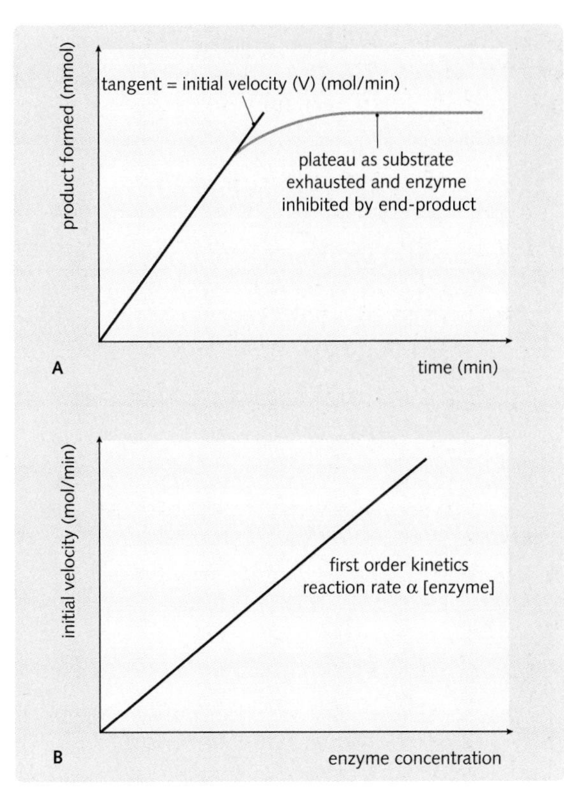

Fig. 5.16 (A) Calculation of initial velocity. (B) Effect of enzyme concentration on reaction rate.

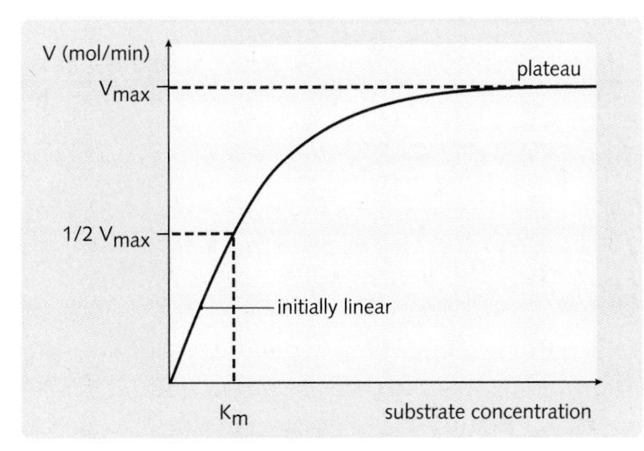

Fig. 5.17 Michaelis–Menten graph showing effect of increasing substrate concentration against reaction rate. K_m, Michaelis constant; V, initial velocity; V_{max}, maximum velocity.

The Michaelis constant (K_m) is the rate of breakdown of the enzyme–substrate complex:

$$K_m = \frac{K_2 + K_3}{K_1}$$

The maximum velocity (V_{max}) is reached under saturating conditions when the substrate concentration is high:

$$V_{max} = K_3[ES]$$

where K_3 is rate of enzyme and product formation and [ES] is enzyme–substrate complex concentration.

The velocity (V) of the Michaelis–Menten reaction is therefore:

$$V = \frac{V_{max}[S]}{K_m + [S]}$$

- V_{max} and K_m are constants for each different enzyme.
- For most enzymes, K_m is the substrate concentration at which the reaction rate is half of V_{max}
- K_m is a measure of the affinity of the enzyme for its substrate, with a low K_m (low enzyme–substrate complex breakdown) corresponding to high affinity (tight binding) and vice versa.
- V_{max} and K_m are difficult to estimate from a Michaelis–Menten graph, so an alternative graph representation is used, the Lineweaver–Burk graph (Fig. 5.18).

Inhibitors

Enzyme inhibitors are substances that lower enzyme activity. False inhibitors include denaturing treatments and irreversible inhibitors (e.g. organophosphorus compounds). True inhibitors can be:

- competitive
- non-competitive
- allosteric.

Competitive inhibitors resemble the substrate, and they compete for the active site (Fig. 5.19A). The K_m is increased, so affinity of the enzyme is decreased; for example, azidothymidine (AZT) used to treat HIV infection resembles deoxythymidine, so it is a competitive inhibitor of HIV reverse transcriptase.

Non-competitive inhibitors (e.g. heavy metals such as lead) do not resemble the substrate, so they

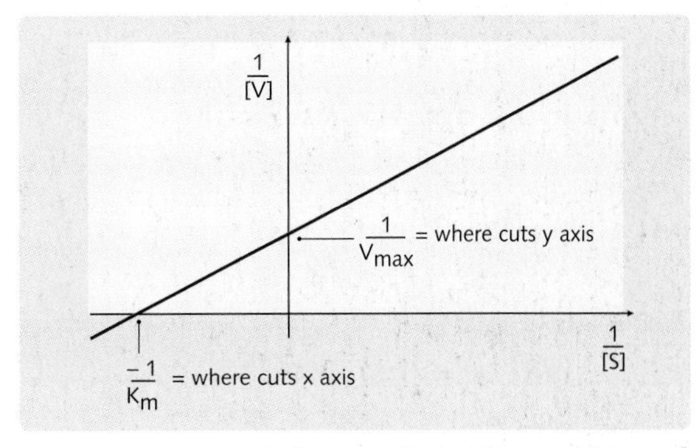

Fig. 5.18 Lineweaver–Burk plot. K_m, Michaelis constant; S, substrate; V, initial velocity; V_{max}, maximum velocity.

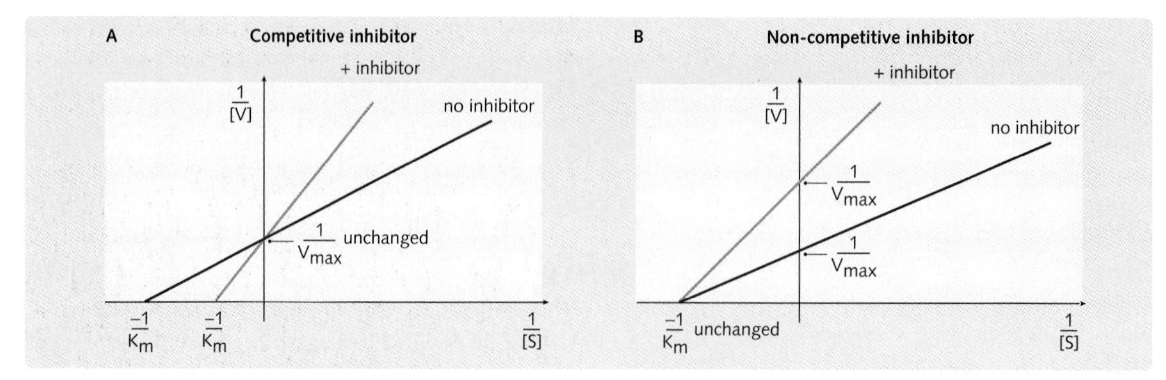

Fig. 5.19 (A) Competitive inhibition of enzyme. The inhibitor increases K_m. V_{max} is unchanged. (B) Non-competitive inhibition of enzyme. The inhibitor decreases V_{max}. K_m remains unchanged. K_m, Michaelis constant; S, substrate; V, initial velocity; V_{max}, maximum velocity.

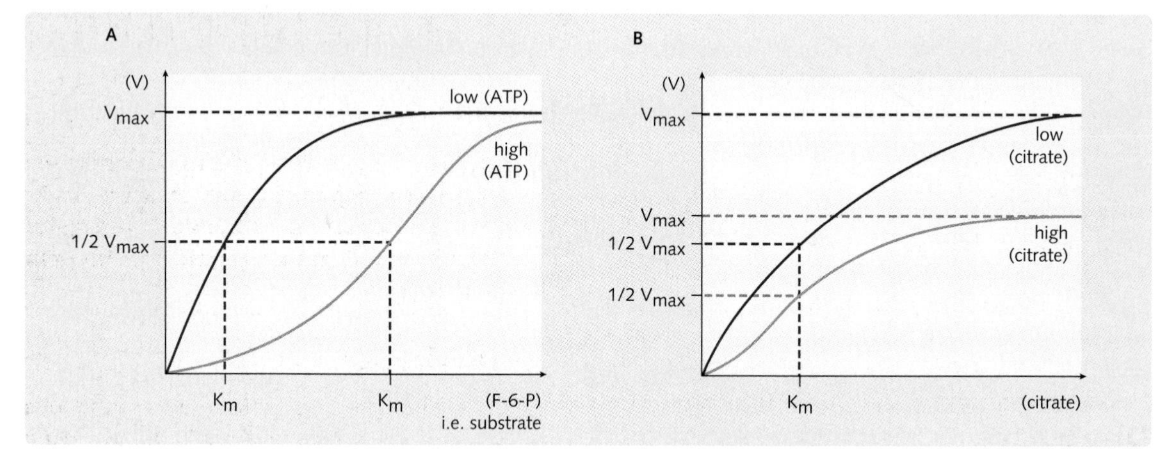

Fig. 5.20 Allosteric inhibition of phosphofructokinase 1. (A) ATP increases K_m (decreases affinity), but it does not alter V_{max}. (B) Citrate decreases V_{max}, but it does not alter K_m. K_m, Michaelis constant; S, substrate; V, initial velocity; V_{max}, maximum velocity.

do not compete for the binding site. They bind to the enzyme and abolish its catalytic activity, although substrate may still bind (Fig. 5.19B). V_{max} is decreased as there is less catalytically active enzyme, but K_m is unchanged as affinity of the enzyme is not altered.

Allosteric inhibitors do not bind to the active site; instead, they bind to a different allosteric site elsewhere on the enzyme. Binding causes a conformational change, which can alter V_{max} and/or K_m; for example, phosphofructokinase I (PFK I) is allosterically inhibited by high concentrations of ATP and by citrate (Fig. 5.20).

CARBOHYDRATES

Carbohydrates are chemical compounds that contain oxygen, hydrogen and carbon atoms, with the general chemical formula $C_n(H_2O)_n$. They have several important biological functions, including:

- metabolism to produce energy (ATP) by glycolysis
- storage in the form of glycogen
- conversion to fatty acids and triacylglycerol for long-term storage
- synthesis of other cellular components, such as the cell membrane.

Monosaccharides

Monosaccharides are the subunits of carbohydrates. They can be distinguished from each other on one of four levels:

1. Chemical nature of the carbonyl group (aldehyde to give an aldose or ketone to give a ketose)

2. Number of carbon atoms
3. Stereochemistry
4. Linear or cyclic nature.

Monosaccharides are polar due to their high proportion of hydroxyl groups, and as such are highly water soluble. With the exception of dihydroxyacetone, all carbohydrates contain at least one asymmetrical (chiral) carbon and, therefore, D and L stereoisomers exist. Unlike amino acids, in humans most carbohydrates exist in the D form, a notable exception being L-ascorbic acid.

Cyclic structures

Monosaccharides with five or six carbon atoms can form cyclic structures. In solution, both the cyclic and the straight chain forms co-exist, with the cyclic form predominating (Fig. 5.21). The aldehyde or ketone group of the straight chain structure may react with a hydroxyl group on a different carbon atom, forming a heterocyclic ring. Rings with five carbon atoms are called 'furanose' rings and those with six, 'pyranose' rings. Each ring structure has one more optically active carbon than the straight-chain form, and so has both an α and a β enatomer, which interconvert in equilibrium (Fig. 5.21).

Disaccharides and oligosaccharides

Disaccharides are formed by the joining of two monosaccharides with the elimination of water. This condensation reaction yields a glycosidic bond between the two subunits (Fig. 5.22), which exists in two forms, the α form and the β form; α glycosidic bonds involve C1 of the α anomer, while β glycosidic bonds involve C1 of the β anomer. Examples of important disaccharides are given in Fig. 5.22.

Lactose is a disaccharide broken down by lactase, an enzyme secreted into the lumen of the small intestine. Most mammals stop producing lactase after infancy; however, a large number of the human population have an autosomal dominant mutation that enables them to continue. Those who do not are said to be lactose intolerant and, as lactose remains in the small intestine, it becomes subject to degradation and fermentation by bacteria, causing abdominal pain and distension and chronic diarrhoea.

Oligosaccharides are short chains of condensed monosaccharides, typically three to seven units long. They have a number of important biological functions, often covalently attached to proteins or to membrane lipids.

Polysaccharides

Polysaccharides are polymers of numerous monosaccharide units, as many as 2500, to yield a polymer with the general formula $C_n(H_2O)_{n-1}$. They can be classified according to whether they are:

- linear or branched
- the nature of glycosidic bond
- homopolysaccharide (repeating units of the same monosaccharide) or heteropolysaccharides (repeating units of different monosaccharides).

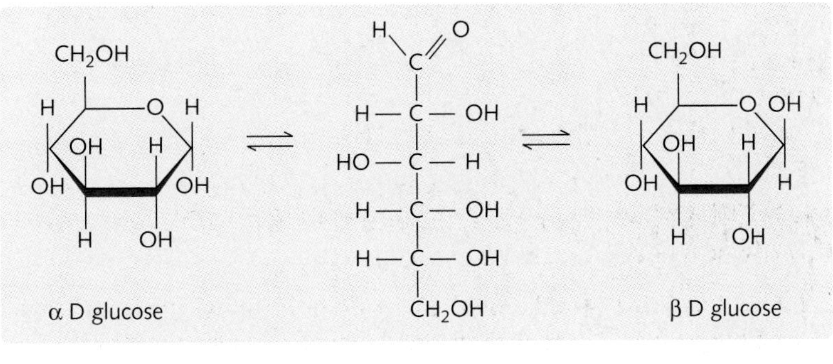

Fig. 5.21 The relationship between straight chain and cyclical monosaccharides.

Fig. 5.22 Some common disaccharides

Disaccharide	Monomers	Linkage	Structure
sucrose	glucose + fructose	alpha 1–2	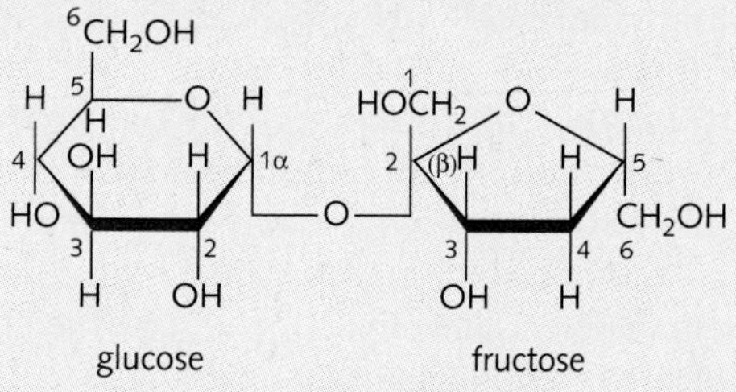 glucose fructose
lactose	galactose + glucose	beta 1–4	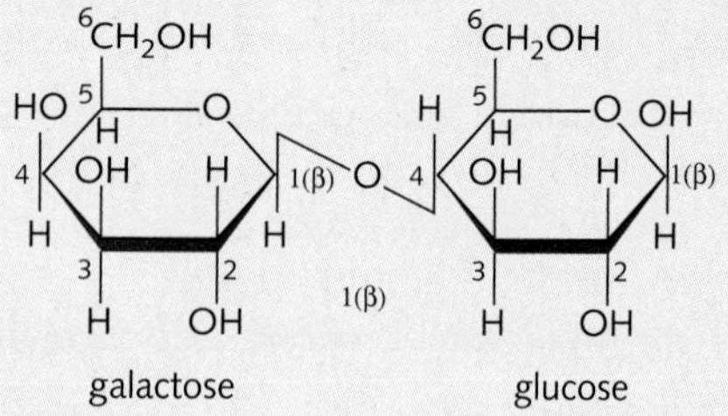 galactose glucose
maltose	glucose + glucose	alpha 1–4	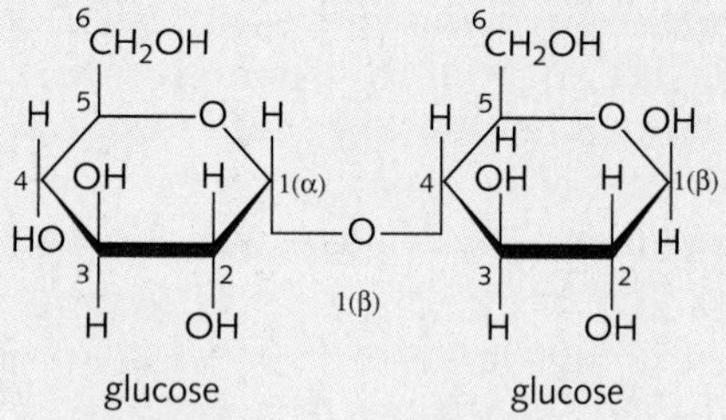 glucose glucose
isomaltose	glucose + glucose	alpha 1–6	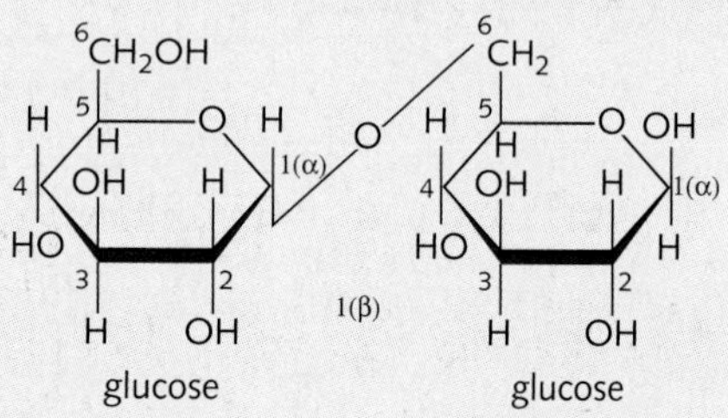 glucose glucose
cellobiose	glucose + glucose	beta 1–4	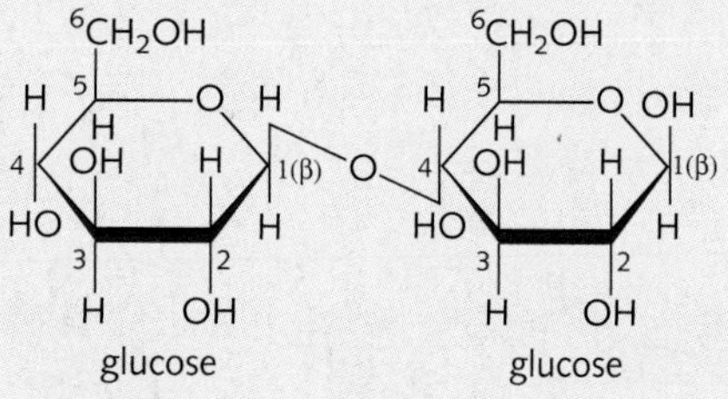 glucose glucose

Fig. 5.22 Common disaccharides.

The main storage polysaccharide in animals is glycogen, a highly branched homopolysaccharide of repeating glucose units. The majority of the subunits are joined by α-1–4 glycosidic bonds. However, at every 10 residues or so falls a branch point, joined to the main molecule by an α-1–6 glycosidic bond. Cellulose, the main structural material of plants, is also a homopolysaccharide of repeating glucose units, but joined by β-1–4 glycosidic bonds. Humans do not have a cellulase capable of digesting such links.

The most abundant heteropolysaccharides in the body are the glycosaminoglycans (GAGs). They are long unbranched molecules consisting of a repeating disaccharide unit of an amino sugar (an N-acetyl-hexosamine) and a sugar or sugar acid (a hexose or a hexuronic acid) (Fig. 5.23).

All cells with the ability to store glucose do so as glucose polymers, such as glycogen. As glycogen is only sparingly soluble in the cell it does not contribute to cellular osmotic pressure and is, therefore, unlikely to lead to cell damage. If a cell were to store free glucose, the osmotic pressure associated would lead to an accumulation of water in the cell and cell lysis.

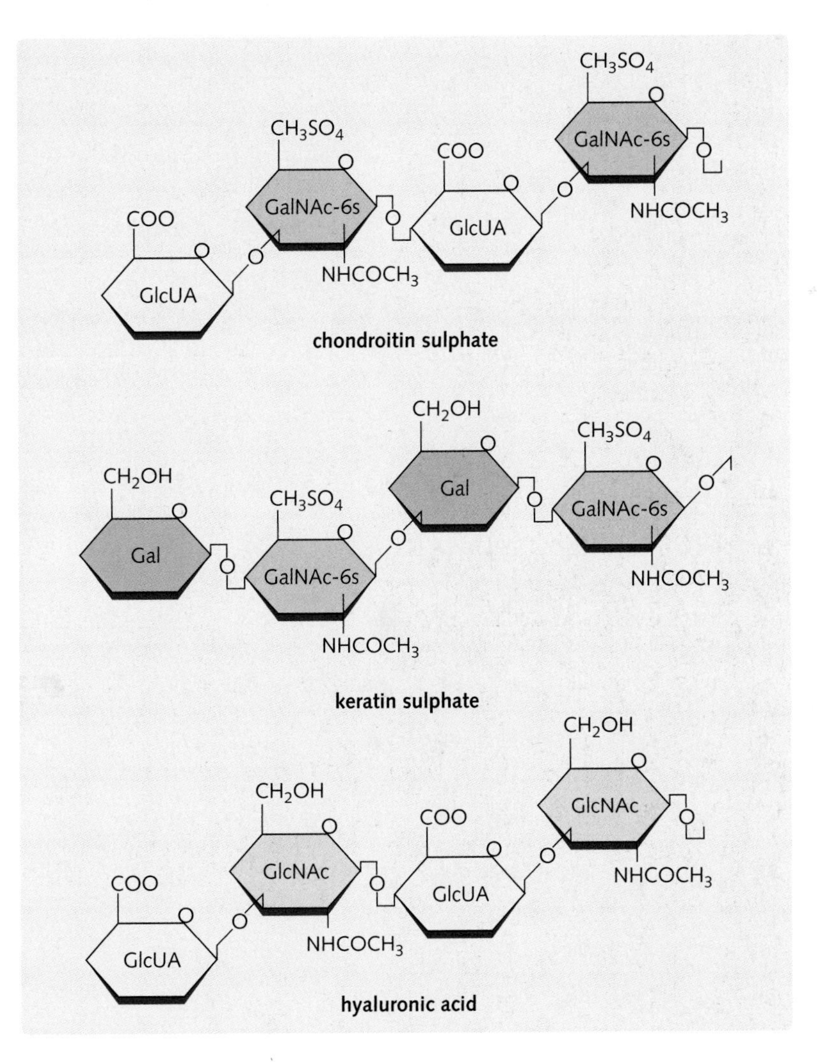

Fig. 5.23 Structures of common glycosaminoglycan units.

Heparin is a glycosaminoglycan. The release of heparin from mast cell granules in response to injury leads to its entry into blood serum and the inhibition of blood clotting. Heparin acts by binding to antithrombin III, exposing its active site and allowing it to inactivate key proteases involved in blood clotting. This property has been manipulated for therapeutic gain, with commercially produced heparin being utilized as an antithrombotic agent for use in a range of venous and arterial clotting disorders.

Sugar derivatives

Various derivatives of sugars exist (Fig. 5.24), including:

- sugar acids. The C1 aldehyde group or the hydroxyl on the terminal carbon is oxidized to a carboxylic acid – e.g. ascorbic acid and glucuronic acid
- sugar alcohols. Formed by the reduction of the carbonyl group of a sugar to a hydroxyl group – e.g. sorbitol, mannitol and ribitol
- amino sugars. One hydroxyl group is substituted for an amino group – e.g. glucosamine and galactosamine
- nucleosides. Specialized amino sugars, in which the amino residue is a pyrimidine or a purine – e.g. adenosine, deoxyadenosine, thymidine and deoxythymidine
- deoxy sugars. One hydroxyl group is substituted for hydrogen – e.g. deoxyribose and fucose.

Fig. 5.24 Sugar derivatives

Class	Example
sugar acids	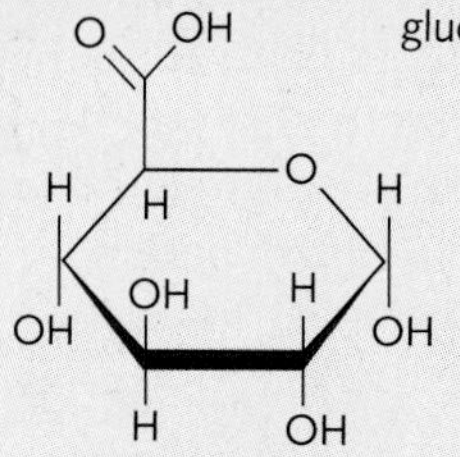 glucuronic acid
sugar alcohols	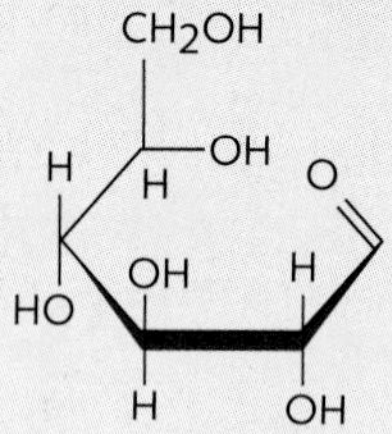 sorbitol
amino sugars	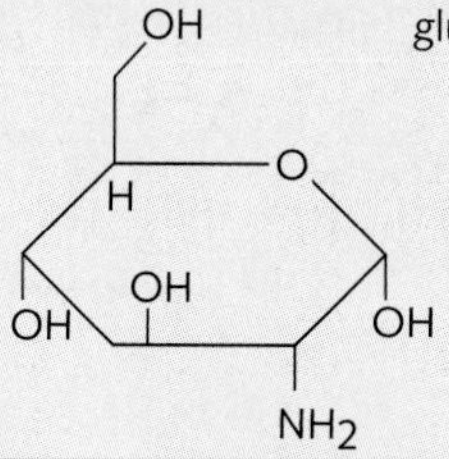 glucosamine
nucleosides	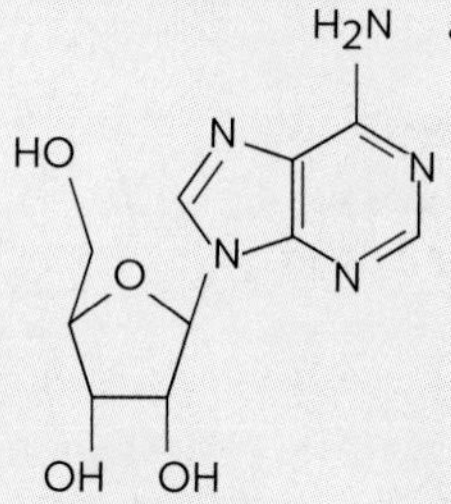 adenosine
deoxy sugars	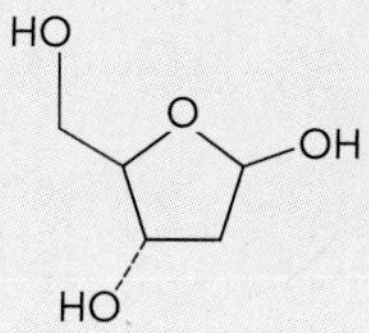 deoxyribose

Fig. 5.24 Sugar derivatives.

Basic molecular biology and genetics

6

Objectives

By the end of this chapter you should be able to:

- Discuss broadly the biosynthesis of purines and pyrimidines, and identify the rate-limiting steps of each.
- Compare and contrast the structures and functions of the different RNA molecules.
- Draw a diagram to show the stages of the cell cycle and explain the role of cyclins in the regulation of the cell cycle.
- Describe the different orders of DNA packaging, from DNA to chromosome.
- Draw an annotated diagram of a eukaryotic replication fork, and contrast it with that of a prokaryotic replication fork.
- Understand the process of eukaryotic transcription and be able to compare and contrast with prokaryotic transcription.
- Summarize the salient features of initiation, elongation and termination of eukaryotic protein synthesis.
- Define mitosis and meiosis, list the stages of each and highlight the differences between them.
- Identify repair mechanisms required to repair single-stranded DNA damage and double-stranded DNA damage.

ORGANIZATION OF THE CELL NUCLEUS

The nucleus is the largest structure in eukaryotic cells. The nucleoplasm is in constant contact with the cell cytoplasm via pores in the nuclear membrane. The nucleus consists of DNA (deoxyribonucleic acid), proteins and RNA (ribonucleic acid), and it plays a vital role in:

- protein synthesis (see pp. 109–115)
- the passage of genetic information from one generation to the next (see pp. 120–122).

Structures of the nucleus

Nuclear envelope

The nuclear envelope encloses the nucleus. It consists of two layers of membrane, the outer being continuous with the endoplasmic reticulum (Fig. 6.1). The space between the inner and outer membranes is called the periplasm, which forms a continuum with the lumen of the endoplasmic reticulum (ER). Ribosomes are attached to the outer layer of the nuclear envelope as well as to the ER.

Nuclear pores

Nuclear pores are found at points of contact between the inner and outer membranes (Fig. 6.2). They are electron-dense structures consisting of eight protein complexes arranged around a central granule. They control the passage of metabolites, macromolecules and RNA subunits between the nucleus and the cytoplasm. Molecules up to 60 kDa pass freely through these pores, but transport of larger molecules is ATP-dependent and requires the receptor-mediated recognition of a nuclear targeting sequence by the pore complex.

Nucleoli

Nucleoli are extremely dense structures in the nucleus that represent the sites of ribosomal RNA synthesis and assembly. There may be one, several, or no visible nucleoli in a cell nucleus (see pp. 25–27).

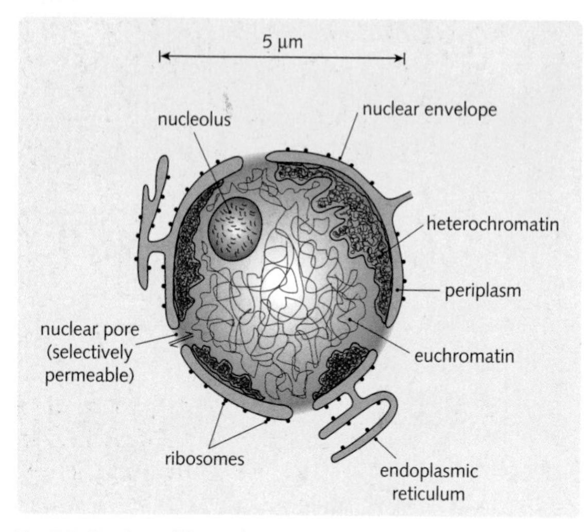

Fig. 6.1 Structure of the nucleus.

Cells that are not active in protein synthesis tend to have nuclei rich in heterochromatin and no nucleoli.

Nuclear matrix

The nuclear matrix consists of DNA, nucleoproteins and structural proteins. Nucleoproteins are proteins closely associated with DNA and can be defined as histone or non-histone. Histones are strongly basic globular proteins around which DNA winds in a regular fashion, like beads on a string, to form chromatin (see pp. 96–97). Chromatin is thought to interact with the lamins, a type of intermediate filament, which are arranged in a lattice forming a thin shell that underlies the inner nuclear membrane. A less regularly organized network of intermediate filaments surrounds the outer membrane, and together these networks provide mechanical support for the nuclear envelope.

NUCLEIC ACIDS

Nucleic acids are produced from the polymerization of nucleotides (Fig. 6.3). During synthesis a series of nucleic acid condensation reactions occur between phosphate and sugar groups to produce strong phosphodiester bonds. Long, unbranching chains form with linkages between C3 and C5 of each sugar, hence the notation 5′ to 3′ or 3′ to 5′ to describe the orientation of nucleotides in a nucleic acid chain. Each sugar is separated from the next by a phosphate group to form a strong and rigid sugar–phosphate backbone from which the bases project (Fig. 6.4). DNA and RNA are nucleic acids that play an integral role in the growth and replication of all living cells.

Nucleotides and nucleosides

A nucleotide is a compound of a pentose sugar residue (deoxyribose in DNA, ribose in RNA; Fig. 6.5) attached to a base (purine or pyrimidine) and a phosphate group (Fig. 6.3) and is the subunit of

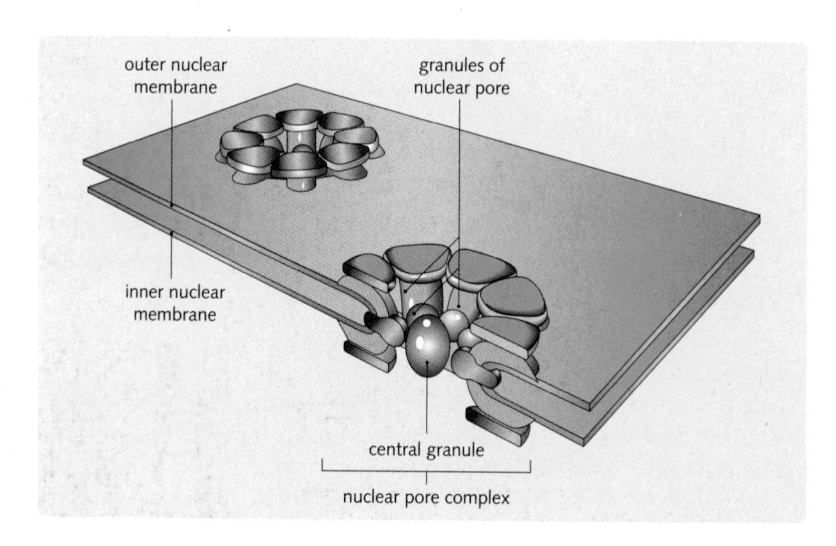

Fig. 6.2 Structure of a nuclear pore. (Adapted from Stevens and Lowe, 1997.)

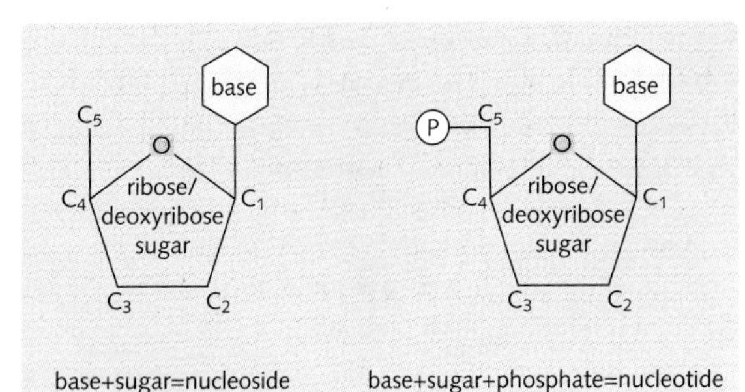

Fig. 6.3 Structure of nucleosides and nucleotides. Deoxyribose has an H group on C2 of the ribose moiety, whereas ribose has an OH group at this position. (Adapted from Molecular Biology of the Cell, 3rd edn, by B Alberts et al, Garland Publishing, 1994. Reproduced by permission of Routledge, Inc., part of The Taylor & Francis Group.)

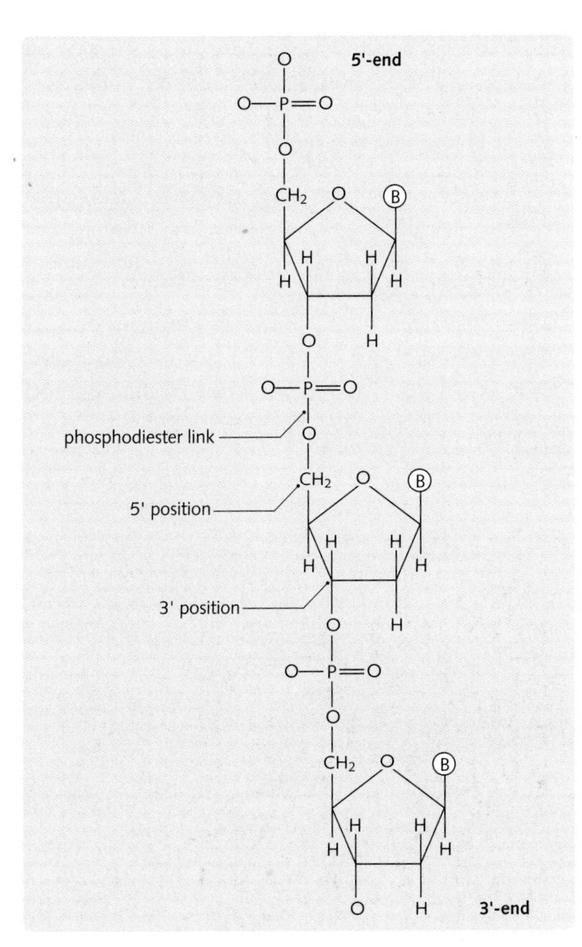

Fig. 6.4 Nucleic acids. The backbone of DNA is formed from deoxyribose sugars linked by phosphodiester bonds. B indicates the position of the base. (Adapted from Norman and Lodwick, 1999.)

Fig. 6.5 Comparison of DNA and RNA

Feature	DNA	RNA
Sugar	Deoxyribose	Ribose
Base pairing	A–T/G–C	A–U/G–C
Structure	Double helix	Single stranded structures

Fig. 6.5 Comparison of DNA and RNA. A, adenine; C, cytosine; G, guanine; T, thymine.

nucleic acids. They are named according to whether the base is a ribonucleotide or a deoxyribonucleotide and the number of attached phosphate groups (e.g. adenosine monophosphate).

A nucleoTide has Three moieties.

A nucleoside is a compound of a sugar residue and a base linked by an N-glycosidic bond between C1 of the sugar and an N atom of the base (Fig. 6.3). The nucleosides of the various bases are named according to whether they are a ribonucleoside or a deoxyribonucleotide, e.g. adenosine or deoxyadenosine respectively.

Purines and pyrimidines

The nucleotide bases in nucleic acids are heterocyclic molecules derived from either purines (adenine, guanine) or pyrimidines (cytosine, thymine, uracil). Uracil is not present in DNA, but takes the place of thymine in RNA. They are planar aromatic rings that contain nitrogen (Fig. 6.6).

Almost all eukaryotic cells are capable of synthesizing purines and pyrimidines *de novo*, which suggests an important role in cell survival.

Pyrimidines are CUT from Purines. The pyrimidines are: cytosine (DNA and RNA), uracil (RNA), thiamine (DNA), and 'single-ringed', hence CUT from the 'double-ringed' purines.

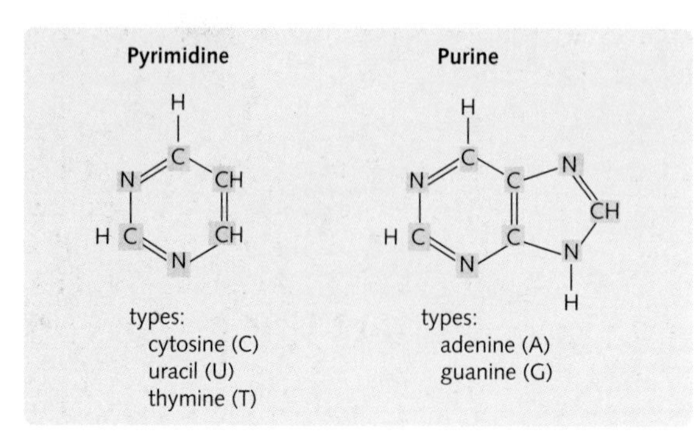

Fig. 6.6 Structure of pyrimidine and purine bases.

Purine biosynthesis

Purines are synthesized in an 11-stage pathway, starting with the formation of 5-phosphoribosyl-1-pyrophosphate (PRPP). Through this pathway, ATP, folate (tetrahydrofolate) derivatives, glutamine, glycine and aspartate are utilized to yield inosine monophosphate (IMP), which is rapidly converted to adenosine monophosphate (AMP) or guanosine monophosphate (GMP). A complex negative feedback control network operates to prevent excessive build-up of AMP and GMP, the rate limiting step lying in the first reaction (the synthesis of PRPP by PRPP synthetase).

> Gout results from abnormal catabolism of purines, either from an excess of purines or partial deficiency in hypoxanthine guanine phosphoribosyl transferase (HGPRT). Clinical manifestations of abnormal purine catabolism arise from the insolubility of the degradation byproduct, uric acid, most of which is usually excreted via the kidneys. Increased formation or reduced excretion of uric acid leads to hyperuricaemia and the formation of sodium urate crystals, which precipitate in the synovial fluid of the joints leading to severe inflammation and arthritis.

Pyrimidine biosynthesis

Pyrimidine synthesis involves a six-step pathway, in which the ribose sugar is incorporated as one of the final steps. From a starting point of carbamoyl phosphate, derived from glutamine and bicarbonate, uridine monophosphate (UMP) is formed. UMP is then phosphorylated to uridine triphosphate (UTP), where it is free to be aminated to form cytidine triphosphate (CTP). Uridine nucleotides are also the precursors for *de novo* synthesis of the thymine nucleotides.

Salvage pathways

In addition to *de novo* synthesis, nucleotides can also be synthesized from the breakdown of endogenous nucleic acids through salvage pathways (Fig. 6.7), in which preformed bases are recovered and reconnected to a ribose unit.

Pentose sugars

These are five-carbon rings (see Fig. 6.4). Deoxyribonucleotides are formed by the reduction of the ribose group of the corresponding ribonucleotide (see p. 90).

DNA double helix

Discovered by Watson and Crick in 1953, the structure of DNA is one of two intertwined polynucleotide strands held together by base pairing to form a double helix. Adenine and thymine pair via two hydrogen bonds between opposing strands, whereas guanine and cytosine pair via three hydrogen bonds. Base pairing results in two complementary polynucleotides, which run antiparallel to each other (i.e. one runs 5′ to 3′, the other runs 3′ to 5′; Fig. 6.8).

RNA molecules

In eukaryotes, all single-stranded RNA is produced from DNA by transcription. It is synthesized

Fig. 6.7 Major salvage pathways for pyrimidine and purine bases.

Fig. 6.7 Major salvage pathways for purine and pyrimidine bases

	Purines	Pyrimidines
Major bases salvaged	Hypoxanthine, guanine	Uracil (thymine)
Enzymes involved	Hypoxanthine-guanine-phosphoribosyltransferase	Uridine phosphorylase, uridine kinase (thymidine phosphorylase, thymidine kinase)
Products	IMP, GMP	UMP (dTMP)

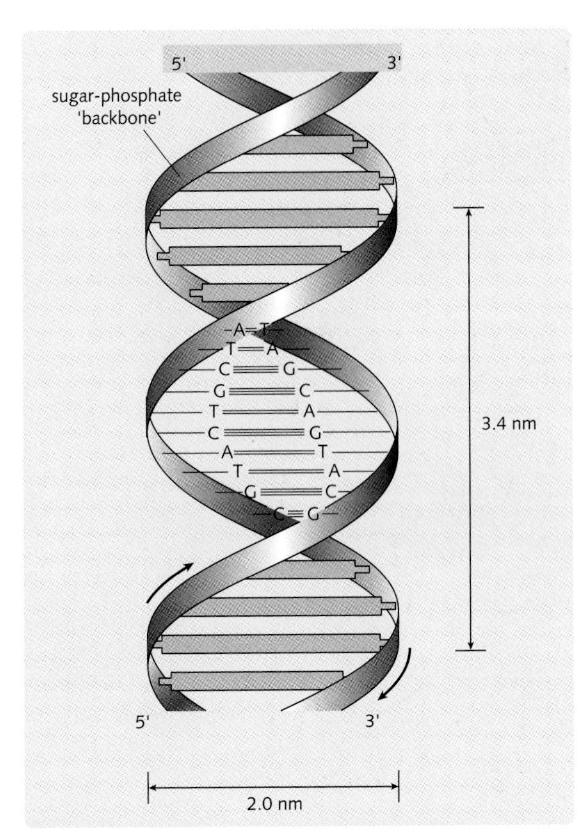

Fig. 6.8 The DNA double helix is a right-handed helix with a common axis for both strands. There are ten base pairs per turn. A, adenine; C, cytosine; G, guanine; T, thymine.

predominantly in the nucleus, moving out into the cytoplasm to carry out its function.

Messenger RNA

Messenger RNA (mRNA) carries genetic information from the nucleus into the cytoplasm. In eukaryotes, it is derived by splicing the initial RNA transcript (heteronuclear RNA). It forms the template upon which polypeptides are manufactured during translation, being a single-stranded molecule without any intramolecular hydrogen bonds (Fig. 6.9).

Heteronuclear RNA

Heteronuclear RNA (hnRNA) is the primary mRNA transcript produced by eukaryotic cells. It is very short lived, as it is processed into mature mRNA. Unlike the final mRNA transcript it contains introns, which are subsequently removed by RNA splicing (see pp. 108–109).

Transfer RNA

Transfer RNA (tRNA):

- carries specific amino acids to the site of protein synthesis
- has two active sites, which allow it to carry out its functions
- is a linear molecule with an average of 76 nucleotides
- exhibits extensive intramolecular base pairing, which gives it a characteristic 'clover-leaf'-shaped secondary structure.

Up to 20% of tRNA bases undergo post-translational modifications, which are postulated to be required for tRNA–protein interactions or in stabilizing the tRNA molecule. Fig. 6.10 shows the arrangement of the secondary structure of tRNA.

Ribosomal RNA

Ribosomal RNA (rRNA) is a component of ribosomes. In a eukaryotic cell each ribosome consists of two unequal subunits, made up of proteins and RNA, called the S (small) and L (large) subunits (Fig. 6.11). The RNA molecules undergo extensive intramolecular base pairing, which is important in determining the ribosomal structure. Ribosomes are capable of self-assembly under physiological conditions with the correct complement of components.

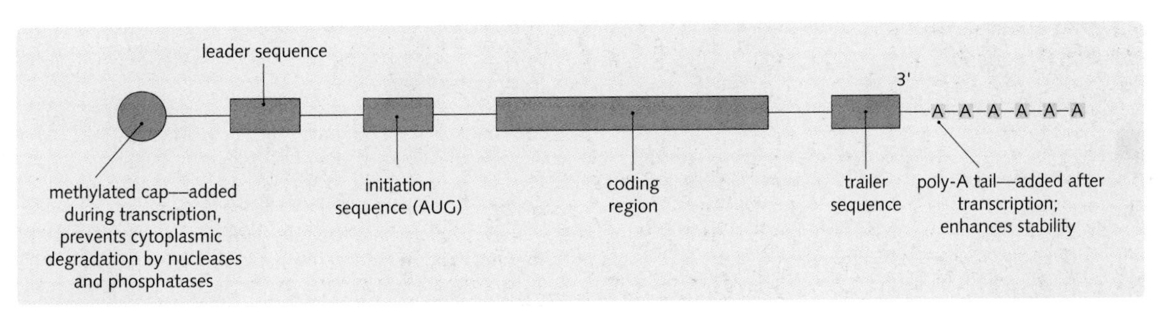

Fig. 6.9 Structure of eukaryotic mRNA. A, adenine; C, cytosine; G, guanine; U, uracil.

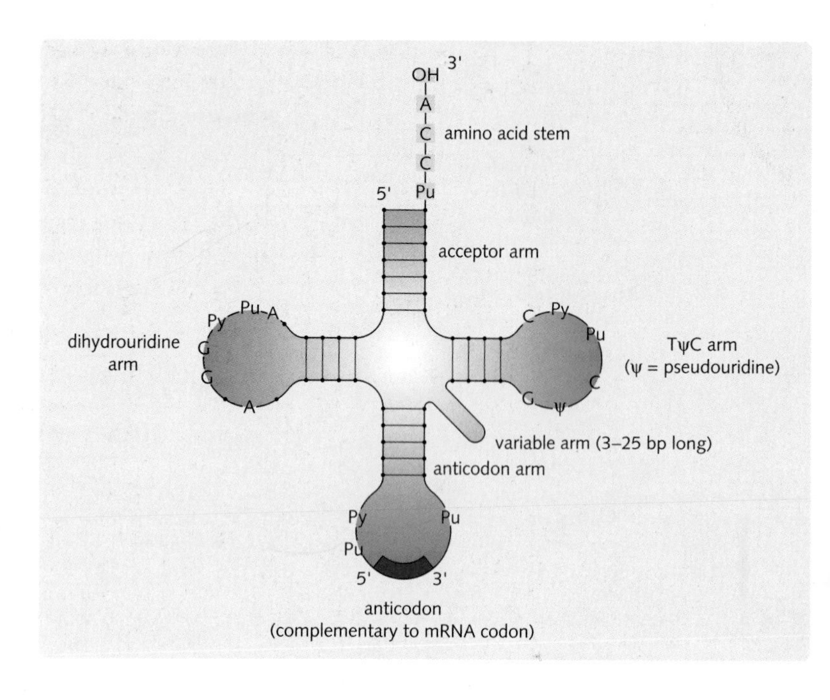

Fig. 6.10 Secondary structure of tRNA. It consists of five arms. The active sites are on the acceptor arm, where the 3′ terminal CCA group can accept a specific amino acid, and the anticodon is on the anticodon arm, which recognizes the corresponding mRNA codon. Specific base pairing within the five arms helps to maintain the secondary structure. A, adenine; C, cytosine; G, guanine; Pu, purine; Py, pyrimidine; U, uridine.

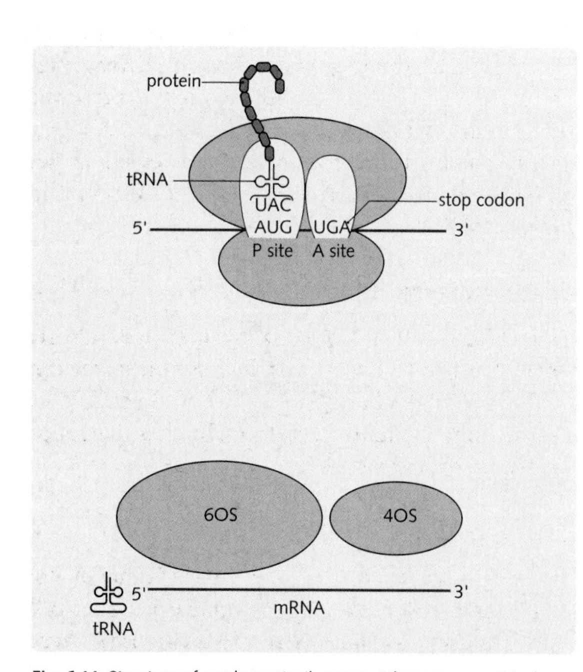

Fig. 6.11 Structure of a eukaryotic ribosome. Ribosomes consist of two unequal subunits, composed of RNA and protein, held together by magnesium ions. In rat cytoplasmic ribosome, the small (40S) unit is composed of 33 polypeptides and 18S rRNA, while the large (60S) unit is composed of 49 polypeptides and three rRNA molecules of 2.8 S, 5.8 S, and 5 S. The ribosome has binding sites for mRNA, the peptidyl tRNA (P-site) and aminoacyl tRNA (A-site). S, Svedberg unit of sedimentation. (Adapted from Baynes and Dominiczak, 1999.)

DNA PACKAGING AND CHROMOSOMES

DNA is found predominantly in the nucleus, but also in mitochondria (see pp. 110–111). It acts as a template during transcription, and is the vehicle of inheritance (i.e. it is passed from one generation to the next). In the nucleus of a normal human cell, there are 46 chromosomes each containing 48–240 million bases of DNA. Watson and Crick's DNA double helix model predicts that each chromosome would have a contour length of 1.6–8.2 cm (i.e. the total length of the DNA would be about 3 m). However, the average nucleus has a diameter of approximately 5 μm! Therefore, an extremely high degree of organization is needed to fit this amount of DNA into the nucleus (Fig. 6.12).

Chromatin

Chromatin is the collective name for the long strands of DNA, RNA and their associated nucleoproteins. During interphase of the cell cycle (see pp. 115–118), chromatin is dispersed throughout the nucleus, becoming more compact during mitosis or meiosis (see pp. 118–122). Two types of chromatin can be seen when a cell is viewed under an electron microscope (see Ch. 2):

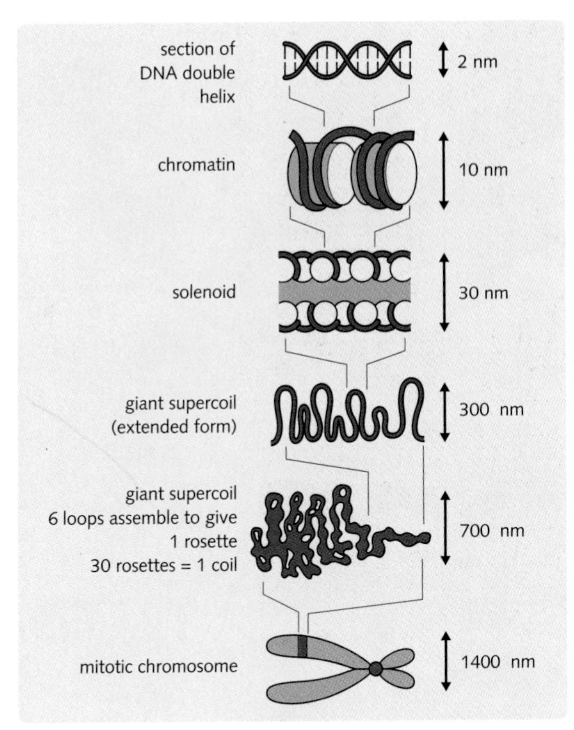

Fig. 6.12 Higher order of chromatin structure. The different levels of DNA condensation, from DNA double helix to mitotic chromosome.

1. Heterochromatin, which is electron dense and is distributed around the periphery of the nucleus and in discrete masses within the nucleus. The DNA is in close association with nucleoproteins, and it is not active in RNA synthesis.
2. Euchromatin, which is electron lucent and represents DNA that is actually or potentially active in RNA synthesis.

In the progressive levels of chromosome packaging:

- DNA winds onto nucleosome spools
- the nucleosome chain coils into a solenoid
- the solenoid forms loops, and the loops attach to a central scaffold
- the scaffold plus loops arrange themselves into a giant supercoil.

Nucleosomes

A nucleosome is formed by 146 bp of DNA wound twice around an octamer of histone proteins (Fig. 6.13). The octamer consists of two copies each of the histone proteins H2A, H2B, H3, and H4. Histone proteins are conserved throughout eukaryotic evolution. They contain a high proportion of positively charged amino-acid residues that can form ionic bonds with the negatively charged DNA. This interaction does not depend on DNA sequence and theoretically histones can bind with any piece of DNA. However, *in vivo*, the position of histone binding is influenced by:

- AT content (AT sequence bends more easily than GC)
- the presence of other tightly bound proteins.

Nucleosome bound regions of DNA are separated by a region of linker DNA that varies from 0 to 80 bp in length. Consequently, on electron micrographs nucleosomes appear as 11 nm 'beads' on a 2 nm DNA 'string'.

DNaseI is an endonuclease that breaks the internal phosphodiester bonds in DNA, irrespective of its base sequence. The regulatory regions of genes are frequently bound by proteins that prevent histones from binding. Since histone binding protects DNA from degradation with DNaseI, the regulatory regions of genes are particularly sensitive to this enzyme, and they are sometimes called 'nuclease-hypersensitive sites'.

Solenoid formation

The second level of DNA packing is mediated by histone HI, binding together adjacent nucleosomes to condense DNA into the supercoiled 30 nm fibre, which is also called the solenoid (see Fig. 6.12, 6.13). The solenoid exhibits six to eight nucleosomes per turn of the spiral. This configuration corresponds to heterochromatin.

Giant supercoil

The third level of organization is less clearly understood, but it is thought to involve the formation of DNA loops radiating from a central scaffold of non-histone proteins. It is thought that the loops form transcriptional units.

Chromosomes

At metaphase (see p. 119), chromatin is maximally condensed, and it forms 1400 nm fibres. After cell staining with biological dyes (i.e. Giemsa), these structures are visible as chromosomes under light microscopy.

Centromeres

Each metaphase chromosome is composed of two identical sister chromatids. Chromatids are

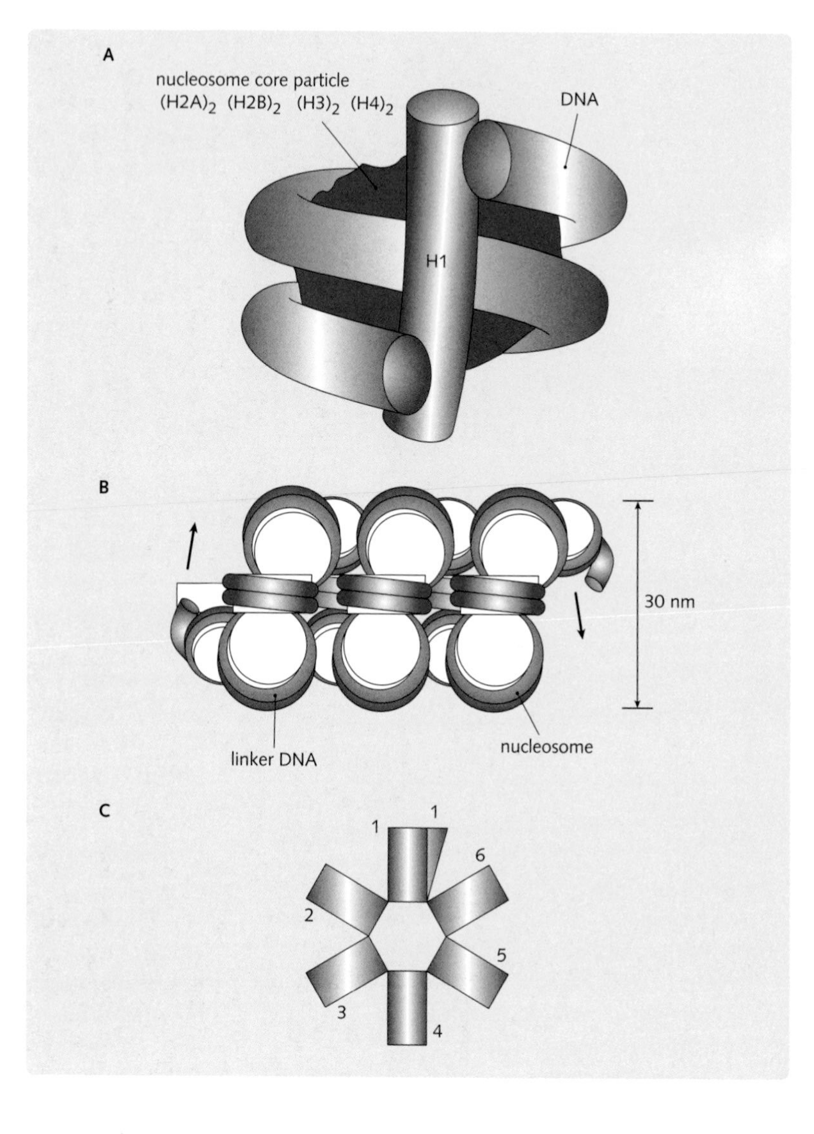

Fig. 6.13 Chromatin fibre organization. (A) The nucleosome core particle is composed of pairs of histones-$(H2A)_2$ $(H2B)_2$ $(H3)_2$ $(H4)_2$. A fifth histone, H1, is also associated with the nucleosome. 166 base pairs of DNA wind around each nucleosome. Linker DNA runs between one nucleosome and the next. It consists of 8–114 base pairs. (B) Chromatin consists of nucleosomes bound together through their H1 proteins (not shown in this part of the figure). (C) Bound nucleosomes form a solenoid, with six nucleosomes per turn. (Adapted with permission from Molecular Biology of the Cell, 3rd edn, by B Alberts et al, Garland Publishing, 1994. Reproduced with permission of Routledge, Inc., part of The Taylor & Francis Group.)

connected at a central region called the centromere, above and below which chromatin strands loop across between chromatids to hold them together (Fig. 6.14). Centromeres consist of hundreds of kilobases of repetitive DNA and are responsible for the movement of chromosomes at cell division. Each centromere divides the chromosome into short and long arms, designated p and q respectively.

Arms of the chromosome, remember: p is petite, queues are long – the p arm is the short arm and the q arm is the long arm.

Centromere position can be used to categorize chromosomes morphologically (see Fig. 6.14):

- Acrocentric chromosomes have centromeres located very close to one end, yielding a small short arm, often associated with small pieces of DNA called satellites, which encode rRNA.
- Metacentric chromosomes have centromeres located in the middle, yielding short and long arms of roughly equal length.
- Submetacentric chromosomes have a centromere that is off-centre, so that one chromosome arm is longer than the other.

The kinetochore is an organelle located at the centromere region (Fig. 6.15). It acts as a microtubule organizing centre and facilitates spindle formation

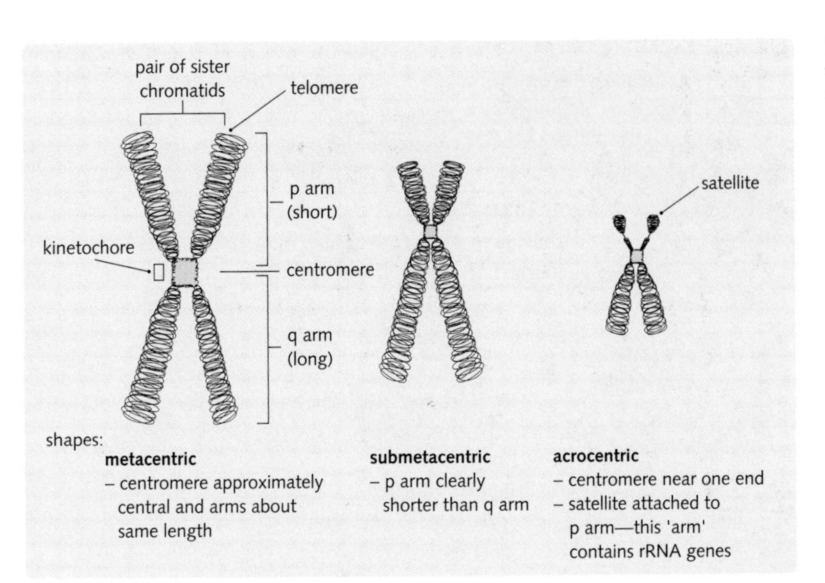

Fig. 6.14 Anatomy of a chromosome showing the three shapes: metacentric, submetacentric, and acrocentric.

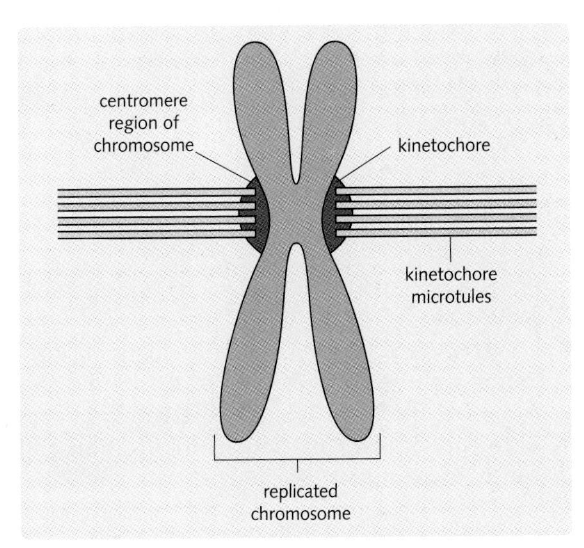

Fig. 6.15 The kinetochore. The kinetochore contains two regions: an inner kinetochore, which is tightly associated with the centromere DNA; and an outer kinetochore, which interacts with microtubules.

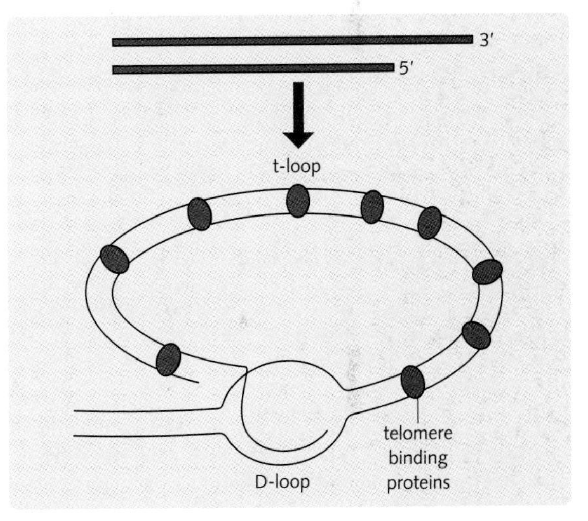

Fig. 6.16 Telomeres. T-loop formation prevents telomerase activity, blocking excessive telomere extension. With each cell division, the telomeric DNA shortens, until the telomere is too short to support a t-loop structure. At this point telomerase can act, lengthening the telomere, allowing the t-loop to reform. This cycle of shortening and lengthening means the t-loop effectively protects the telomere end.

by polymerization of tubulin dimers to form microtubules during the early stages of mitosis (see Ch. 4).

Telomeres

The ends of chromosomes are protected by DNA structures called telomeres. Telomeres are tandem repeats of the hexameric sequence 'TTAGGG', ending in a 3′ single-stranded overhang that ranges in length from about 50–400 nucleotides and loops back on itself to form the T-loop (Fig. 6.16). Telomeres have several functions in preserving chromosome stability, including:

- preventing abnormal end-to-end fusion of chromosomes
- protecting the ends of chromosomes from degradation
- ensuring complete DNA replication
- having a role in chromosome pairing during meiosis.

DNA polymerases require an RNA primer (see pp. 102–105). This poses a potential problem at the ends of eukaryotic chromosomes because there is nowhere for the lagging strand primer to bind to facilitate replication of the terminal sequence. Thus, potentially the chromosome could become progressively shorter after successive rounds of replication, resulting in a loss of genetic information. In order to avoid this, the cell employs the addition of non-coding DNA to the end of the 3′ tail via the enzyme telomerase (see p. 105).

Nuclear genes

Genes are sequences of DNA that encode proteins, and are composed of a transcribed region and a regulatory sequence. The 'one gene, one polypeptide' hypothesis states that the base sequence of DNA determines the amino-acid sequence in a single corresponding polypeptide. By convention, gene sequences are described in the direction 5′–3′, as this is the direction of *in vivo* nucleic acid synthesis.

Genes lie within expanses of 'non-coding DNA' which, until recently, were not believed to serve any function. Emerging research suggests that as much as 50% of non-coding DNA is transcribed and produces non-coding RNA (ncRNA) with postulated critical roles in regulating DNA structure, RNA expression, protein translation and protein functions.

Eukaryotic gene structure

The typical eukaryotic gene consists of a number of conserved features (Fig. 6.17).

Promoters

The promoter region lies in the 5′ DNA immediately preceding a gene and is sometimes called the 'upstream flanking region'. Sequence analysis studies have revealed 'consensus' sequences in both prokaryotes and eukaryotes, which are vital for promoter function. The most conserved sequence in the *Escherichia coli* promoter is the TATAAT box, which forms the initial binding site for the transcription enzyme RNAP (Ch. 1). The situation is more complex in eukaryotes, where multiple consensus sequences have been detected, including the GC box (5′-GGGCGGG-3′); the TATA box (5′-TATAAAAA-3′), located about 25 bp upstream of the transcription start site; and the CAAT box (5′-GGCCAATCT-3′), located 75 bp upstream of the start site, which act as binding sites for specific transcription factors.

Eukaryotic genes may also require enhancer sequences for physiological expression that may be situated many kilobases from the start of transcription (see pp. 105–109).

Introns and exons

The typical eukaryotic gene contains both exons and introns (Fig. 6.17). Exons are transcribed sequences (i.e. represented in final mRNA) – the majority of exons will code for amino acids, but 5′ or 3′ UTRs are also included in exons. Introns are the non-coding intervening sequences. The length of a typical exon is a few hundred base pairs whereas introns tend to be several kilobases long. Introns are:

- rare in prokaryotic genes
- uncommon in lower eukaryotes, such as yeast
- abundant in higher eukaryotes (vertebrate structural genes rarely lack introns).

3′ sequences

The 3′UTR is defined as the sequence extending from a coding region stop codon up to the point at its 3′ end where the transcript is cleaved, and is thought to influence mRNA translation, localization and stability. In particular, the polyadenylation signal (5′-AATAAA-3′) determines the site of polyadenylation of the resultant mRNA molecule, which in turn protects the molecule from the action of enzymes (exonucleases) and is important for transcription

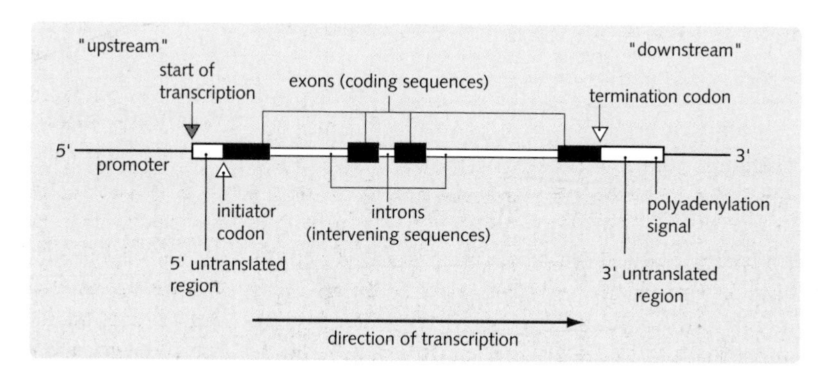

Fig. 6.17 Structure of a typical eukaryotic gene. The first and final exons include sequence that is transcribed and present in the mature mRNA, but not translated. These are called the 5′ and 3′ UTR (untranslated regions) respectively. Note that this diagram is not to scale and in reality the average intron is much longer than the average exon. (Adapted from Nussbaum, McInnes and Willard, 2001.)

termination, for export of the mRNA from the nucleus, and for translation.

Active chromatin

If nuclei isolated from different vertebrate cell types are treated with the enzyme DNaseI, they show different patterns of degradation. This is attributed to actively transcribed genes (which differ between cell types) being more sensitive to the enzyme, which suggests that active chromatin and inactive chromatin are packaged differently. In contrast to inactive chromatin, active chromatin shows the following features:

- Histone H1 is less tightly bound.
- Nucleosomal histones are highly acetylated.
- Histone H2B is less phosphorylated.
- There is enrichment of a variant of H2A.

The differences between active and inactive chromatin suggest that chromatin structure is important in the regulation of gene expression. It is thought that large areas of the genome may be transcriptionally silenced during differentiation as a result of specialized packing. Although the precise details are not understood, it is thought that methylation status may influence chromatin structure.

Methylation

The CpG dinucleotide is generally underrepresented in the genome. However, it is found at expected levels in promoter regions. In the promoters of inactive genes the 5′ position of cytosine is frequently methylated, whereas it is generally unmethylated in active genes. Methyl-CpG binding proteins bind to methylated regions and recruit histone deacetylase, which may trigger the formation of inactive chromatin. In this way, methylation may act to mark chromatin for inactivation, although the factors that prompt it to do so are unclear.

Several neurodevelopmental disorders, namely Rett, fragile X and ICF (immunodeficiency, centromeric instability and facial anomalies) syndromes, are thought to arise as a result of defective methylation.

Gene evolution

Intron shuffling

Eukaryotic genes contain introns, whereas prokaryotic genes do not. Exons often encode functional protein domains. If these are separated by long introns, then any chromosome break is likely to be in a non-coding region. Such breakage may lead to recombinational exchange between genes, 'shuffling' new combinations of exons together, to produce new, potentially advantageous proteins. If this process were to occur in the absence of introns, it is likely that the new protein would be out of frame, and the functional domains would be lost.

Gene duplication and multigene families

Gene duplication is a rare consequence of normal recombinational events. Duplicated genes are free to mutate and evolve new functions and patterns of expression, since they are not required for the survival of the organism. The mechanism of gene duplication events means that some functionally related proteins, such as the family of β-globin genes and the human leucocyte antigen (HLA) complex, are each clustered close together on the same chromosome.

HLA complex

The human leukocyte antigen (HLA) complex (also known as the major histocompatibility complex, MHC) on the short arm of chromosome 6 is an example of a gene cluster. The HLA genes:

- are mostly members of the immunoglobulin superfamily – a family of hundreds of genes, which occur throughout the human genome and are involved in cell surface recognition
- are characterized by the presence of immunoglobulin 'domains' (or folds) consisting of 110 amino-acid residues stabilized by a disulphide bridge
- share sets of related exons that they have evolved from an ancestral gene by duplication.

There are three classes of HLA genes (Fig. 6.18A). Classes I and II encode proteins that are involved in the presentation of peptide antigens to T cells. This is usually in the context of foreign antigens derived from infective pathogens, but includes 'non-self' antigens introduced through organ transplantation, hence their role in organ rejection (Fig. 6.18B). Class III genes encode proteins that modulate or regulate immune responses in other ways. These include tumour necrosis factor (TNF) and complement proteins (C2, C4) (Fig. 6.18B).

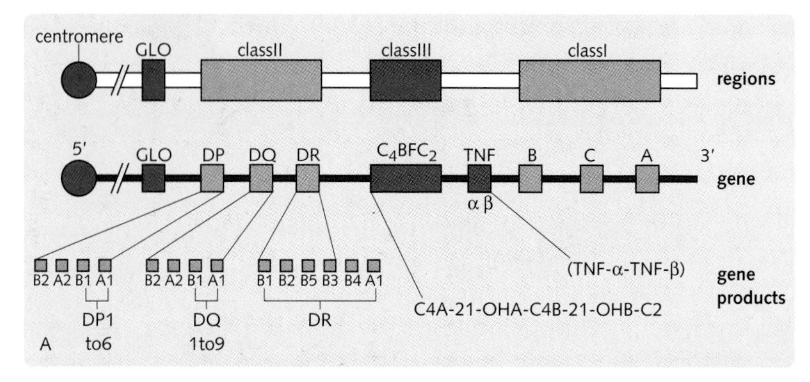

Fig 6.18 (A) The major histocompatibility complex (HLA region), showing the regions, genes and gene products on the short arm of chromosome 6 (6p21.1 to p21.3). GLO, glyoxalase; TNF, tumour necrosis factor. (Adapted from Kumar and Clark, 2005.) (B) Features of different classes of HLA genes.

B **Classification of the HLA haplotype genes**

	Class I	Class II	Class III
Main products	A, B, C code for α_1, α_2, α_3 proteins, which complex with β_2 microglobulin coded for on chromosome 15	DR, DP, DQ	Complement components C4a, C4b, TNF, etc
Expressed	On surface of all cells except erythrocytes	By antigen-presenting cells (e.g. macrophages and B lymphocytes)	
Structure		Each has an α and a β domain	Sequence homologies with class I and class II

The term haplotype is used to describe a cluster of alleles that occur together on a DNA segment and are inherited together. Children get one HLA haplotype from their mother and one HLA haplotype from their father. In the case of renal transplantation, the degree of matching for HLA loci A, B, and DR (Fig. 6.18A) is assessed to reduce the risk of organ rejection.

Pseudogenes

Pseudogenes are DNA sequences that closely resemble structural genes, but that cannot be translated into a functional protein. The presence of pseudogenes within the human genome gives an insight to gene evolution, especially as many functional genes have pseudogene equivalents. They are thought to have arisen in a number of ways, including incomplete duplication events, the silencing of functionally redundant genes and as a result of the insertion of complementary DNA sequences produced by the action of reverse transcriptase on a naturally occurring mRNA transcript (retrotransposition).

DNA REPLICATION

As discussed in Chapter 1, DNA replication is the process whereby a cell duplicates its nuclear DNA in preparation for cell division. Replication proceeds in the 5′ to 3′ direction with new nucleotides being attached to the 3′ OH of the growing molecule. One strand, known as the leading strand, is formed continuously, moving in the direction of the replication fork. The other strand, the lagging strand, is formed in short sequences of 1000–5000 nucleotides known as Okazaki fragments, which are then joined together enzymatically (see Fig. 6.21).

Eukaryotic DNA replication

Eukaryotic DNA synthesis is remarkably similar to that seen in prokaryotes (see Ch. 1). However, eukaryotes have many more origins of replication (Fig. 6.19), which are activated simultaneously during the S phase of the cell cycle (see pp. 115–118), enabling rapid replication of entire chromosomes.

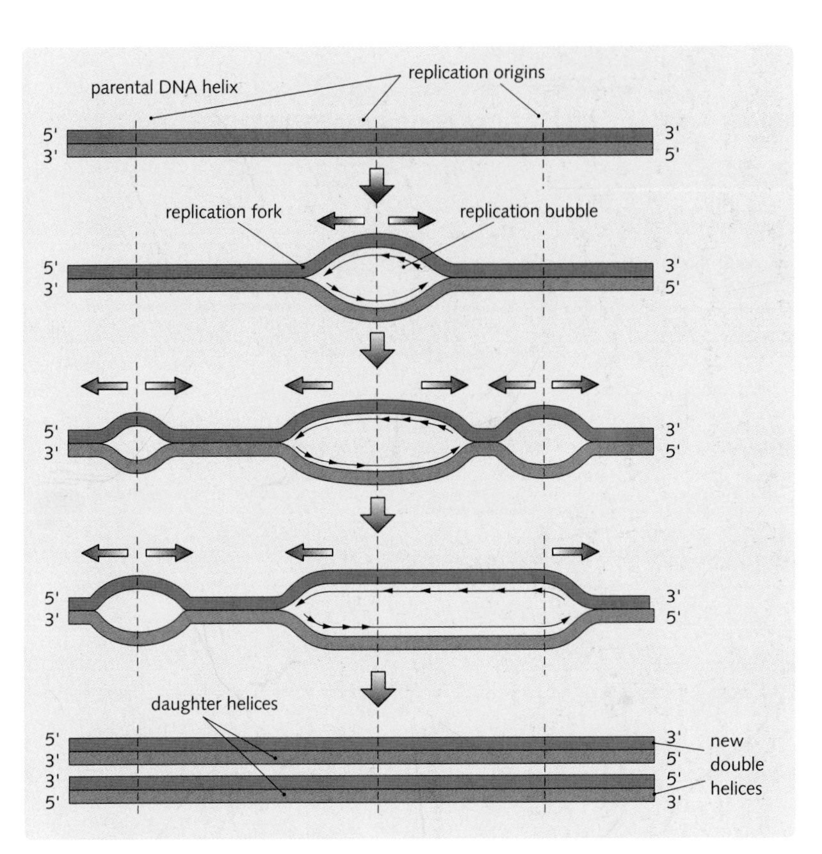

Fig. 6.19 Eukaryotic DNA replication. There are multiple origins of replication, but replication is initiated at specific points at specific times to ensure that the entire genome is replicated once, and once only. (Adapted from Jorde et al, 1997.)

The origin recognition complex, along with licensing factors, provides the trigger for an origin of replication to begin DNA replication, and prevents over replication from occurring. Replication proceeds in two directions from each origin of replication and continues until neighboring forks fuse. The rate of synthesis of DNA in eukaryotic cells is much slower than that of prokaryotic cells at only 75 nucleotides/s; about one-tenth the rate of bacterial DNA synthesis.

Heterochromatin replicates later in S phase than euchromatin. The genome is replicated once only, and it is thought that the chromatin is marked after replication to prevent it being replicated a second time, possibly by DNA methylation.

DNA polymerases

More than 19 polymerases have been identified in eukaryotes, with α, β, γ, δ and ε being the key enzymes required to maintain the integrity of the eukaryotic genome. Of these, two (α and δ) are especially important for the replication of chromosomes.

Replication forks

After the replication fork passes, chromatin structure is reformed by the addition of new histones. The presence of histone proteins in eukaryotic chromosomes is thought to account for the slower rate of eukaryotic replication and shorter Okazaki fragments, as compared to prokaryotic replication (Fig. 6.20).

Additional proteins involved in DNA replication

Proliferating cell nuclear antigen

Proliferating cell nuclear antigen (PCNA) is the eukaryotic counterpart of the regulated sliding clamp protein of *E. coli.* (see Figs 1.8, 6.21). It acts as a co-factor for DNA polymerase δ in S phase and also during DNA synthesis associated with DNA damage repair mechanisms.

Replication protein A

Replication protein A (RPA) is the eukaryotic equivalent to the single-strand binding proteins (see Fig. 1.8). RPA molecules facilitate the unwinding of the helix to create two replication forks. Experiments *in vitro* have shown that replication is 100 times faster when these proteins are attached to the single-stranded DNA. RPA is also required for nucleotide excision repair, and homologous recombination (Fig. 6.21).

Replication factor C

Replication factor C (RFC) is required, in the presence of ATP, to load the PCNA sliding clamp onto DNA,

Fig. 6.20 Comparison of prokaryotic and eukaryotic DNA replication

	E.coli (prokaryote)	Mammalian (eukaryote)
Site	Cytoplasm	Nucleus (and mitochondrion)
No. of proteins involved	30	100s
DNA polymerase	Pol I—fidelity and repair Pol III—DNA synthesis	Four enzymes identified, α polymerase is the principal nuclear polymerase
Initiation	Single origin of replication (OriC)	Multiple origins of replication (spatially and temporally separated during DNA replication)
Rate of replication	10^3 nucleotides/s	10^2 nucleotides/s
Post-replication	RNA primers removed from lag strand by Pol I 5′–3′ exonuclease	RNA primers removed by 5′–3′ exonuclease (NOT associated with α polymerase)
Timing of replication	Continuous DNA synthesis between cell divisions	DNA synthesis and cell division separated by G_1 and G_2 (gap) phases
Okazaki fragments	Large (1000 – 2000 bp)	Small (100 – 200 bp)
DNA polymerase	Same for leading and lagging strands	Different for leading and lagging strands

Fig. 6.20 Prokaryotic and eukaryotic DNA replication.

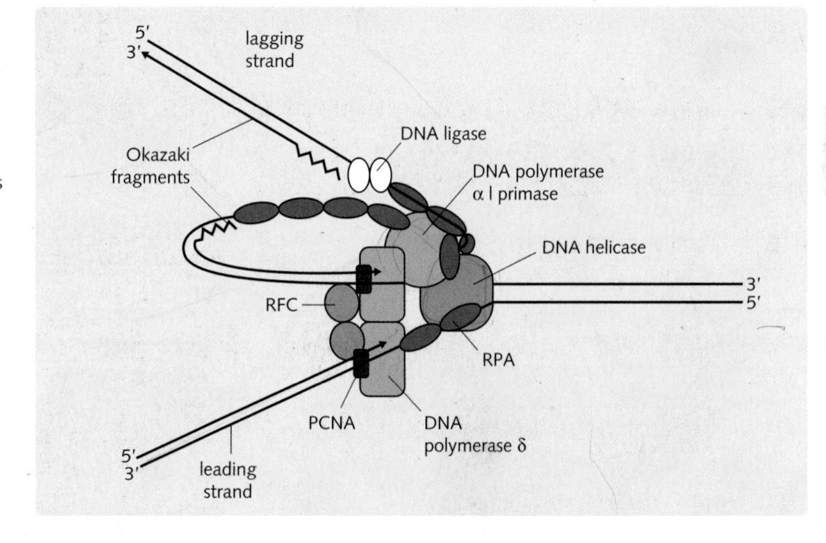

Fig. 6.21 Eukaryotic DNA replication. Leading strand synthesis starts with the primase activity of DNA polymerase α to lay down an RNA primer, to which it adds a stretch of DNA. RFC assembles PCNA at the end of the primer, followed by displacement of DNA polymerase α. DNA polymerase δ binds to PCNA at the 3′ ends of the growing DNA strand to carry out highly processive DNA synthesis. Lagging strand synthesis begins in the same manner as leading strand synthesis. RNA primers are synthesized by DNA polymerase α every 50 nucleotides. Polymerase switching generates Okazaki fragments, which are ultimately ligated together by DNA ligase.

thereby recruiting DNA polymerases to the site of DNA synthesis (Fig. 6.21).

Leading strand synthesis

Leading strand synthesis (Fig. 6.21) progresses as follows:

- DNA polymerase α has primase activity which lays down an RNA primer.
- DNA polymerase α adds a stretch of DNA to the RNA primer.
- RFC assembles PCNA at the end of the primer.
- PCNA displaces DNA polymerase α.
- DNA polymerase δ binds to PCNA to carry out highly processive DNA synthesis.

Lagging strand synthesis

Lagging strand synthesis (Fig. 6.21) commences in a similar fashion to leading strand synthesis:

- DNA polymerase α synthesizes RNA primers every 50–75 nucleotides.

- PCNA mediated polymerase switching extends the RNA-DNA primers to generate Okazaki fragments.
- DNA polymerase δ extends its activity towards the RNA primer of the downstream Okazaki fragment, at which point enzymes (RNases and exonucleases) remove the RNA primer.
- DNA polymerase δ fills in the gap as the RNA primer is being removed.
- DNA ligase joins the Okazaki fragment to the growing strand.

Telomerase

Telomeres are predisposed to becoming progressively shorter with each round of DNA replication (see pp. 99–100). Telomerase circumvents this problem by adding protective sequences to the ends of each chromosome. Although active in germ-line cells, telomerase is not normally active in somatic cells. It has been proposed that the progressive shortening of telomeres in somatic tissue is an important component of ageing. Moreover, immortal cancer cells frequently show regained telomerase activity.

EUKARYOTIC TRANSCRIPTION AND RNA SYNTHESIS

Definition of transcription

Transcription is the process by which RNA is synthesized according to a DNA template. The process is catalysed by DNA-dependent RNA polymerases, which unwind dsDNA to expose unpaired bases upon which DNA–RNA hybrids form. RNA is synthesized 5′–3′. The DNA template displays polarity in that only one strand can act as a template (template strand). The non-transcribed strand is called the coding strand as it has the same base composition as the RNA except that thymines (Ts) are substituted for uracils (Us).

The template strand is transcribed. It is identified by RNA polymerase, which binds to the specific DNA sequences that comprise the promoter. Transcription may be influenced and regulated by both *cis* and *trans* acting factors:

- *cis* acting factors are specific sequences of DNA that lie on the same molecule of DNA as the gene they regulate
- *trans* acting factors are proteins that bind to *cis* acting elements. They are transcribed from genes distinct from the ones they regulate.

Eukaryotic transcription

Eukaryotes have three chromosomally encoded RNA polymerases, which recognize different promoters and, therefore, transcribe different types of RNA molecules. The TATA (Hogness-Goldberg) box is found in the promoters of genes transcribed by RNA polymerase II. These polymerases can be identified because they differ in their sensitivity to a mushroom toxin, α-amanitin (Fig. 6.22).

Gilbert's syndrome (GS) is a benign, mildly symptomatic, unconjugated hyperbilirubinaemia, and affects ~10% of the population. Bilirubin UDP glucuronosyltransferase 1 (UGT1A1) activity in patients with GS is decreased to about 30% of normal, leading to a failure of uptake of albumin-bound bilirubin into hepatocytes. The large majority of mutations found to be associated with GS involves an expansion of the TATA box in UGT1A1 leading to reduced expression and, ultimately, reduced activity of the enzyme product.

There are three different RNA polymerases in eukaryotes:

1. RNA polymerase I transcribes ribosomal RNA in the nucleolus
2. RNA polymerase II transcribes messenger RNA in the nucleoplasm
3. RNA polymerase III transcribes transfer RNA (and one rRNA species) in the nucleoplasm.

Sequence of events

Initiation

Transcription in eukaryotes is more complicated than in prokaryotes, and it requires the presence of several transcription factors (proteins that are required to initiate or regulate eukaryotic transcription). All genes that are transcribed and expressed via mRNA are transcribed by the RNA polymerase II complex. It contains at least six basal transcription factors (TFII)

Fig. 6.22 Eukaryotic RNA polymerases. These differ with respect to their template specificity, localization and sensitivity to α-amanitin (the active ingredient in *Amanita phalloides*, a poisonous mushroom). (Adapted from Biochemistry, 3rd edn, by L Stryer, W.H. Freeman and Company, 1988.)

Fig. 6.22 Eukaryotic RNA polymerases

RNA polymerase	Localization	Cellular transcripts	Effect of α-amanitin
I	Nucleolus	18S, 5.8S, and 28S rRNA	Insensitive
II	Nucleoplasm	mRNA precursors and hnRNA	Strongly inhibited
III	Nucleoplasm	tRNA and 5S rRNA	Inhibited by high concentrations

A, B, D, E, F and H, and is believed to be assembled, and transcription initiated, as follows (Fig. 6.23):

- TFIID recognizes and binds to the TATA box. TFIID consists of TBP (TATA binding protein) and numerous TAFs (TBP associated factors).
- TFIIA binds TFIID and DNA, stabilizing the interaction.
- TFIIB binds to TFIID, recruiting TFIIF-RNA Pol II.
- TFIIF, having helicase activity, is believed to help expose the template strand of the DNA.
- TFIIE binds and recruits TFIIH to the complex.
- TFIIH phosphorylates Pol II.
- Pol II is released from the complex and begins transcription.

If the transcription complex contains only the six basal transcription factors highlighted above, the level of transcription is low. Physiological levels of transcription require the presence of enhancers. These are *cis* acting elements that may be located many kilobases away from the start of transcription in a 5′ or 3′ direction. Enhancers are bound by *trans* acting transactivating proteins, which are then able to associate with the polymerase complex to increase the level of transcription (Fig. 6.24).

Since the expression of transactivators is tissue specific, enhancers facilitate tissue-specific control of gene expression in multicellular organisms. Similarly, transcription from a gene may be reduced in a tissue-specific manner by the presence of silencers (the *cis* acting element) and repressor proteins (the *trans* acting element).

Elongation

The overall equation is the same as for prokaryotic elongation:

$$(RNA)_n + XTP \rightarrow (RNA)_{n+1} + PP_i \rightarrow 2P_i$$

However, the transcript is modified while it is still being transcribed by the addition of a 5′ cap (see below), which does not occur in prokaryotes.

The length of the primary RNA transcript (hnRNA), at an average of 7000 nucleotides, is very much larger than the average prokaryotic transcript. Moreover, it is larger than the predicted 1200 nucleotides needed to code for an average protein of 400 amino-acid residues. This discrepancy reflects the presence of introns, which are not found in prokaryotic transcripts.

Termination

Unlike prokaryotes, eukaryotic genes have no strong termination sequences. Instead, RNA polymerase II continues transcribing up to 1000 to 2000 nucleotides beyond where the 3′ end of the mature mRNA will be. The actual 3′ end is determined by the cleavage of the transcript at a highly conserved AAUAAA sequence, potentially generating the 3′ end of the message to which a polyA tail is added (see p. 108). This sequence is thought to be important as, when the corresponding sequence is deleted from the DNA template, no mature mRNA is made.

Eukaryotic post-transcriptional modification

Addition of a 5′ cap

This is a very early modification that occurs soon after transcription initiation. The cap structure, a 7-methylguanosine residue, is enzymatically added to the 5′ end of the hnRNA molecule by a unique 5′–5′ linkage (Fig. 6.25). It is thought to have four main functions:

1. It protects the mRNA from enzymatic attack.
2. It aids in splicing (i.e. the removal of introns from hnRNA).
3. It enhances translation of the mRNA.
4. It regulates nuclear export (via the cap binding complex).

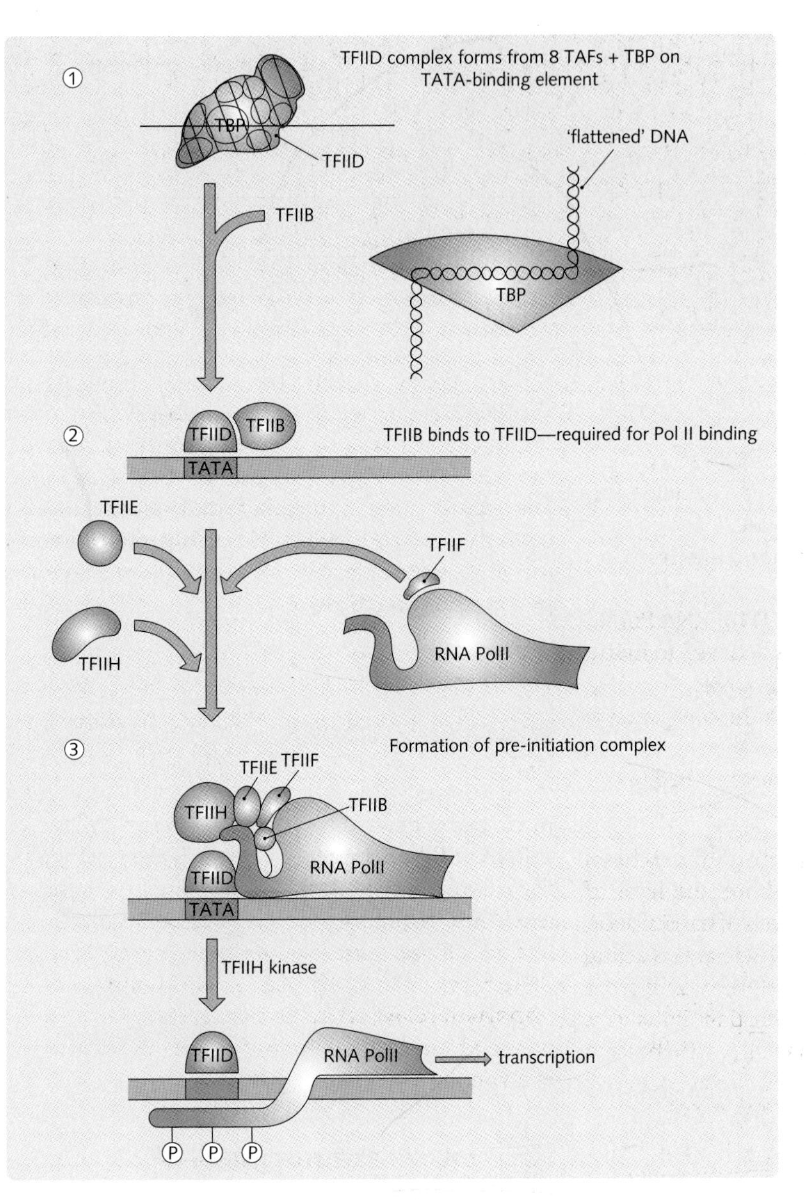

Fig. 6.23 Initiation of transcription with eukaryotic RNA polymerase II. In the final stage, TFIIH phosphorylates amino acids in the tail of RNA Pol II, which reduces its affinity for TAFs and releases RNA Pol II for transcription. A, adenine; T, thymine; TAF, TATA-associated factor; TBP, TATA-binding protein–a saddle-shaped protein that unwinds the DNA helix; TF, transcription factor.

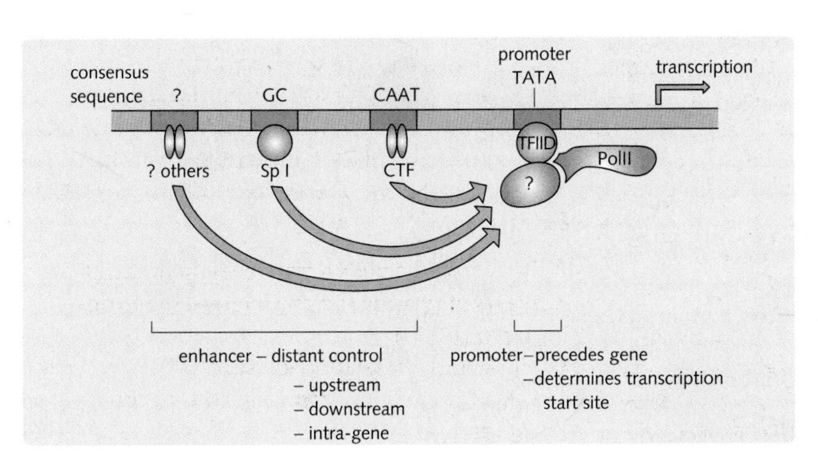

Fig. 6.24 Hierarchical control over gene expression in eukaryotes. Physiological levels of expression depend on the interaction between polymerase and transactivating and repressor proteins. There is sequence homology between TATAAT and eukaryotic TATA. Both form the site of the transcription initiation complex. A, adenine; T, thymine; C, cytosine; G, guanine.

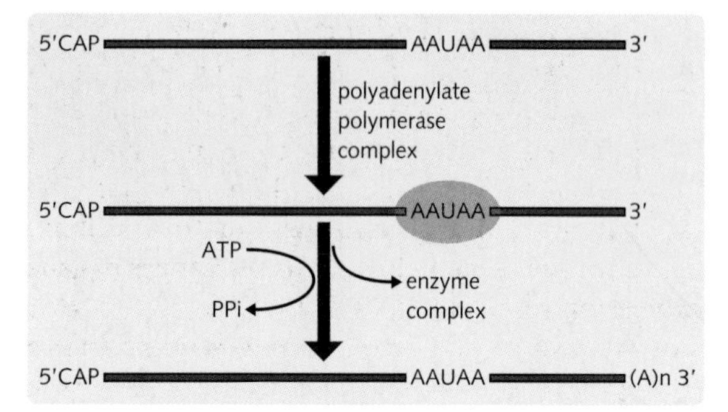

Fig. 6.25 Polyadenylation of mRNA. Cleavage occurs at the polyA site, which is usually found 10–35 nucleotides 3′ of the upstream polyA signal (AAUAAA). PolyA polymerase then adds several hundred A nucleotides to the 3′ end of the mRNA.

Polyadenylation

Although RNA polymerase continues to transcribe the DNA, the transcript is cleaved by the endonuclease activity of the polyadenylate polymerase complex approximately 10 to 30 nucleotides downstream from the polyadenylation signal (AAUAAA). The cleavage gives the 3′ end of the transcript a well-defined end. The 50–250 nucleotide poly(A) tail is generated from adenosine triphosphate (ATP) by the polyadenylate polymerase complex (Fig. 6.25). The degree of polyadenylation correlates with the half life of the mRNA molecule, generally the longer the poly(A) tail, the more stable the mRNA and the longer its half life.

Splicing

HnRNAs are the primary transcripts from genomic DNA. The production of mature eukaryotic mRNAs from hnRNA involves a process called gene splicing. This is the removal of non-coding introns and the joining together of the intervening exons, a process facilitated by a ribonucleoprotein complex called the spliceosome, which assembles immediately after the intron sequence has been transcribed. Spliceosomes consists of:

- a core structure made up of a number of subunits called small nuclear ribonuclear particles (snRNPs), which consist of both RNAs and proteins. There are four major classes of snRNPs, named according to the snRNA they contain: U1, U2, U5 and U4/U6
- non-snRNP splicing factors and an hnRNA.

Autoantibodies against snRNP proteins have been implicated in several autoimmune conditions. Components of the U1, U2 and U4–U6 snRNPs are antigenic targets in the anti-Smith antigen (Sm) response, present in about 30% of all patients with systemic lupus erythematosus. Autoantibodies against U1 snRNP-specific proteins (anti-nRNP) are specifically seen in mixed connective tissue disease. Both the Sm and nRNP antigens are required for the normal post-transcriptional, pre-messenger RNA processing to excise introns.

The splicing process depends on the existence of consensus sequences within the hnRNA intron, which are recognized by components of the spliceosome (Fig. 6.26):

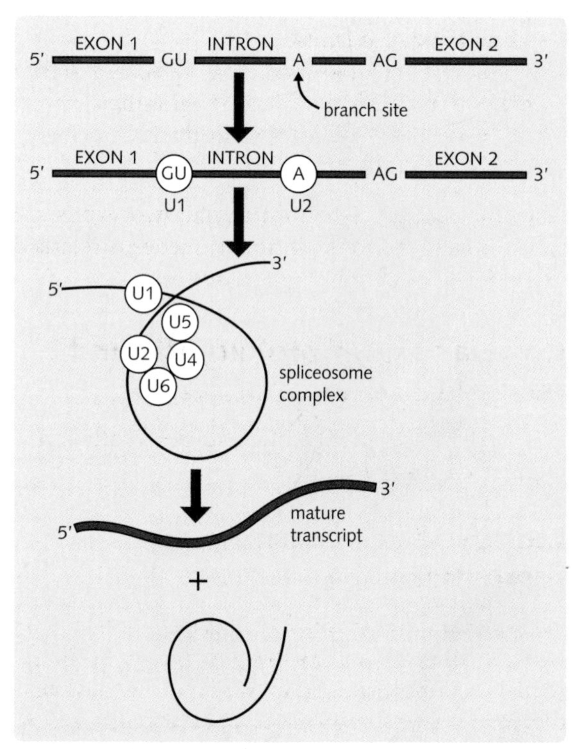

Fig. 6.26 Spliceosome mediated RNA splicing. The first stage in the reaction results in the breaking of the phosphate bond at the 5′ exon/intron boundary and its joining to the adenine at the branch site. In the second step the phosphate bond at the 3′ exon/intron boundary is cleaved, which is followed by the reformation of a phosphate bond between the terminal nucleotide of the first exon and the first nucleotide of the second exon.

- The first two nucleotides of the intron are always GU, to which binds U1 forming the splice donor site.
- An 'A' nucleotide approximately 30 nucleotides from the 3′ end of the intron binds U2 and forms the branch-point.
- The last two nucleotides of the intron are always AG, which binds U5 and forms the splice acceptor site.

It should be noted that RNA sequences contain many GU and AG sequences that are not used in splicing. The reason for this is poorly understood.

The binding of the snRNPs to the intron causes it to form a loop. The splicing reaction then proceeds in two stages, with the first stage releasing the first exon and the second stage the second exon (Fig. 6.26):

- The nucleotide at the branch site attacks the donor site and cleaves it, such that the 5′ end of the intron becomes covalently attached to the adenine nucleotide at the branch site, forming a 'lariat'-shaped structure.
- The 3′-OH end of the first exon (generated in the previous reaction) adds to the beginning of the second exon sequence, cleaving the RNA at the splice acceptor site.

After the reaction is complete, the two exons are joined together and the intron sequence is released as a lariat.

Comparison of prokaryotic and eukaryotic transcription

A comparison of prokaryotic and eukaryotic transcription is shown in Fig. 6.27.

Eukaryotic transcriptional regulation

Transcriptional regulation in eukaryotes is much more complex than in prokaryotes (see Ch. 1). Although the DNA complement in all eukaryotic cell types is the same, the genes expressed from it vary immensely. Differences in chromatin structure between cell types determine whether specific transcriptional activators and repressors are expressed. These bind to enhancers and silencers respectively and influence the levels of transcription of individual genes in a tissue-specific manner.

Even within active euchromatin, the expression of genes is subject to regulation according to the needs of the whole organism. Thus, signalling cascades initiated by hormones in response to changes in the external environment will ultimately alter gene expression. Similarly, changes in the cellular internal environment may result in changes in gene expression (for example, an accumulation of DNA damage will trigger the expression of apoptosis genes).

Epigenetic mechanisms

When a specific cell type replicates, the daughter cells retain the structural characteristics of the parent cell. This suggests that the changes in chromatin structure initiated in differentiation can be transmitted. Epigenetic mechanisms are those that exert a heritable influence on gene activity that is not accompanied by a change in DNA sequence. Examples include histone modification and methylation (see p. 101).

EUKARYOTIC TRANSLATION AND PROTEIN SYNTHESIS

Translation is the mRNA-directed biosynthesis of polypeptides. It is a complex process that involves several hundred macromolecules.

The genetic code

In the translation process from mRNA to protein, amino acids are coded for by groups of three bases called codons. Since nucleic acids contain four bases, there are 4^3 (64) possible codons. The same, non-overlapping, genetic code (Fig. 6.28) is seen in most living organisms, so it is said to be universal. Out of 64 possible codons, the genetic code consists of 61 amino-acid coding codons and three termination codons, which stop the process of translation.

Since there are 61 amino-acid coding codons, but only 20 amino acids that are commonly used in polypeptide synthesis, a large proportion of the codons could be considered redundant. Instead, the code is considered to be degenerate, that is more than one codon exists for each amino acid. For example, the codons GGU, GGC, GGA and GGG all encode the amino acid glycine.

Codons in the mRNA transcript are recognized by the 3 nucleotide 'anticodon' tRNA molecules

Fig. 6.27 Comparison of prokaryotic and eukaryotic transcription. A, adenine; C, cytosine; G, guanine; T, thymine; U, uracil.

Fig. 6.27 Comparison of prokaryotic and eukaryotic transcription

	Prokaryotic	**Eukaryotic**
Site of transcription	Cytoplasm	Nucleus: 5.8S/18S/28S rRNA in nucleolus 5S rRNA, tRNA, mRNA in nucleoplasm
RNA polymerase	Single species: RNAP (*E. coli*); large holoenzyme; core enzyme consists of four subunits: $\alpha_2\beta\beta'$	Three RNA polymerases in nucleus: Pol I transcribes rRNA; Pol II, mRNA; Pol III, rRNA and tRNA; consist of two large subunits with homology to the prokaryotic β subunits and a complex array of approximately 12 small subunits (e.g. Pol II assembly initiated by TFIID binding to promoter)
Initiation	S subunit associates with the core enzyme and facilitates binding to the promoter	Complex assembly of proteins
Termination	RNA forms stable hairpin loop between A–T-rich and G–C-rich region, then weak base pairing between poly(U) RNA and DNA encourages dissociation; the ρ factor is an enzyme that facilitates transcription termination with or without the hairpin loop	Termination is imprecise and signal is unknown
Post-trans-criptional modification	mRNA requires little or no modification	hnRNA = mRNA precursor: 5′ methyl cap added during transcription; 3′ end cleaved at AAUAAA site and poly(A) tail added; introns removed by splicing
mRNA	Polycistronic: each transcript can code for more than one polypeptide	Monocistronic: can code for only one polypeptide

charged with the appropriate amino acid. Codons that differ in the third base may be recognized by the same tRNA, while those that differ in the first or second bases are not. The 'wobble hypothesis' seeks to explain this by suggesting that the third base in tRNA anticodons allow for a certain amount of play (or 'wobble'), and so it may bind a variety of bases.

Three codons (UAA, UAG, and UGA) are not recognized by tRNAs, and these are termed stop codons. They mark the end of a polypeptide and signal to the ribosome to stop synthesis.

Mitochondrial DNA and the genetic code

Mitochondria contain their own DNA, which differs from that in the rest of the cell. Human mitochondrial DNA consists of 16 kb of circular dsDNA. It codes for:

- 22 mitochondrial (mt) tRNAs
- two mt rRNAs
- 13 proteins synthesized by the mitochondrion's own protein-synthesizing machinery. All are subunits of the oxidative phosphorylation pathway.

Fig. 6.28 Standard genetic code

First base (5')	Second base U	C	A	G	Third base (3')
U	UUU phe	UCU ser	UAU tyr	UGU cys	U
	UUC phe	UCC ser	UAC try	UGC cys	C
	UUA leu	UCA ser	UAA stop	UGA stop	A
	UUG leu	UCG ser	UAG stop	UGG trp	G
C	CUU leu	CCU pro	CAU his	CGU arg	U
	CUC leu	CCC pro	CAC his	CGC arg	C
	CUA leu	CCA pro	CAA gln	CGA arg	A
	CUG leu	CCG pro	CAG gln	CGG arg	G
A	AUU ile	ACU thr	AAU asn	AGU ser	U
	AUC ile	ACC thr	AAC asn	AGC ser	C
	AUA ile	ACA thr	AAA lys	AGA arg	A
	AUG met	ACG thr	AAG lys	AGG arg	G
A	GUU val	GCU ala	GAU asp	GGU gly	U
	GUC val	GCC ala	GAC asp	GGC gly	C
	GUA val	GCA ala	GAA glu	GGA gly	A
	GUG val	GCG ala	GAG glu	GGG gly	G

Fig. 6.28 Standard genetic code. A, adenine; C, cytosine; G, guanine; T, thymine; U, uracil. To find out which amino acid a particular codon codes for, first select the 5′ end base (left column). Then read across to select the column corresponding to the second position base. Read down to find the 3′ end base (right column) and locate the row on which the corresponding amino acid lies. (Adapted from Nussbaum, McInnes and Willard, 2001.)

The differences between mitochondrial DNA code and nuclear DNA code are shown in Fig. 6.29. Codon/anticodon pairings show more 'wobble' pairings than in the process originating in the nucleus. This is made possible by unusual mt tRNA sequences such as mt $tRNA^{ser}$, which lacks a D arm.

Eukaryotic protein synthesis

Protein synthesis is very similar to that seen in prokaryotes (see Ch. 1), but more associated factors are involved. Fig. 6.30 compares prokaryotic and eukaryotic translation.

Fig. 6.29 Variations between mitochondrial and standard genetic code

	Standard	Mammalian mitochondrion
UGA	Stop	Trp
AUA	Ile	Met (initiation signal)
CUN	Leu	–
AGA/AGG	Arg	Stop
CGG	Arg	–

Fig. 6.29 Summary of variations between mitochondrial and standard genetic code. N, one of four nucleotides.

Initiation

The mechanism behind initiation of translation in eukaryotic cells is poorly understood. In general, there are two ways by which ribosomal subunits can reach the initiator AUG codon:

1. Cap-dependent initiation by a scanning ribosome (Fig. 6.31)
2. Cap-independent initiation by the presence of an internal ribosome entry site (IRES) within the 5′-UTR (Fig. 6.31).

The majority of eukaryotic translation initiation is believed to be cap-dependent, and involves ribosomal scanning of the 5′ untranslated region (5′-UTR) for an initiating AUG start codon. Both the proximity to the cap and the nucleotides surrounding the AUG start codon can influence the efficiency of the start site recognition during the scanning process. If the recognition site is of poor quality, the scanning (40S) ribosomal subunit will ignore and skip potential starting AUGs, a phenomenon called leaky scanning. As such, initiation is not always restricted to the AUG codon nearest the 5′ end. Like prokaryotes, the eukaryotic initiator tRNA carries a methionine residue. However, unlike prokaryotes, this methionine residue is not formylated.

Fig. 6.30 Comparison between prokaryotic and eukaryotic translation

	Prokaryotes	Eukaryotes
Ribosome	Large subunit 50S, small subunit 30S, whole ribosome 70S	Large subunit 60S, small subunit 40S, whole ribosome 80S
Initiation	Three initiation factors called IF-1, 2, 3; initiator tRNA carries f-Met (formylated methionine); start codon AUG; Shine–Dalgarno sequence precedes the start site on the mRNA; binds to a complementary sequence on the ribosome's S subunit	Over 10 initiation factors with multiple subunits called eIFs ('e' for eukaryote); initiator tRNA carries Met (not *N*-formylated); start codon AUG; no Shine–Dalgarno sequence; mRNA 5′-methylated cap may have a binding site on ribosome S subunit to guide translation complex to start site
Type of mRNA code	Polycistronic (mRNA often codes for more than one protein)	Monocistronic (mRNA always codes for a single protein)
Elongation	Elongation factors called EF-Tu, EF-Ts and EF-G	EF-Tu and EF-Ts are replaced by a single factor, eEF-1; EF-G replaced by eEF-2
Termination	Three release factors, RF-1, 2, 3; RF-3 is bound to GTP and the RF-3–GTP complex stimulates ribosomal binding of RF-1 and 2; GTP hydrolysis triggers complex to disassemble	Single release factor, eRF, which binds to the ribosome with GTP; GTP hydrolysis triggers eRF release from ribosome

Fig. 6.30 Comparison between prokaryotic and eukaryotic translation.

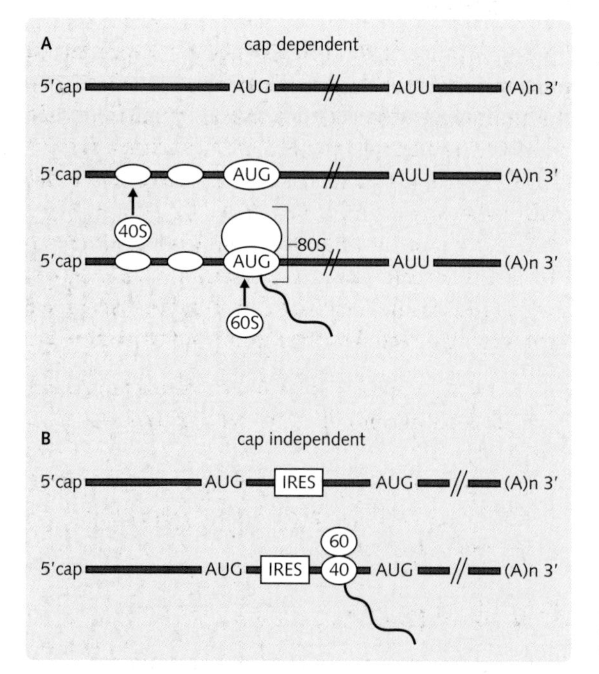

Fig. 6.31 Initiation of translation. (A) Cap dependent. The 40S ribosomal subunit binds the mRNA in a 5′ cap dependent manner, and scans the mRNA until it recognizes the relevant start codon, or AUG sequence. The 60S ribosomal subunit then binds the 40S in a GTP dependent manner. (B) Cap independent. Initiation is mediated by binding of the 40S ribosomal subunit to internal ribosome entry site (IRES) elements.

Elongation

As is the case with prokaryotes, the translation elongation cycle adds one amino acid at a time to a growing polypeptide according to the sequence of codons found in the mRNA. This process requires eukaryotic elongation factors (eEFs).

> Diphtheria toxin is a bacterial exotoxin, encoded by a bacteriophage gene. It is secreted as a single polypeptide and cleaved into two fragments. One fragment is required for toxin binding to the membrane of the host cell, while the other contains enzymatic activity (ADP-ribosylation) for the inhibition of elongation factor-2 (EF-2), inactivating the transfer of amino acids from tRNA to the growing polypeptide chain and inhibiting protein synthesis. Diphtheria toxin is very potent in its action; a single molecule within a cell is lethal, inactivating millions of EF-2 molecules.

Termination

All three stop codons, UAA, UGA and UGC, are recognized by the same specific eukaryotic releasing factor (eRF), which is composed of two proteins, eRF1 and eRF3. Binding of eRF to a stop codon results in the release of the mRNA from the ribosome. The ribosome then dissociates into its constituent 40S and 60S subunits ready to reassemble on another molecule to begin a new round of protein synthesis (Fig. 6.30).

CONTROL OF GENE EXPRESSION AND PROTEIN SYNTHESIS

Control of protein synthesis is synonymous with control of gene expression. In the prokaryote, control can be at the level of transcription or translation (see Ch. 1). The situation is more complex in eukaryotes and a total of six control points have been identified (Fig. 6.32).

Constitutive, inducible and repressible enzymes

Constitutive enzymes are present at fixed concentrations in the cell, irrespective of changes in the cell environment. They are examples of products of housekeeping genes in multicellular organisms.

The level of expression of inducible/repressible enzymes is altered by the cellular environment. A certain chemical may induce the expression of a gene, while another chemical may repress its expression.

Many hepatic drug-metabolizing enzymes are induced or inhibited by therapeutic drugs and other compounds. The P450 enzyme subfamily CPY3A is the most abundant of the hepatic cytochrome enzymes. It can be induced (e.g. by phenytoin and rifampicin) or inhibited (e.g. by erythromycin and fluoxetine). Induction of these enzymes can lead to accelerated drug breakdown and a reduced therapeutic window, while inhibition leads to prolonged drug activity with the potential of toxicity.

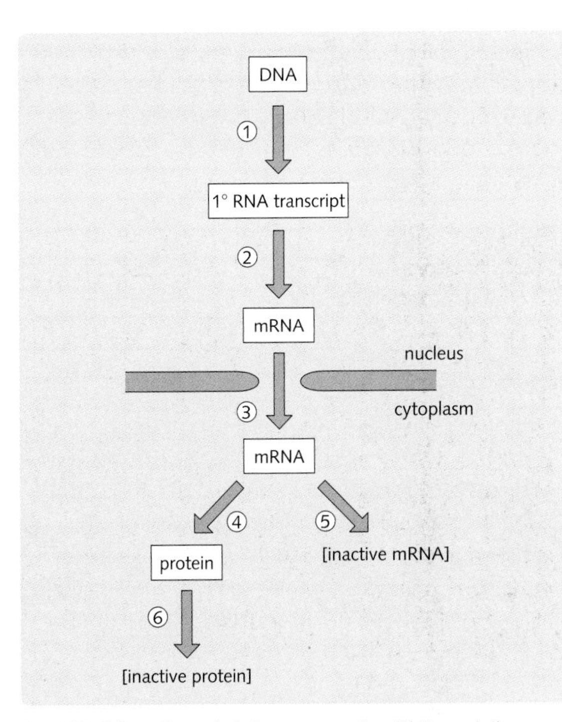

Fig. 6.32 Eukaryotic control of gene expression. (1) Transcription control. (2) Processing of transcript. (3) Transport control. (4) Translational control by selection of ribosomes by mRNA. (5) mRNA degradation control. (6) Protein activity control and post-translational modification.

POST-TRANSLATIONAL MODIFICATION OF PROTEINS

Concepts

Post-translational modification is the alteration of proteins after translation. These modifications give the mature protein functional activity and include peptide cleavage and covalent modifications, such as glycosylation, phosphorylation, carboxylation and hydroxylation of specific residues.

A newly synthesized protein may be destined for extracellular secretion, the cytoplasm or organelles, such as the plasma membrane, nucleus or lysosomes. Proteins are directed to the appropriate location by:

- conserved amino-acid sequence motifs (e.g. the signal peptide and the nuclear targeting signal)
- moieties added by post-translational modification (e.g. mannose-6-phosphate for lysosomal delivery).

Signal peptide

The signal peptide (or leader sequence) is a characteristic hydrophobic amino acid sequence of 18–30 amino acid residues near the amino terminus of the polypeptide that directs non-cytoplasmic polypeptides into the ER lumen as they are translated (Fig. 6.33).

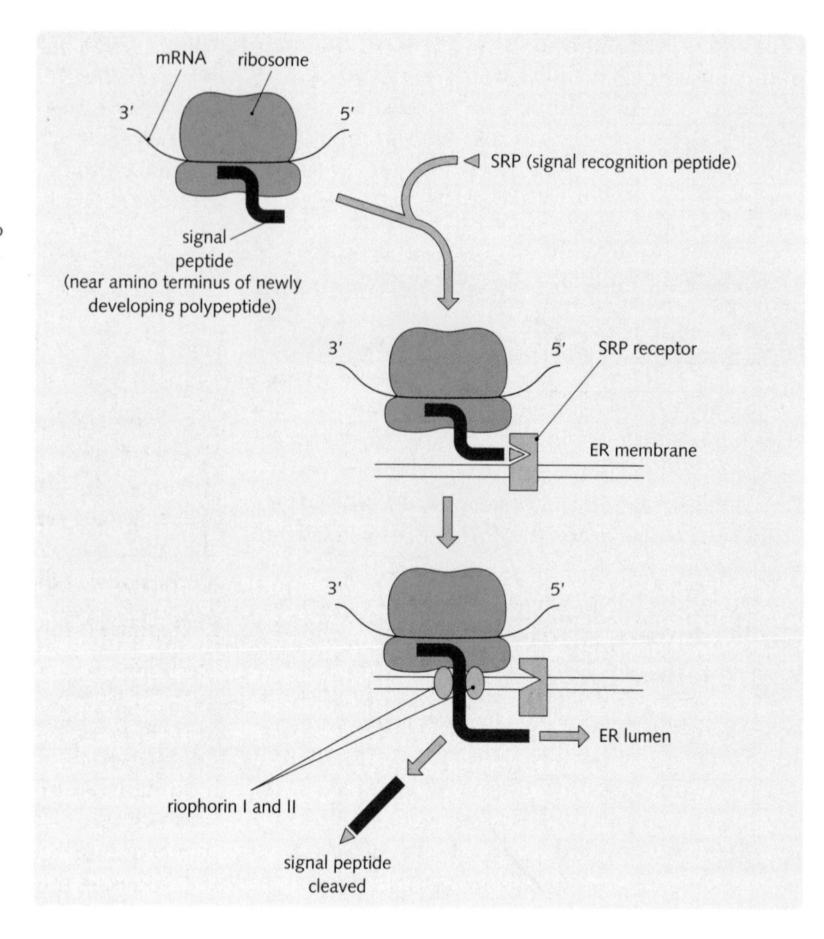

Fig. 6.33 Role of the signal peptide in translocation into the endoplasmic reticulum (ER) lumen. The signal peptide is near the amino terminus of the newly developing polypeptide. It associates with a cytoplasmic signal recognition peptide (SRP) and then with an SRP receptor ('docking protein') on the ER membrane. The ribosome then interlocks between two membrane-associated proteins, riophorin I and II, which drive the developing polypeptide into the ER lumen.

Inside the ER lumen the polypeptide can undergo post-translational modifications specific to each protein. Proteins destined for extracellular secretion are transported to the Golgi apparatus in vesicles that bud off the ER. Further modification may take place here and the secreted protein is packaged into vesicles that fuse with the cell plasma membrane, releasing their contents to the exterior.

I-cell disease (mucolipidosis type II) is a lysosomal storage disease resulting from a deficiency of mannose-6-phosphate glycosyltransferase, which is responsible for glycosylation of enzymes destined for lysosomes. As the cell's lysosomal enzymes lack their lysosomal uptake recognition markers, they are secreted into the extracellular matrix, and undigested substrates accumulate within the lysosomes with severe pathological consequences, including rapidly progressive growth failure and severe developmental delay, skeletal deformation and early death.

Glycosylation of proteins

Glycosylation occurs in the ER lumen or the Golgi apparatus, and it involves the addition of an oligosaccharide to specific amino-acid residues. There are two types of glycosylation, designated N-linked and O-linked, that employ specific glycosyltransferases:

- N-linked – the oligosaccharide is added to the polypeptide by a β-N-glycosidic bond to an aspartate residue
- O-linked – the oligosaccharide is an α-O-glycosidic bond to a serine or threonine residue.

Glycosylation is important for the functioning of some proteins and the correct compartmentalization of others:

- O-linked glycosylation is involved in the production of blood group antigens.
- N-linked glycosylation is involved in the transfer of acid hydrolase to lysosomes and the production of mature antibodies.

Other modifications of protein

Proteins may be modified by (Fig. 6.34):

- phosphorylation, which targets Ser or Tyr residues and tends to regulate enzyme activity (Ser) or protein activity (Tyr). Kinases transfer phosphate groups from ATP onto the target residue
- sulphation, which targets Tyr. It is important in compartmentalization (e.g. marking proteins for export) and biological activity
- hydroxylation, which targets Lys and Pro residues. It is very important in the production of collagen (and extracellular matrix protein). The hydroxylation of Lys and Pro residues occurs during translation, and it is essential for the formation of the collagen triple helix
- lipidation of Cys and Gly residues, which is necessary for anchoring proteins, such as antibody receptors, into the membrane
- acetylation of Lys, which can change the charge of the residue. In histone H4, this alters its binding properties to DNA
- cleavage, which activates some enzymes and hormones.

CELL CYCLE

Concept of the cell cycle

The cell cycle is a controlled set of events, culminating in cell growth and division into two daughter cells. The events of the cell cycle are ordered into pathways in which the initiation of late events, such as cell division, is dependent on the successful completion of early events, such as DNA synthesis. The cell cycle consists of four distinct phases (Fig. 6.35):

1. G_1 phase is the gap between mitosis of the preceding round of the cell cycle and the DNA synthesis phase of the current cycle. It contains the restriction point, which denotes the start of the cell cycle.
2. S phase is the DNA synthesis phase, during which the cell's chromosomes are replicated in preparation for cell division.
3. G_2 phase is the gap between the completion of DNA synthesis and the decision to divide.

Fig. 6.34 Summary of some post-translational modifications

Destination	Protein function	Modification	Residue	Example
Secreted	Structural	Hydroxylation	Lys/Pro	Collagen
	Enzyme	Hydrolytic cleavage	(peptide bond)	Pepsinogen → pepsin
	Hormone	Hydrolytic cleavage	(peptide bond)	Proinsulin → insulin
	Clotting factor	Carboxylation	Glu	Prothrombin → thrombin
	Antibody	Glycosylation (N-linked)	Asp	IgG
Membrane	Receptor	Lipidation	Gly/Cys	Antibody receptors
	Cell recognition	Glycosylation (O-linked)	Ser	Blood group antigens
	Receptor activation	Phosphorylation	Tyr	Growth factor receptor
Cytoplasm	Enzyme	Phosphorylation	Tyr/Ser	
Lysosome	Hydrolytic enzymes	Glycosylation (N-linked)	Asp	Acid hydrolases

Fig. 6.34 Summary of some post-translational modifications.

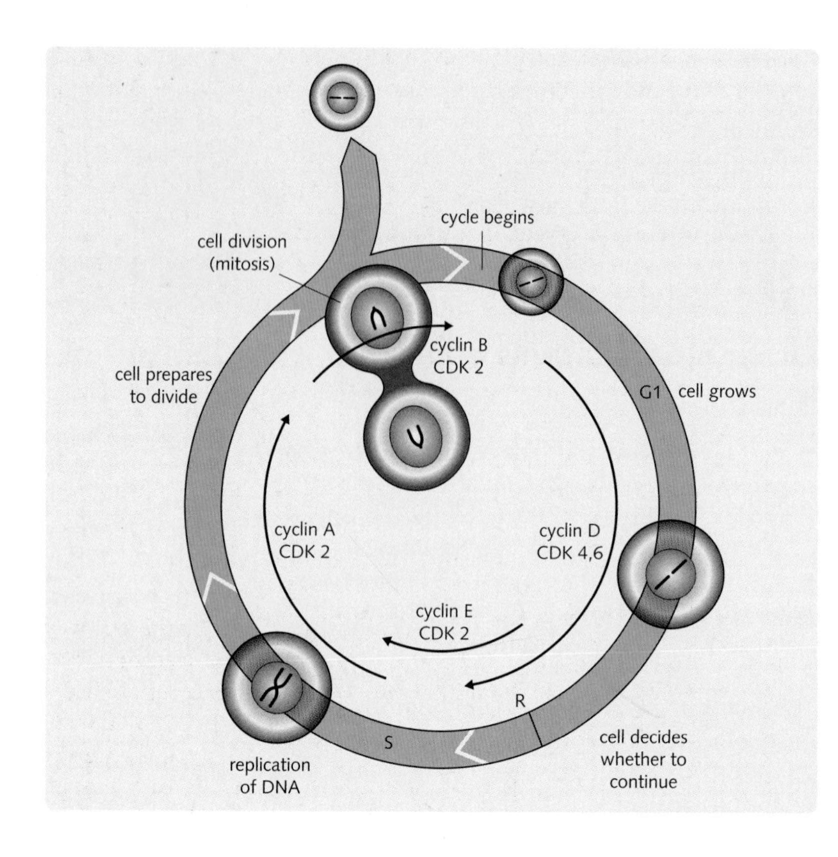

Fig. 6.35 The cell cycle – see text for explanation of each phase.

4. M phase, or mitosis, is the cell division phase and results in the production of two daughter cells.

Together, G_1, S and G_2 phases are known as interphase, the interval between divisions during which the cell undergoes its functions and prepares for mitosis. Non-dividing cells, such as neurons, are quiescent, i.e. are not cycling, and they remain in a resting state called G_0.

Regulation of the cell cycle

Progression through the cell cycle is controlled by the activity of cyclins, which are proteins that govern the transition from one phase to another (Fig. 6.35). The activity of the cyclins is modulated by changes in their intracellular concentrations, which is regulated both at the level of mRNA expression and at the protein level. In the presence of DNA damage and structural problems proteins known as the 'checkpoint proteins', function to stall the cell cycle and allow DNA repair to occur, preventing the propagation of mutations.

Cyclin-dependent kinases

Forward progression though the cell cycle is controlled by a set of protein kinases, the activity of which rises and falls at different cycle phases. These protein kinases are noncovalent complexes of an activating protein called a cyclin, and a catalytic subunit called a cyclin-dependent kinase (CDK). Activated CDKs are serine/threonine kinases which stimulate cell cycle progression by phosphorylating specific proteins in the cell required for transition to the next stage. For example:

- at the beginning of prophase of mitosis, nuclear membrane breakdown is initiated by phosphorylation of lamins, which form part of the nuclear skeleton
- chromosome condensation at the beginning of mitosis is initiated by the phosphorylation of H1 histone, a nuclear-associated protein
- G_1–S transition is initiated by CDK-dependent phosphorylation of Rb protein. Unphosphorylated Rb protein forms a

complex with the transcription factor E2F. On phosphorylation of Rb, E2F is released and activates transcription of genes required for the transition from G_1 into S phase.

Human cyclins are typically designated A, B, D and E. Each one accumulates at a different time in the cell cycle (Fig. 6.36).

Maturation-promoting factor (MPF) is an example of a cyclin–CDK complex. It initiates transition from G_2–M, under the control of cyclin B. A small increase in cyclin levels produces a big increase in MPF kinase activity, promoting chromosomal condensation.

CDKs are negatively controlled by CDK inhibitors (CKIs). When cyclin levels rise above a threshold, CKIs can no longer exert their effect. They are also utilized to temporarily arrest the cell cycle in response to DNA damage and unfavourable environmental conditions. Examples of CDK inhibitor families include:

- INK4 – inhibits CDK4 and CDK6
- CIP/KIP – inhibits G1/S CDKs and S phase CDKs, activates cyclin D-CDK4.

The activity of CDKs, and thus progression through the cell cycle, is further influenced by several extracellular signalling pathways that facilitate coordinated cell division in the multicellular organism.

Extracellular regulation of the cell cycle

A mitogen is an agent that induces mitosis. In addition to the activity of cyclins, the cell cycle is influenced by a multitude of factors that activate the mitogen activated protein kinase (MAPK) pathways, including:

- growth factors
- hormones
- cell–cell interactions.

Fig. 6.36 Cyclins and CDKs involved in cell cycle progression

Cyclin	Kinase	Function
D	CDK4, CDK6	Progression past restriction point at G_1/S boundary
E, A	CDK2	Initiation of DNA synthesis in early S phase
B	CDK1	Transition from G_2 to M

Fig. 6.36 Cyclins and cyclin-dependent kinases (CDKs) involved in the cell cycle. Cyclin C has recently been discovered and, along with CDK8, is thought to regulate RNA transcription during the cell cycle.

The huge variety of factors involved allows cell growth and replication to be finely controlled, responding to changes in the environment. Growth factors are soluble substances that can act locally or over long distances to affect cell growth. In common with the effects of hormones and cell–cell interactions, growth factors exert their action by binding to specific target receptors on the cell surface. This in turn initiates the phosphorylation of target proteins within the cell in a cascade that ultimately results in changes in gene expression.

Checkpoints

There are numerous points of control in the cell cycle where progression through the cycle may be regulated. These include the following:

- The restriction point, which occurs during G_1. At this point the cell becomes independent of external mitogenic stimuli and becomes committed to completing a cycle of division. The cell will not proceed beyond this point if there are inadequate nutrients or growth factors available.
- The G_1–S DNA integrity checkpoint ensures that the previous cycle of division has been completed and any resultant damage repaired before DNA synthesis occurs. This checkpoint is the main site of p53 action (see p. 118).
- The G_2–M DNA integrity checkpoint ensures that DNA synthesis and any DNA damage has been repaired before mitosis occurs.
- The spindle-assembly checkpoint ensures that the prerequisites for chromosome segregation have been met before executing chromosome segregation.

The cell cycle and cancer

Cancer is characterized by uncontrolled cellular growth. Normal cells are in equilibrium, with a balance between proliferation, quiescence and death. Malignant cells may have the ability to grow autonomously and are not subject to the normal controls that regulate cell division.

Malignancy occurs as a result of DNA mutations that result in either increased or decreased expression of genes associated with cell cycle control. The most common causes of altered gene expression are:

- chemical damage (e.g. by benzene, nitrosamines)
- radiation (e.g. ultraviolet light)
- integration of viral DNA into the host genome
- inherited defects.

Cancer represents clonal expansion of a cell in which there has been sufficient change to the genomic DNA to transform the cell's phenotype from a normal to a malignant cell. Usually in the progression to cancer there is an accumulation of mutations that together cause malignant transformation (see Ch. 8).

The genes that, when mutated, are associated with cancer can be grouped into four categories:

1. Oncogenes (e.g. *Ras, Fos, Myc*) are mutated or up-regulated versions of normal cellular genes (proto-oncogenes) that induce uncontrolled growth. Proto-oncogenes are frequently components of cell signalling pathways (see Ch. 3).
2. Tumour suppressor genes (e.g. *p53, Rb, GAP*) are genes expressed in normal cells, with loss of their activity resulting in uninhibited growth. Many of these genes code for proteins that are normally involved in the regulation of cell division and differentiation.
3. Apoptotic regulatory genes (e.g. *BCL2*, caspases) are genes expressed in normal cells with the function of controlling the apoptotic process. Inappropriately activated anti-apoptotic genes can drive cancer, as can the loss of pro-apoptotic genes.
4. DNA repair genes (e.g. *XPA, BRCA1*) are required for genome integrity. Inability to repair DNA damage allows mutations to be passed on to subsequent cell generations, leading to immortalization. Genomic instability is a key feature of tumour cells.

Tumour suppressor genes are responsible for stalling the cell cycle if there is DNA damage, in order to allow time for repair. They can be thought of as like the brakes on a car.

Proto-oncogenes are generally responsible for telling a cell it may divide. When mutated or deranged, they inappropriately promote cell division and survival. They can be thought of as the accelerator.

For cancer to arise you need to remove the brake (i.e. mutate a tumour suppressor gene, such as *p53*), and then apply the accelerator (i.e. turn on/up a mitogen or an anti-apoptotic gene, or turn off/down a pro-apoptotic one).

p53

p53 is an important tumour suppressor that has been dubbed the 'guardian of the genome'. Its basic function is to restrict the entry of cells with damaged DNA into S phase (i.e. it regulates progression of the cell cycle past the restriction point). Cells with mutant p53 are not arrested in G_1 and they progress through the cell cycle and division with damaged DNA. The *p53* gene lies on chromosome 17 and codes for a nuclear phosphoprotein of 53 kDa. Three major roles of p53 have been identified:

1. Transcription activator – regulating certain genes involved in cell division.
2. As a G_1 checkpoint for DNA damage – if there has been excess DNA damage (e.g. ultraviolet damage) it arrests cell division, allowing time for repair.
3. Participation in the initiation of apoptosis (programmed cell death).

MITOSIS AND MEIOSIS

Overview of cell division

Cell division is the process by which a cell, including the nucleus, undergoes replication and splits to produce two daughter cells. Many somatic cell types are continually replenished by cell division. In order to be viable, each daughter cell must contain a complete set of genetic material so that all its proteins can be expressed at the appropriate levels.

In addition to its role in directing protein synthesis, DNA enables the passage of genetic information from one generation to the next. Therefore, in the sexually reproducing multicellular organism, cells must have two mechanisms of cell division, resulting in both diploid and haploid daughter cells:

- Mitosis is the type of cell division that occurs in somatic cells and results in the production of two genetically identical daughter cells.
- Meiosis occurs in gamete formation (e.g. sperm and ova). Each daughter cell contains half the genetic information of the parent cell and crossing-over ensures a reassortment of genetic material between homologous ('paired') chromosomes.

Mitosis

Two identical diploid daughter cells are formed as a result of mitosis.

Mitosis: 2n→2n

There are six distinct phases in mitosis (Fig. 6.37).

Prophase

The cell's chromatin condenses into the classical chromosome structure with each duplicated chromosome identifiable as a pair of sister chromatids joined by the duplicated, but unseparated, centromere. The centrioles duplicate and migrate towards opposite poles of the cell. A spindle of microtubules is formed simultaneously. The nucleoli disperse.

Prometaphase

It is characterized by the breakdown of the nuclear envelope (in eukaryotes), and the formation of kinetochores, points of attachment between the chromosome and the spindle, at the centromeres of the chromosomes.

Metaphase

The chromosomes become attached to the spindle at the kinetochore. The chromosomes become arranged along the spindle, forming the equatorial plate. At this point the chromosomes are maximally condensed and are at their most visible.

Anaphase

The centromeres separate allowing the chromatids to be pulled to opposite poles by the spindle. The end of anaphase is marked by the clustering of a complete set of chromosomes at each pole of the cell.

Telophase

The chromosomes begin to uncoil, assuming the extended state characteristic of interphase. The nuclear membrane re-forms and nucleoli reappear.

Cytokinesis

A cleavage furrow forms around the mid-region between the poles, dividing the cytoplasm into two and ultimately leading to the formation of two daughter cells each with a complete diploid chromosome complement.

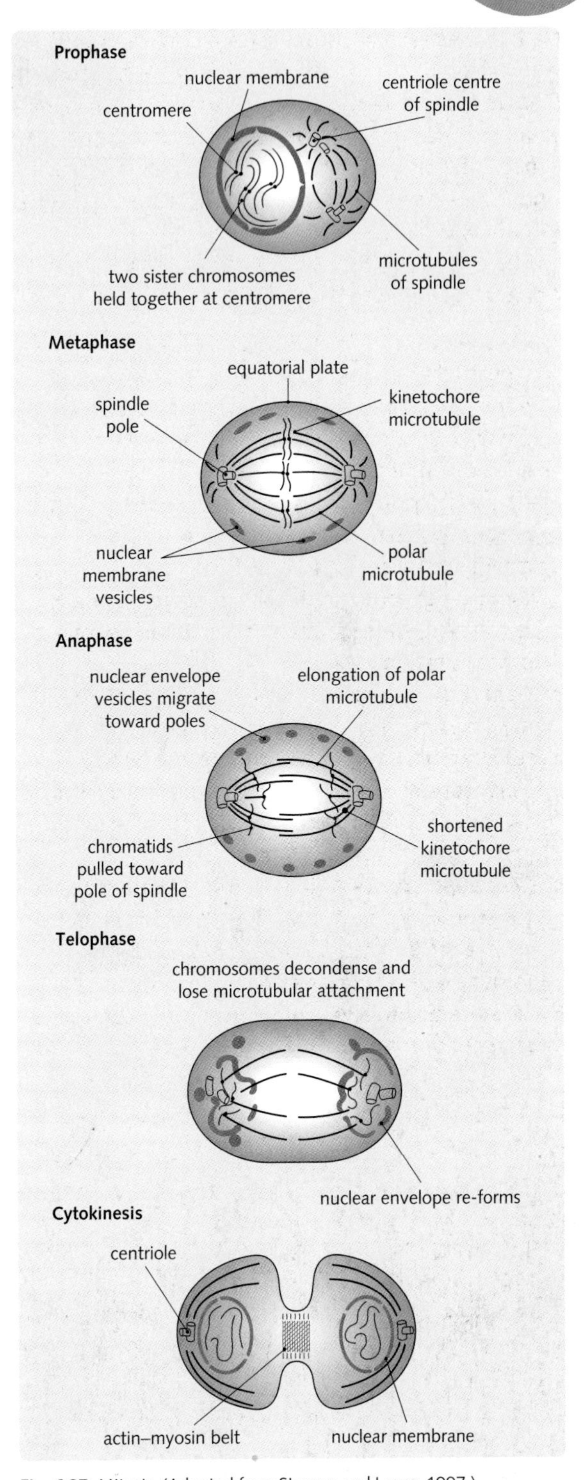

Fig. 6.37 Mitosis. (Adapted from Stevens and Lowe, 1997.)

Each sister chromatid of a metaphase chromosome is a double helix, because both strands were replicated in the preceding S phase. Thus, after separation in mitotic anaphase each daughter cell receives a complete copy of the genome. (It is a common misconception that the strands of the double helix are separated at anaphase, in which case the genome would only be complete after S phase.)

Meiosis

In the first division of meiosis two genetically different haploid cells are formed (Fig. 6.38). In the second division, each haploid cell is duplicated.

Meiosis: 2n→n

Prophase I

There are five stages during which homologous chromosomes come together and exchange segments in homologous recombination:

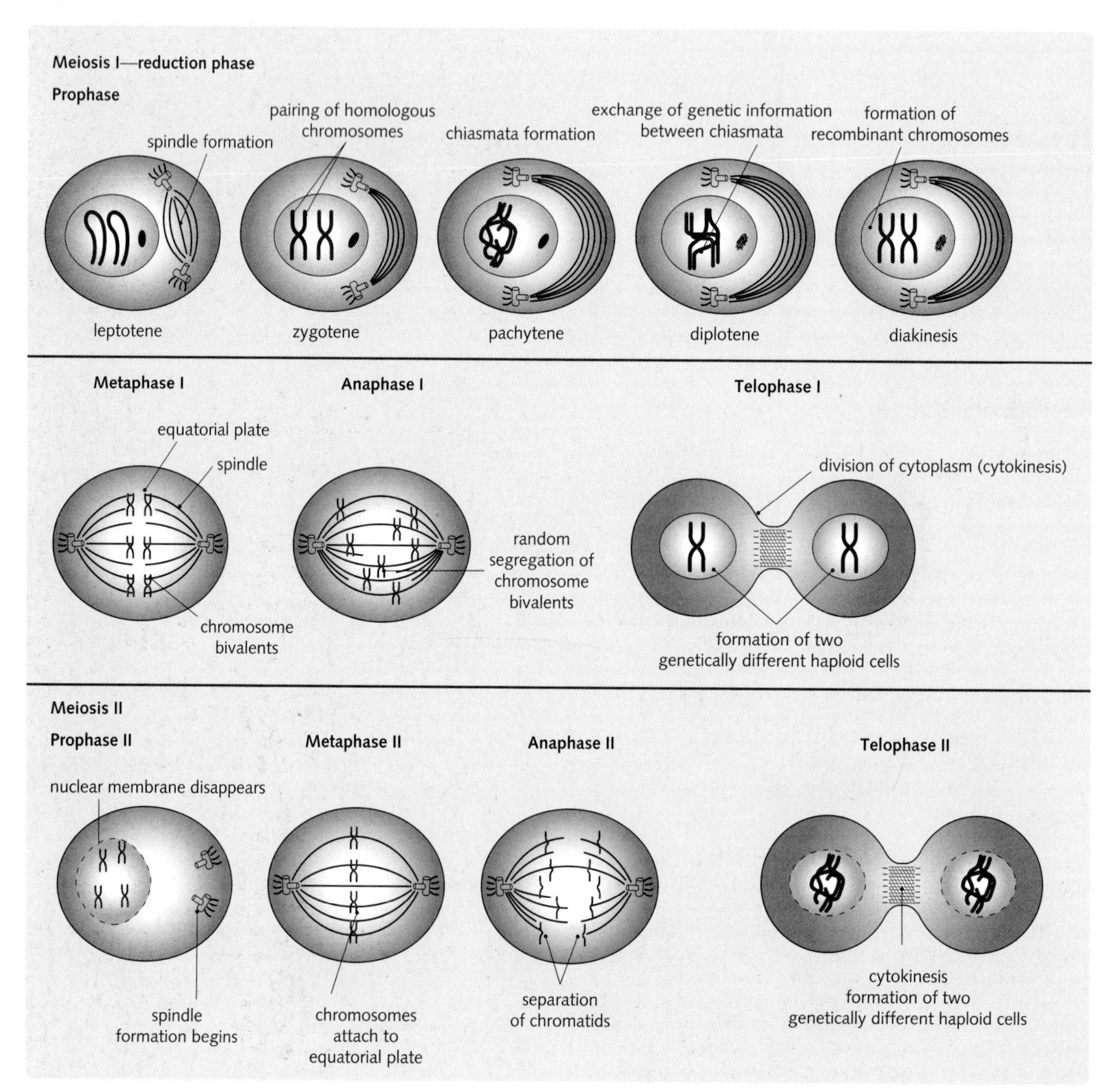

Fig. 6.38 Meiosis – see text for explanation of each phase.

- Leptotene – spindle forms
- Zygotene – homologous chromosomes pair, shorten and thicken, and form bivalents (pairs of homologous chromosomes)
- Pachytene – chiasmata begin to form. These are points at which non-homologous chromatids become associated with each other via base pairing. These become points of 'crossing-over' between the chromatids
- Diplotene – exchange of genetic material in chiasmata and nuclear membrane disappears
- Diakinesis – recombinant chromosomes are formed.

Metaphase I

Chromosomes become attached to a spindle (see mitotic metaphase).

Anaphase I

The chromatids do not separate and whole chromosomes migrate to opposite poles of spindle, hence 'reduction division'.

Telophase I

Two genetically different haploid cells are formed.

Second division

The second division is like mitosis, but a haploid number of chromosomes are involved. The chromatids separate in anaphase II.

Genetic diversity and gametogenesis

Two processes in meiosis are vital in the generation of genetic diversity:

1. Chiasmata formation ('crossing-over'), which allows random exchange of genetic material between homologous chromosomes
2. Independent segregation of homologous chromosomes.

During anaphase I, homologous chromosomes segregate independently of each other. Since humans possess 23 pairs of homologous chromosomes, there are 2^{23} possible ways that the chromosomes can segregate to form a haploid set.

In humans, meiosis begins during gametogenesis. In females this occurs in the ovaries. The first ('reduction') meiotic division begins during the 5th month of embryonic life, but it is arrested at prometaphase and completed just before ovulation. Meiosis II takes place after ovulation. Therefore:

- there is a fixed number of oocytes
- there is a period of arrest between the start of meiosis I and the completion of meiosis II of 12–45 years.

In oogenesis, meiosis is arrested in prometaphase of meiosis I, with a long latent period (dictyotene phase). It is probable that this lengthy interval leads to an accumulation of wear and tear effects on the primary oocyte, damaging the cell's spindle formation and repair mechanisms, and predisposing to a failure of chromosome separation at meiosis I and chromatid separation at meiosis II (non-disjunction). Non-disjunction at meiosis I has been implicated in trisomies such as Down syndrome (trisomy 21), Edward's syndrome (trisomy 18) and Patau's syndrome (trisomy 13).

Oogenesis produces a single oocyte and two polar bodies (Fig. 6.39).

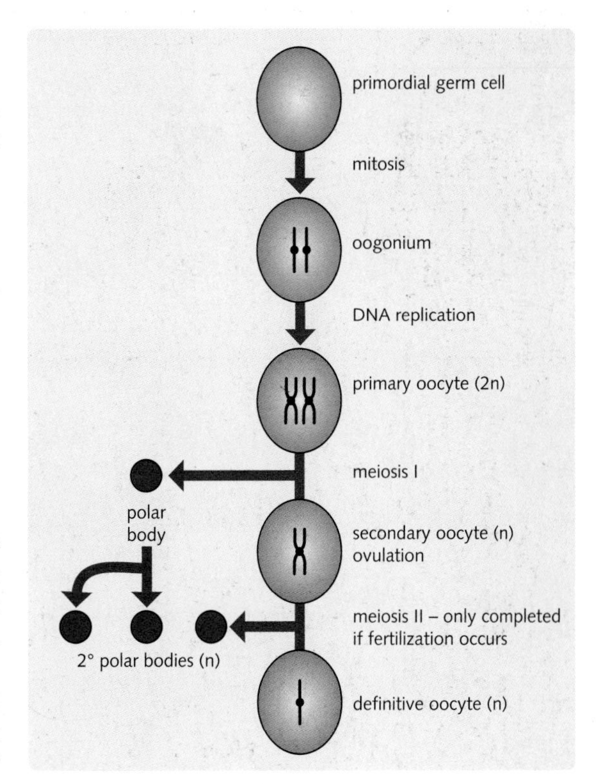

Fig. 6.39 Oogenesis.

Male spermatogenesis occurs in the seminiferous tubules of the testes. After sexual maturity the spermatogonia continuously multiply by mitosis, subsequently undergoing meiosis to produce unlimited numbers of spermatocytes (Fig. 6.40).

Endoreduplication

Endoreduplication, also known as endomitosis, is the process of repeated DNA replication in the absence of nuclear division and cytokinesis (Fig. 6.41). It can generate huge nuclei with up to 16 copies of the DNA. Endoreduplication occurs in the formation of megakaryocytes, which are cells in the bone marrow from which anucleate platelets bud off.

DNA DAMAGE AND REPAIR

DNA damage

DNA damage is the introduction of structural changes to the DNA molecule, including conformational

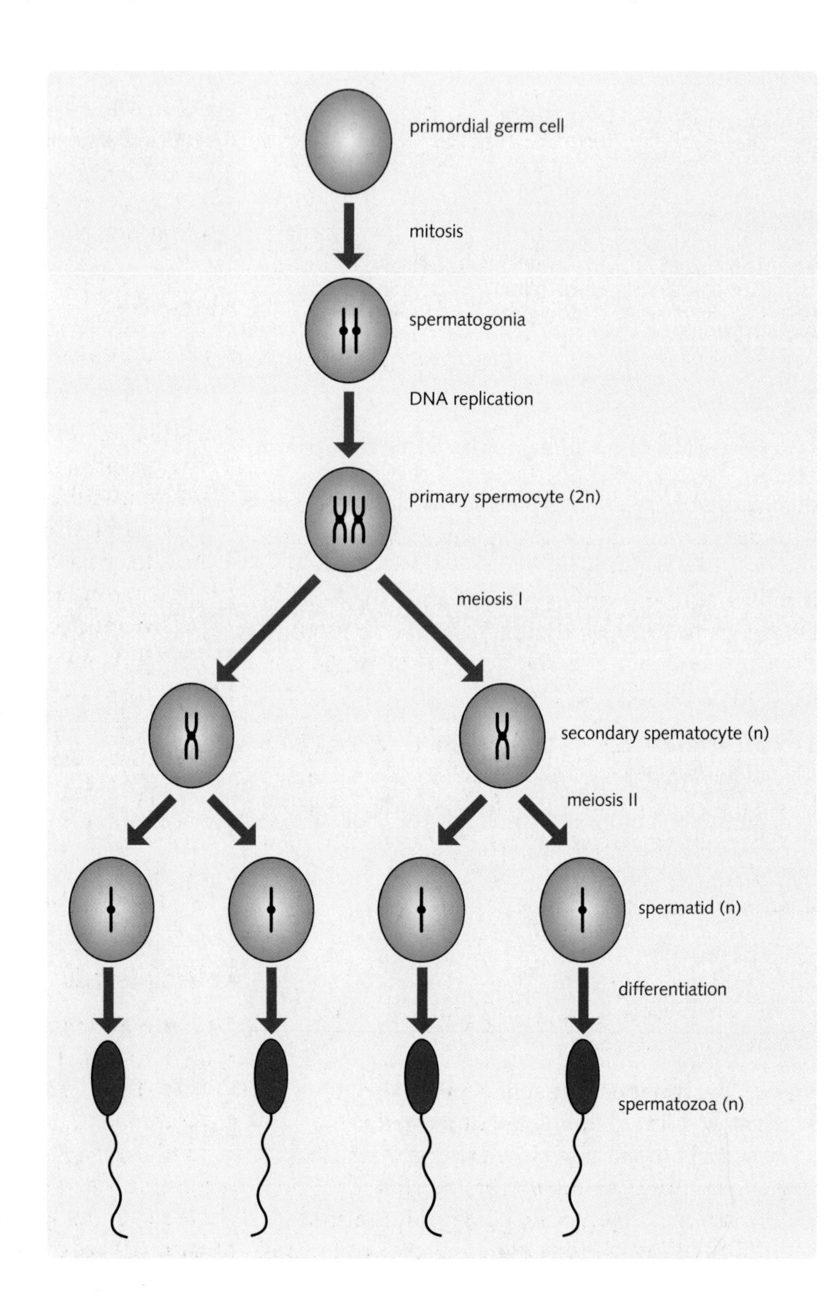

Fig. 6.40 Spermatogenesis.

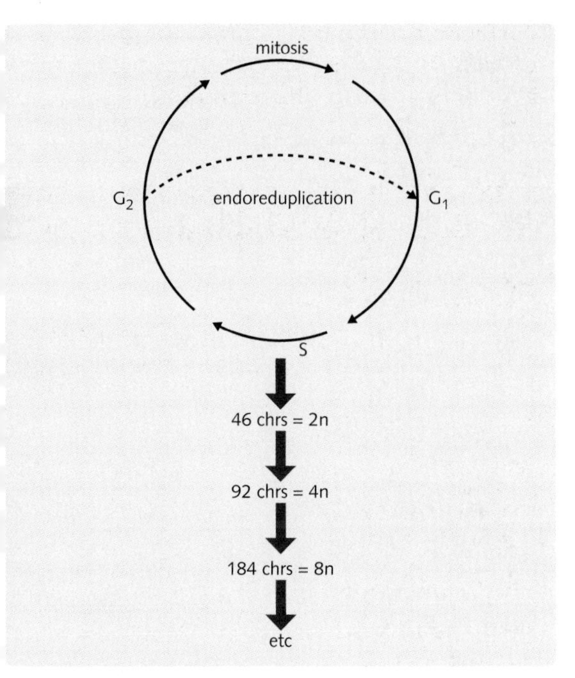

Fig. 6.41 Endoreduplication. Chromosomal duplications without intervening mitoses, resulting in increases in DNA content and cell enlargement. Megakaryocytes undergo endoreduplication, with an average ploidy of 16n (range 4n–64n).

distortion of the double helix, which interferes with replication and transcription, and sequence changes, which may disrupt base pairs or lead to the incorporation of an incorrect base into the replicating DNA strand. Damage occurs at a rate of 1000 to 1 000 000 molecular lesions per cell per day and can be divided into two main types:

1. Endogenous – caused by the products of normal metabolism, from errors in DNA replication and spontaneously arising. For example, adenine and cytosine can undergo spontaneous deamination to produce hypoxanthine and uracil residues.
2. Exogenous – caused by external agents.

Mutagens

Agents that cause exogenous DNA damage are known as mutagens, and include:

- ionizing radiation, such as γ-rays and X-rays
- ultraviolet light, which promotes chemical cross-linking between two adjacent thymine residues on a DNA strand, resulting in the formation of a pyrimidine dimer. These distort the DNA double helix in the region of the dimer
- chemical mutagens, which can be of three types: base analogues (e.g. 5-bromouracil, which resembles thymine and base pairs with adenine), which become incorporated into the DNA and cause misreading; chemical modifiers (e.g. hydroxylamine and compounds containing and propagating free radicals – such as those formed during the metabolism of polycyclic aromatic hydrocarbons in tobacco smoke), which react with bases to form derivatives that cause misreading; and intercalators (e.g. some antibiotics and heavy metals), which slip between adjacent bases and inhibit RNA transcription
- viral genomes, which can become incorporated into eukaryotic chromatin. This can disrupt coding regions or promoting regions, or it can affect levels of expression of existing genes.

Mutations

A mutation is a change in the base sequence of DNA:

- If this is in a coding region and there is a change in the number of bases, for instance due to insertion or deletion, it may result in a 'frameshift' error (see Ch. 8).
- An altered base or point mutation may result in a misreading error, which can result in an altered protein product (see Ch. 8).
- Double-strand DNA breaks may result in a chromosomal level rearrangement, such as a translocation (see Ch 8).
- The mutation may disrupt a regulatory region and affect the level of expression of a particular gene (see pp. 105–106).

DNA damage has two fates; the cell may repair it or, if the damage is extensive, trigger apoptosis.

DNA repair

DNA repair mechanisms act as a major defence system against environmental and cellular damage to DNA. They act to minimize cell killing, mutations, replication errors, persistence of DNA damage and genomic instability, and abnormalities in these processes have been implicated in cancer and ageing. There are several systems for the detection and repair of various types of DNA damage (Fig. 6.42).

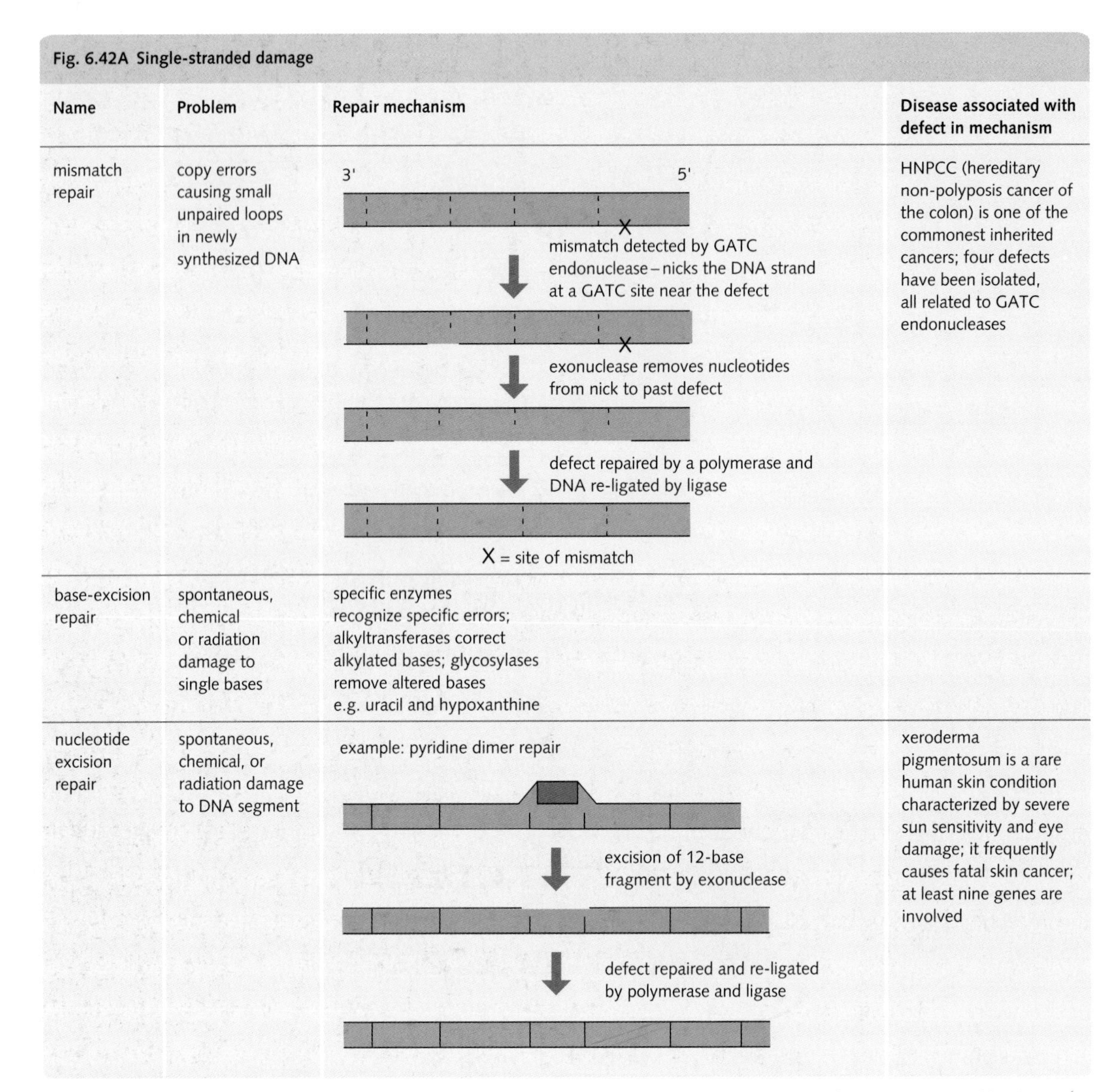

Fig. 6.42A Single-stranded damage

Name	Problem	Repair mechanism	Disease associated with defect in mechanism
mismatch repair	copy errors causing small unpaired loops in newly synthesized DNA	3' 5'; mismatch detected by GATC endonuclease – nicks the DNA strand at a GATC site near the defect; exonuclease removes nucleotides from nick to past defect; defect repaired by a polymerase and DNA re-ligated by ligase; X = site of mismatch	HNPCC (hereditary non-polyposis cancer of the colon) is one of the commonest inherited cancers; four defects have been isolated, all related to GATC endonucleases
base-excision repair	spontaneous, chemical or radiation damage to single bases	specific enzymes recognize specific errors; alkyltransferases correct alkylated bases; glycosylases remove altered bases e.g. uracil and hypoxanthine	
nucleotide excision repair	spontaneous, chemical, or radiation damage to DNA segment	example: pyridine dimer repair; excision of 12-base fragment by exonuclease; defect repaired and re-ligated by polymerase and ligase	xeroderma pigmentosum is a rare human skin condition characterized by severe sun sensitivity and eye damage; it frequently causes fatal skin cancer; at least nine genes are involved

Fig. 6.42 Mammalian DNA defect repair mechanisms. Guanine, adenine, thymine, and cytosine (GATC) exonuclease recognizes and binds to a GATC base sequence near to the DNA defect and nicks the DNA at this site.

Direct reversal of DNA damage

One of the most frequent causes of point mutation in the human genome is the alkylation (especially methylation) of bases at specific sites. If unrepaired, such lesions can lead to the incorporation of incorrect bases at the next round of replication. For example O^6-alkylated guanine base pairs with thymine. Such damage is reversed by a class of enzymes known as the DNA alkyltransferases, suicide enzymes, which are consumed by the process of de-alkylation.

Single-stranded damage

Base excision repair

Base excision repair (BER) is a process by where a damaged or inappropriate base is removed from its sugar linkage and replaced (Fig. 6.42). Repair involves the following stages:

- Removal of the damaged base by an enzyme called a glycosylase. This leaves an abasic or AP site. There are at least eight different human DNA glycosylases, each enzyme being responsible for identifying and removing a specific kind of base damage.

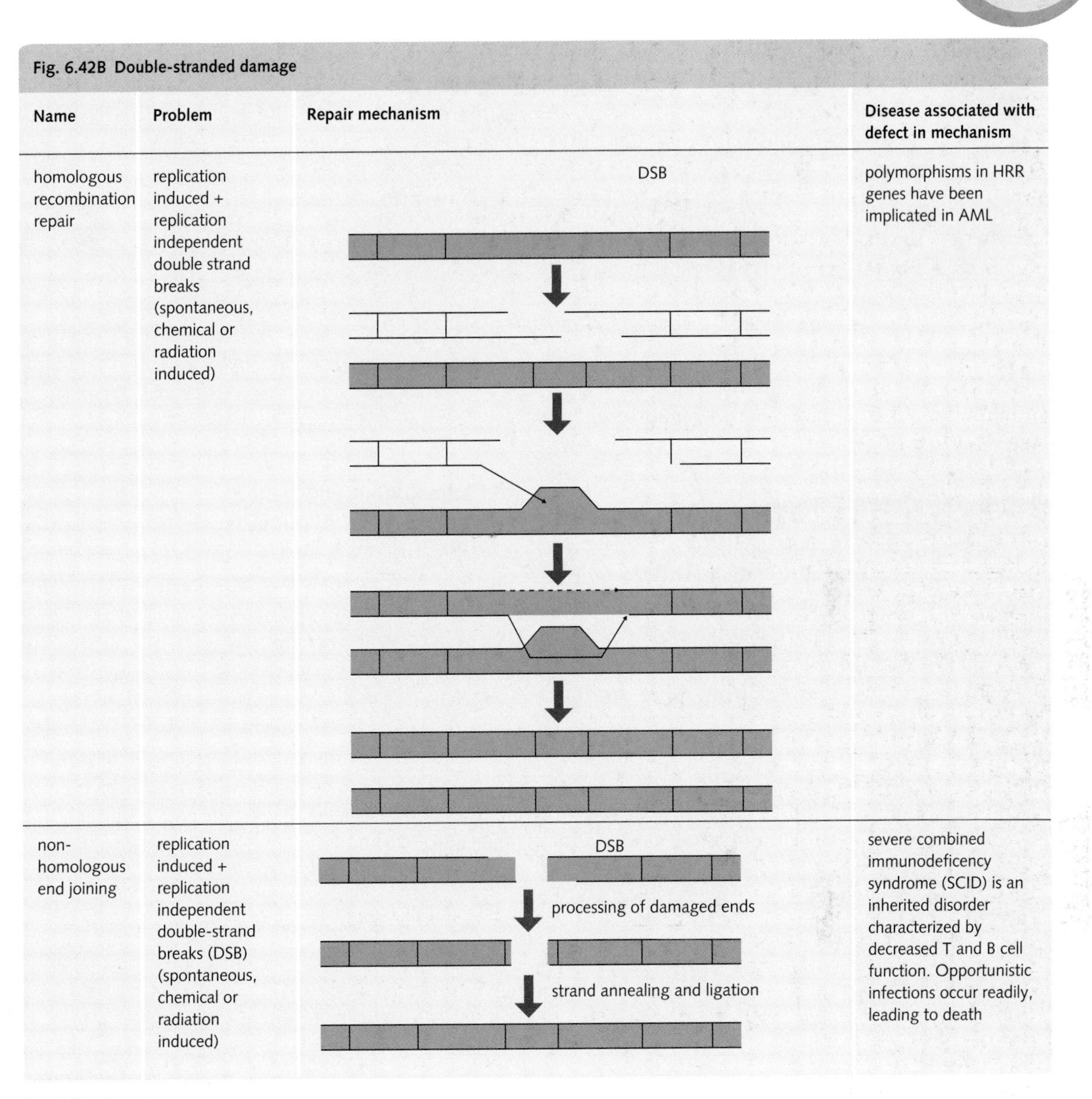

Fig. 6.42B Double-stranded damage

Name	Problem	Repair mechanism	Disease associated with defect in mechanism
homologous recombination repair	replication induced + replication independent double strand breaks (spontaneous, chemical or radiation induced)	DSB	polymorphisms in HRR genes have been implicated in AML
non-homologous end joining	replication induced + replication independent double-strand breaks (DSB) (spontaneous, chemical or radiation induced)	DSB processing of damaged ends strand annealing and ligation	severe combined immunodeficency syndrome (SCID) is an inherited disorder characterized by decreased T and B cell function. Opportunistic infections occur readily, leading to death

Fig. 6.42—Cont'd

- An endonuclease and phosphodiesterase recognize the AP site and cut the sugar phosphate DNA backbone on the 5′ of the AP site, creating a 3′-OH terminus.
- Using the complementary DNA strand as a template, DNA polymerase β then extends the DNA from the free 3′-OH to replace the nucleotide of the damaged base.
- The nicked strand is sealed by a DNA ligase.

Variations of this general mechanism are used to repair up to 10 damaged bases, sugar backbone and single-strand DNA breaks.

Nucleotide excision repair

Nucleotide excision repair (NER) is utilized to repair more bulky and complex DNA damage, such as thymidine dimers and chemically modified bases over a longer stretch of DNA (Fig. 6.42).

Steps involved in NER include:

- the damage is recognized by a protein complex and the DNA unwound by a helicase to produce a 'bubble'
- cuts are made on both the 3′ side and the 5′ side of the damaged area and the damage containing oligonucleotide is removed

- using the complementary DNA strand as a template, the resulting gap is filled by DNA polymerase δ and ε
- a DNA ligase covalently binds the fresh piece into the backbone.

Mismatch repair

Mismatch repair (MMR) primarily repairs undamaged, but mismatched, base pairs and small insertions or deletions.

The overall process of MMR is similar to the other excision repair pathways. The DNA lesion is recognized, a patch containing the lesion is excised, and the strand is corrected by DNA repair synthesis and re-ligation (Fig. 6.42).

Double-stranded damage

Homologous recombination repair

Homologous recombination repair (HRR) typically occurs between DNA sequences with extended homology. On detection of a DNA double-stranded break (DSB), each of the 5′ ends of the break are resected by exonucleases, to leave long sections of 3′ ended single-stranded DNA (ssDNA) tails. These 3′ ends are recombinogenic and can invade a suitable homologous template molecule (such as the sister chromatid or homologous chromosome). The invading ssDNA ends then act as primers for DNA synthesis using the invaded homologous molecule as a template. Thus, synthesis results in the restoration of the degraded single-strands (Fig. 6.42).

Mutations in BRCA1 and BRCA2 genes are associated with familial breast cancer. The proteins coded by these genes are thought to be important in the repair of double-strand DNA damage by homologous recombination repair.

Non-homologous end joining

Non-homologous end joining (NHEJ) is regarded as an illegitimate repair pathway. Unlike HRR, NHEJ does not require an undamaged partner molecule and does not rely on extensive homologies between the two recombining ends. In most cases, direct end-ligation of a DSB is not possible, due to the presence of damaged bases and sugar moieties flanking the DSB and DNA. DSBs will require some processing before they can be ligated. This means NHEJ is rarely error-free and sequence deletions of various lengths are usually introduced around the area of the original DSB (Fig. 6.42).

MEDICAL GENETICS

Tools in molecular medicine

Objectives

By the end of this chapter you should be able to:

- Explain how restriction endonucleases are used in the preparation of nucleic acids in the laboratory.
- Understand the basis of PCR and its applications.
- Compare and contrast Southern, Northern and Western hybridization techniques.
- Understand the differences between single nucleotide polymorphisms and variable number tandem repeats.
- Describe three different cytogenetical methods for examining chromosome structure.
- Appreciate the important role of bioinformatics in the postgenomic era.
- Describe the main aims of the human genome project.
- Understand the principles of genetic linkage and its role in identifying human disease genes.
- Define the terms genome, proteome and metabolome.
- Appreciate the difference between somatic and germ-line gene therapy.

MOLECULAR TECHNIQUES

The increase in our understanding of the molecular basis of disease since the 1950s follows the discovery of the structure of DNA and the development of new technologies that permit the detailed analysis of normal and abnormal genes. In addition to providing the basis for theoretical advances, the techniques of molecular genetics permit the detection and laboratory diagnosis of genetic disease. It is a fast-moving field in which new techniques are introduced on a regular basis. Rapid access to the most up to date information is often necessary and the best place to access it is through the Internet.

The National Center for Biotechnology Information (www.ncbi.nlm.nih.gov) is a gateway into a range of public databases, including scientific literature (e.g. PubMed and Online Mendelian Inheritance in Man (OMIM) – a catalogue of all human single gene defects) as well as molecular databases and access to bioinformatics tools.

Molecular genetics is the study of the structure and function of genes at the molecular level. The basic techniques of molecular genetics concern:

- the separation of nucleic acids from the other components of the cell
- the characterization of DNA sequence
- the study of gene expression
- the manipulation and modification ('engineering') of DNA
- gene cloning and mapping.

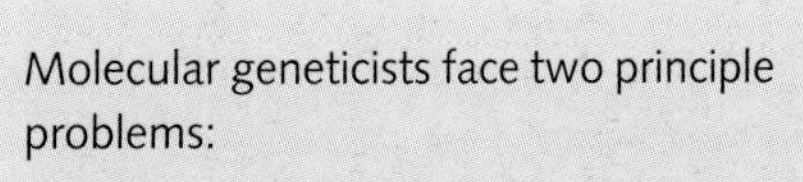

Molecular geneticists face two principle problems:

1. Obtaining sufficient quantities of nucleic acid to work with
2. Identifying specific sequences within a complex mixture of sequences.

Most of the techniques of molecular genetics address one or both of these issues.

Isolation and preparation of nucleic acids

The major constituents of any cell type are protein, lipid, carbohydrate, DNA and RNA. There are many different methods of separating these components from one another, depending on the type of cells used, how much material is required, and the purpose for which they are to be used. Most strategies for extracting nucleic acids from cells exploit their differential solubilities as compared with the other cellular constituents or rely on synthetic resins that reversibly bind them.

Restriction enzymes

Restriction enzymes are a class of endonucleases that are generated by microorganisms and cleave DNA after recognizing a specific sequence.

Restriction enzymes are characteristically named according the bacterial species from which they were isolated, followed by a number. For example EcoR1 was isolated from *E. coli* RY13, while SmaI was isolated from *Serratia marcescens*.

Restriction enzymes function by scanning the length of a DNA molecule, until they encounter their specific recognition sequence, at which point it binds to the DNA double helix and makes one cut in each of the two sugar-phosphate backbones of the double helix.

There are three types of restriction enzymes, those employed for recombinant DNA technologies are type II restriction enzymes, which cut at or near a short, and often palindromic, recognition sequence to generate either blunt or sticky ends (Fig. 7.1). Most type II restriction enzymes have recognition sites consisting of four, six, or eight base pairs. Assuming a random distribution of bases in a piece of double-stranded DNA, a sequence of eight specific base pairs will occur much less frequently than a sequence of six base pairs (Fig. 7.2).

Gel electrophoresis

The fragments produced by digestion with restriction enzymes are separated by gel electrophoresis. The products of digestion are loaded into the well of an agarose gel, across which an electric field is applied:

- DNA is negatively charged, so it migrates towards the positive electrode.
- Small pieces of DNA migrate more quickly than large pieces, so fragments become sorted according to their size.
- Small DNA fragments end up furthest away from the wells at the end of electrophoresis.

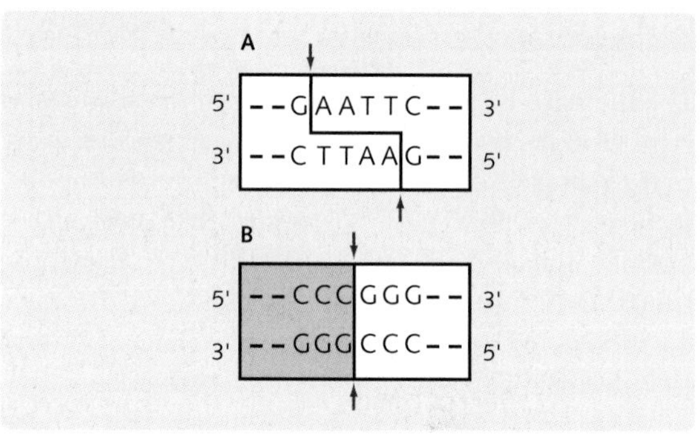

Fig. 7.1 Sticky and blunt ends after restriction enzyme digestion. (A) EcoRI cuts to produce staggered termini ('sticky ends'). (B) SmaI cuts to produce blunt ends. (Adapted from Mueller and Young, 2001.)

Fig. 7.2 Fragment size produced and cutting frequency of restriction enzymes

Restriction enzyme recognition sequence	Average size of restriction fragment	Average number of cuts in 10 kb plasmid	Average number of cuts in *S. cerevisiae* (1.4×10^4 kb)	Average number of cuts in haploid human genome (3×10^6 kb)
4 bp	4^4 bp (0.2 kb)	50	7.0×10^4	1.1×10^7
6 bp	4^6 bp (4.1 kb)	2	3.4×10^3	7.3×10^5
8 bp	4^8 bp (65.5 kb)	0	2.1×10^2	4.5×10^4

Fig. 7.2 Fragment size produced and cutting frequency for the digestion of human, *Saccharomyces cerevisiae*, and plasmid DNA with restriction enzymes that recognize four, six and eight base pair sequences.

DNA is visualized on agarose gels under ultraviolet light after staining with ethidium bromide. The size of the fragments produced by restriction enzyme digestion can be determined by reference to a 'marker' lane that contains fragments of known size.

Electrophoresis of digested DNA

The digestion of human DNA with a six base pair cutter produces multiple fragments with an average size of 4 kb (Fig. 7.2). However, since the sequence of bases in much of the genome is random, the size of fragment varies about this mean, and many different sizes of fragment are produced. These appear as a 'smear' on an appropriately treated agarose gel (Fig. 7.3A).

Regular agarose gels are limited by the size of fragment that they can separate, with a maximum resolution of around 30 kb. Larger fragments can be resolved using a technique called pulsed field gel electrophoresis (PFGE). By alternating the direction of the electric field, PFGE allows fragments well over 10 Mb to be separated out (Fig. 7.3B). However, this technique requires a gel run time in the region of days, rather than the hours of standard gel electrophoresis.

Digestion of a 10 kb plasmid with a six base pair cutter will cut the plasmid twice on average, generating two bands (Fig 7.2). An enzyme that cleaves once within the circular plasmid will linearize it, producing a single band. Thus, the products of digestion of plasmid DNA give distinct bands that can be visualized directly on appropriately treated agarose gels (Fig. 7.4).

Polymerase chain reaction

Polymerase chain reaction (PCR) is a means of amplifying short segments of DNA (2–3 kb using standard methods), and the technique has revolutionized molecular genetics. Its many uses include:

- probe preparation
- cloning (see pp. 142–147)
- genotyping polymorphic markers (see pp. 137–139)
- diagnostic detection of mutations.

PCR hinges on the manipulation of conditions so that a DNA polymerase enzyme repeatedly replicates a specific sequence of DNA (Fig. 7.5). There are five reagents essential for PCR:

Target DNA

The target DNA is the sequence that is to be amplified, which acts as a template for the first round of replication.

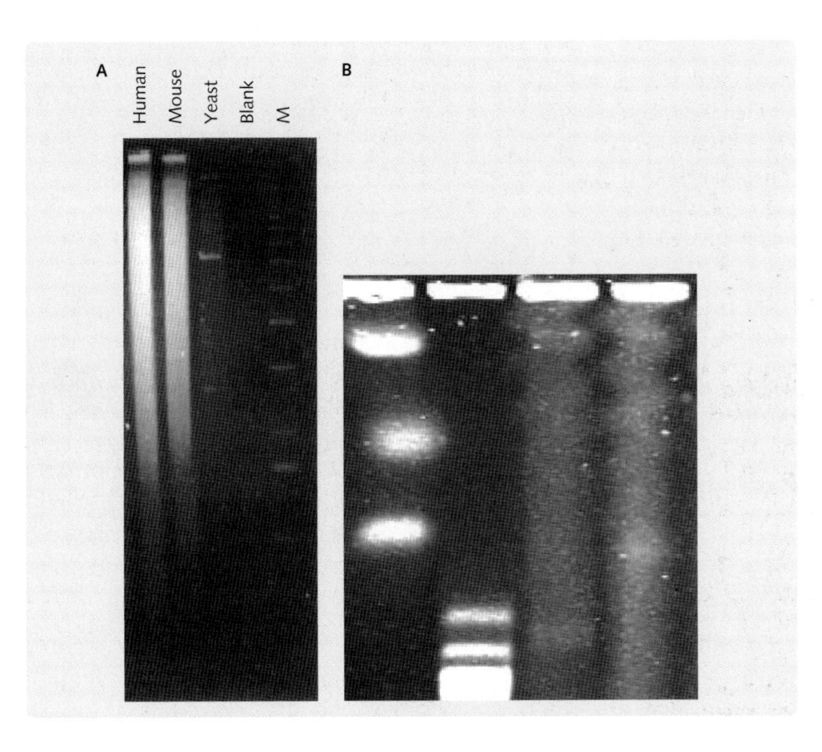

Fig. 7.3 Electrophoresis of human DNA. (A) Human, mouse, and yeast (*S. cerevisiae*) genomic DNA cut with a restriction enzyme that recognizes a six base pair sequence (*Hin*dIII). The marker lane contains DNA fragments of known size. (Courtesy of Dr Steve Howe.) (B) Resolution of large fragments of DNA by pulsed field gel electrophoresis.

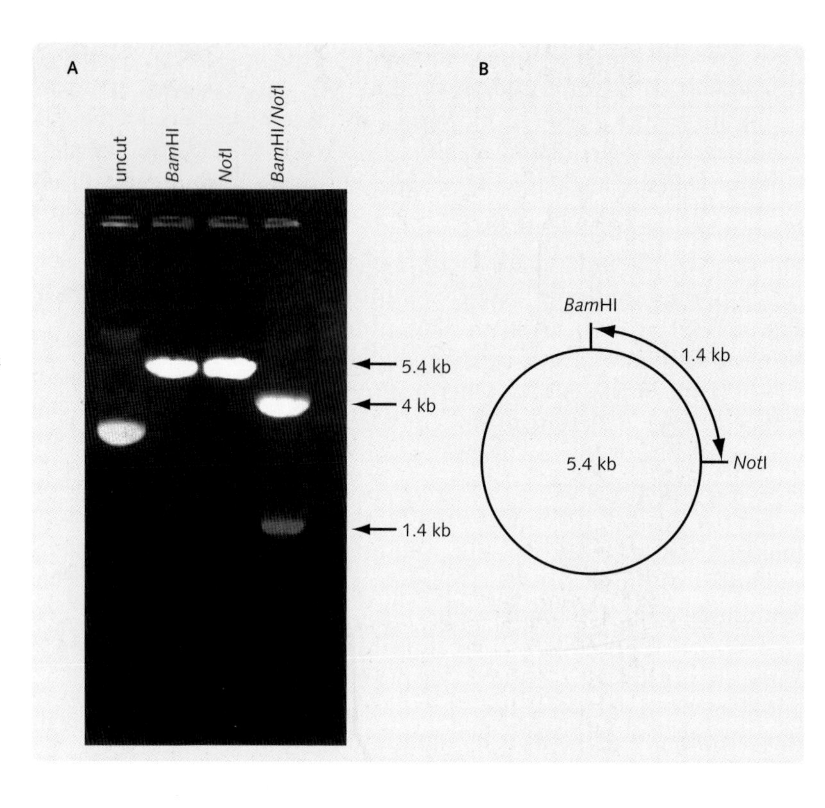

Fig. 7.4 (A) Plasmid DNA cut with restriction enzymes. The wells are labelled with the enzyme used to digest the plasmid. *Bam*HI and *Not*I each cut the plasmid once and linearize it. Digestion with both enzymes produces two bands. The lane labelled 'uncut' contains plasmid DNA that has not been digested. Note this contains two bands that correspond to supercoiled and open circular forms of the plasmid, and that these forms migrate differently to the linearized plasmid. The restriction fragments are sized by comparison to a lane containing fragments of known size (not shown). (B) Restriction map of the plasmid based on the restriction enzyme digestion. (Courtesy of Dr Ajay Mistry.)

Very little DNA is required for a PCR reaction. If the purpose of the reaction is to determine whether a patient carries a specific mutation, sufficient DNA can be extracted from buccal cells, which can be painlessly scraped from the inside of the cheek. Alternatively, DNA may be prepared from peripheral leukocytes that are obtained by venepuncture.

PCR polymerases

Taq polymerase is a heat-stable DNA polymerase enzyme that was originally isolated from *Thermus aquaticus*, a thermophilic bacterium that naturally inhabits hot springs. Like all DNA polymerases (see also p. 102) Taq polymerase:

- requires a primer to initiate synthesis
- synthesizes DNA in a 5′ to 3′ direction.

As Taq polymerase lacks 3′ to 5′ exonuclease activity, having a mutation rate in the region of 1 in 10 kb, if sequence fidelity is key, a polymerase with 3′ to 5′ exonuclease activity, such as Pfu polymerase (extracted from *Pyrococcus furiosus*) can be used.

Deoxynucleotide triphosphates

Deoxynucleotide triphosphates (dNTPs) are the substrates for Taq polymerase, from which the new strands of DNA are synthesized. The reaction should include equal amounts of dATP, dTTP, dCTP and dGTP.

Primers

Each reaction includes 5′ and 3′ primers that together flank the target sequences and anneal to complementary sequences on opposing DNA strands. Primers are usually synthetically produced oligonucleotides that are about 17–26 bases in length. Their design is key, as it is the primer sequence that gives PCR its specificity for amplification of target regions. It is also important that both of the oligonucleotide primers should be designed such that they have similar melting temperatures, and a base composition of between 45% and 55% GC, but lacking PolyG or PolyC runs, as these encourage non-specific annealing.

Buffer

The buffer maintains the optimum pH and chemical environment for the polymerase enzyme.

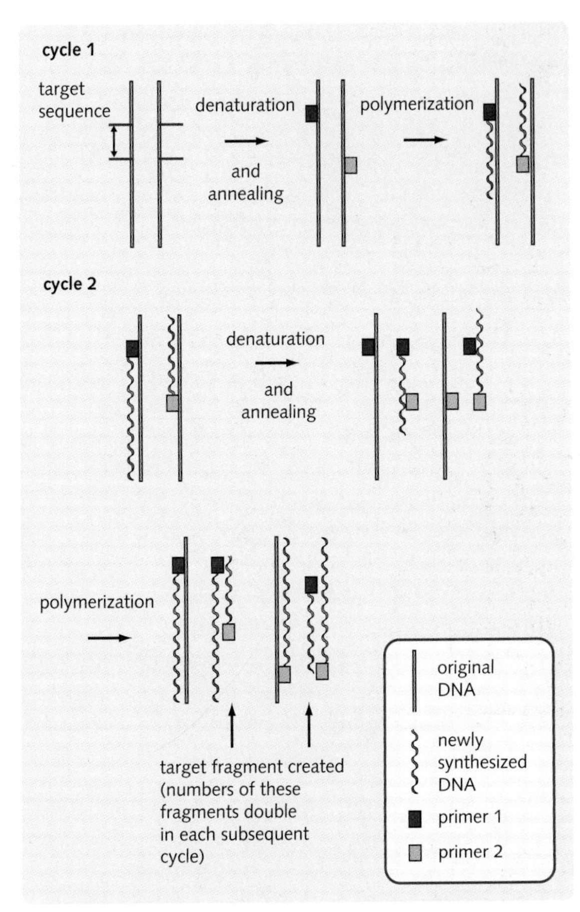

Fig. 7.5 Mechanism of action of the polymerase chain reaction (PCR). Each primer is complementary to sequence on opposing strands, flanking the target sequence. *Taq* polymerase catalyses the addition of dNTPs onto the 3′ ends of both primers so that each round of PCR doubles the amount of target DNA.

PCR in practice

Tubes containing the various reagents are placed in a 'thermocycling' machine. This rapidly heats and cools the reaction tubes in a cyclical manner, each cycle consisting of heating and cooling to three distinct temperatures.

Each cycle theoretically doubles the amount of DNA, although the reaction becomes less efficient when residual amounts of reagents such as dNTPs eventually become limiting. Typically 30 cycles are performed, after which about 10^5 copies of the target sequence are present. This is sufficient DNA to be visualized on an agarose gel. The steps involved in each PCR cycle are summarized in Fig. 7.6.

Visualizing the products of PCR

The products of the PCR reaction can be visualized directly on an agarose gel under ultraviolet light after staining with ethidium bromide (Fig. 7.7). They appear as distinct bands, the size corresponding to the number of base pairs between the 5′ and 3′ ends of the two primers. The original template DNA is not visible because such a small amount is used in the PCR reaction.

Controls

Compared with Southern blotting (see pp. 134–135), PCR is a quick method of genotyping polymorphic markers or identifying disease-causing mutations that is well suited to automation (Fig. 7.8). However, it is extremely sensitive to contamination, and small amounts of template DNA in any of the reagents will generate spurious results. Furthermore, for reasons that are often difficult to determine, artefacts (e.g. unexpected bands) are common or the amplification may fail completely. For this reason every experiment should include a minimum of two controls. The negative control (or 'blank') contains all the reagents except for the target DNA. A band in the lane that corresponds to this reaction suggests contamination.

Fig. 7.6 A summary of a typical PCR reaction, with an approximate time scale

Polymerase chain reaction		
Step 1	95 °C for 4 minutes	Initial denaturation step
Step 2	94 °C for 1 minute	Short denaturation period
Step 3	55 °C for 1 minute	Primer-annealing step
Step 4	72 °C for 1 minute	Elongation step
Step 5	Go back to Step 2, 30 times	Cycling steps
Step 6	72 °C for 10 minutes	Final elongation step
Step 7	Hold at 4 °C	

Fig. 7.6 A summary of a typical PCR reaction, with an approximate time scale.

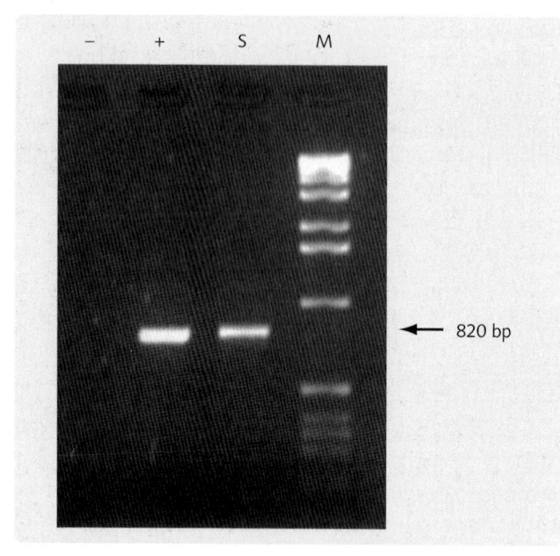

Fig. 7.7 The products of a PCR reaction. A 820 bp band results because the primers are separated by this amount of sequence. The lane marked '–' is the negative control that contains all the components necessary for PCR except DNA. The lane marked '+' is the positive control, which is known to contain the sequence to which the primers anneal. The 820 bp band is seen in the sample lane (S), suggesting that this clone also includes the sequence to which the primers anneal. 'M' represents the marker lane that contains DNA fragments of known size. (Courtesy of Dr Steve Howe.)

The positive control contains DNA known to contain the target sequence, and which has ideally produced the expected band in the same PCR reaction on a previous occasion. If the lane that corresponds to this reaction contains no band or unexpected bands, the experiment should be repeated.

Reverse transcriptase PCR

This is a modification of PCR that is used to study gene expression. RNA cannot be used directly in PCR because Taq polymerase is unable to recognize it as a template. However, the viral enzyme reverse transcriptase (RT) can use RNA as a template for the production of a strand of DNA. In the first step of RT-PCR, RNA is incubated with this enzyme in the presence of dNTPs and an appropriate primer to allow the synthesis of a strand of cDNA (the 'first strand reaction'). The products of this reaction are then used directly in a PCR reaction as described above (the 'second strand reaction').

Nucleic acid hybridization

Hybridization techniques take advantage of the ability of two complementary single-stranded DNA molecules to form a double-stranded DNA molecule. This enables the identification of individual sequences of DNA (or RNA) within a mixture that contains many sequences, using specific probes. It requires both the target and probe sequences to be first denatured (i.e. made single stranded). They are then mixed together under stringent conditions that only permit pairing of complementary strands of DNA. Thus, the labelled probe sequence binds specifically with complementary target sequence to form a hybrid, but it does not bind to non-complementary sequences. The presence of such hybrids can be detected:

- by autoradiography, using radioactively labelled probes (Fig. 7.9)
- by chromogenic detection, using fluorescent probes developed more recently to circumvent the health and safety risks associated with the use of radioactive compounds.

DNA

Southern blotting

Although DNA within agarose gels can be subjected to hybridization directly, it is only possible to do so once, because gels are fragile. Southern blotting is a technique by which DNA is transferred from the agarose gel to a nylon or nitrocellulose membrane (Fig. 7.10). Having the products of a restriction enzyme digestion on

Fig. 7.8 The relative advantages and disadvantages of PCR and Southern hybridization.

Fig. 7.8 PCR compared with Southern hybridization

Consideration	PCR	Southern hybridization
Amount of DNA required	Small amount (100 ng)	Larger amount (5 μg)
Typical time taken	1–2 days	4–5 days
Suitability for automation	Yes	No
Spurious results	Frequent	Infrequent
Contamination problems	Common	Rare

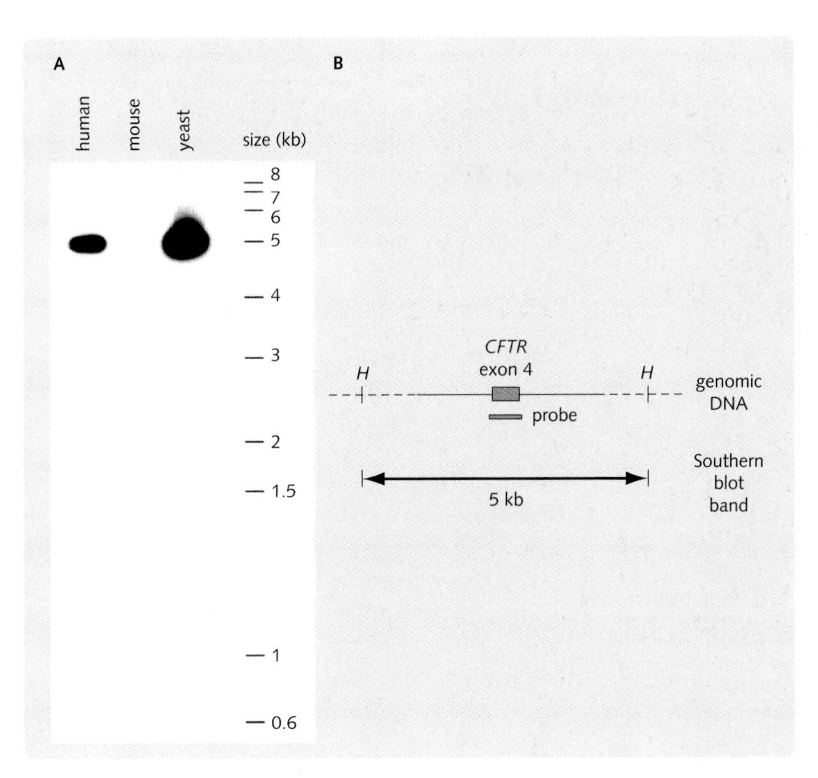

Fig. 7.9 (A) Autoradiograph obtained after the DNA digest pictured in Fig. 7.3A is hybridized with a single copy probe that hybridizes to exon 4 of the human *CFTR* (cystic fibrosis) gene. The probe is specific for human sequence, so hybridization to a 5 kb fragment occurs with human DNA, but there is no hybridization with mouse DNA. The probe anneals to a fragment in the yeast lane because this clone contains a yeast artificial chromosome (YAC) that includes the human *CFTR* gene. (B) Map depicting probe annealing on human genomic DNA. A 5 kb fragment results because exon 4 is flanked by two *Hin*dIII fragments (H) that are 5 kb apart. (Courtesy of Dr Steve Howe.)

such a membrane is highly desirable, since it can be used in many consecutive hybridization experiments without the need to repeat the tedious digestion and electrophoresis steps. Although Southern blotting is used in research settings, it is rarely used clinically. While historically, small polymorphic variations in DNA sequence (RFLPs – see pp. 137–139) were analysed by Southern analysis of restriction enzyme digested genomic DNA, they are now most commonly typed using PCR (see pp. 131–134).

RNA

Northern blotting

Northern blotting is a technique used to study the expression of genes. It is analogous to Southern blotting, except that molecules of RNA, as opposed to DNA, are separated by electrophoresis and transferred to a nylon or nitrocellulose membrane, which can then be used in hybridization experiments.

Protein

Western blotting

Although not concerned directly with nucleic acids, Western blotting is the protein equivalent of the above hybridization techniques. Gel electrophoresis separates out denatured proteins by mass, which are then transferred to a nitrocellulose membrane and probed for using specific fluorescently labelled antibodies to the protein under investigation.

Southern blotting was named after its developer Ed Southern. Northern blotting and Western blotting were named as a pun on the points of a compass. A handy mnemonic for remembering which molecules each hybridization technique detects is SNOW DROP:

Southern	**D**NA
Northern	**R**NA
O	**O**
Western	**P**rotein

The O's are zeros, since there is no eastern blot.

DNA sequencing

Sequencing is the determination of the order of bases in a piece of DNA. There are a variety of methods that may be used for sequencing:

Fig. 7.10 Restriction map (A) and Southern blotting (B) showing a clinical example of diagnosis of sickle-cell disease: a point mutation in the gene for β-haemoglobin that destroys a restriction site for *Mst*II. Genomic DNA is digested with *Mst*II and run on an electrophoresis gel. The gel separates the DNA fragments, making the small ones travel further. The DNA is stained with ethidium bromide and viewed under ultraviolet light. The DNA can be transferred onto nitrocellulose film by Southern blotting. (1) Gel soaked in NaOH to denature DNA. (2) The filter paper under gel soaks up NaOH solution. (3) The filter paper sucks NaOH upward through gel and membrane and so transfers DNA from gel to membrane. (4) DNA forms a chemical bond with nitrocellulose membrane and is firmly fixed. Radiolabelled probes on the Southern blot can be used to detect the β-globin gene. The mutation is detected by a single large *Mst*II fragment. (Normal genes have two smaller *Mst*II fragments because the probe spans this restriction site.)

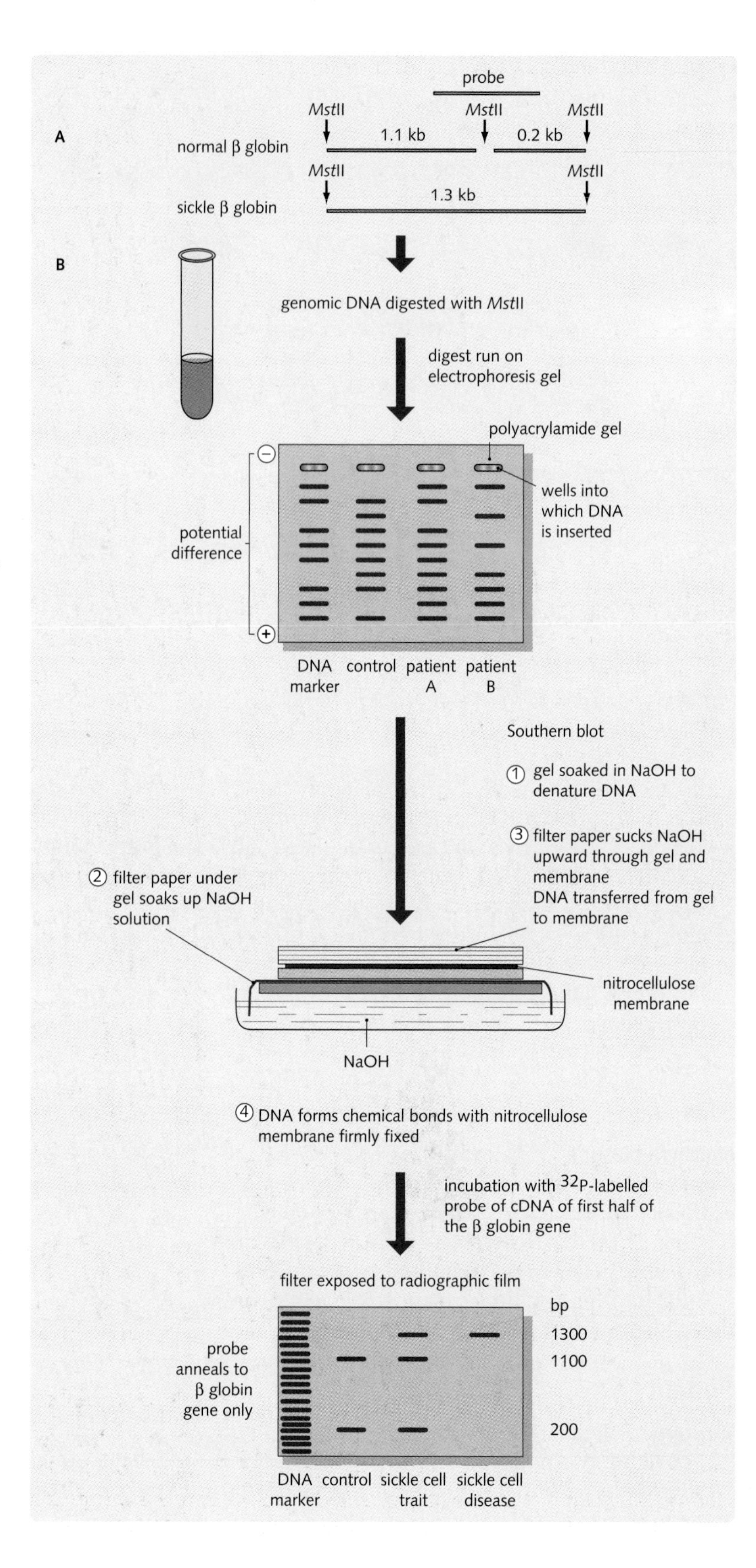

- chemical cleavage (Maxam and Gilbert method)
- chain termination (Sanger sequencing)
- thermal cycle sequencing.

Automated Sanger sequencing is the most widely used approach.

Sanger sequencing

The principle of Sanger sequencing hinges on DNA polymerase being unable to extend a growing DNA strand once a nucleotide analogue (a ddNTP) that lacks the 3′ hydroxyl group has been incorporated (Fig. 7.11).

Historically, Sanger sequencing was achieved using radiolabelled dNTPs. Using this method, for each piece of DNA sequenced, four parallel reactions are set up differing with respect to which of the four dideoxynucleotide triphosphates (ddNTPs) is included:

- template DNA (the DNA that is to be sequenced)
- the sequencing primer
- dNTPs (dATP, dTTP, dGTP and dCTP, one of which is radioactively labelled)
- ddNTP (one of ddATP, ddTTP, ddGTP or ddCTP)
- DNA polymerase.

Extension of the DNA chain continues until a ddNTP, known as a 'chain terminator', is incorporated. Since each reaction contains all four dNTPs (onto which bases can be added), a population of newly synthesized DNA strands result that vary in length according to how quickly a ddNTP was encountered. These can be separated by electrophoresis on a polyacrylamide gel, and the radioactively labelled bands can be visualized by exposure to X-ray film (Fig. 7.11).

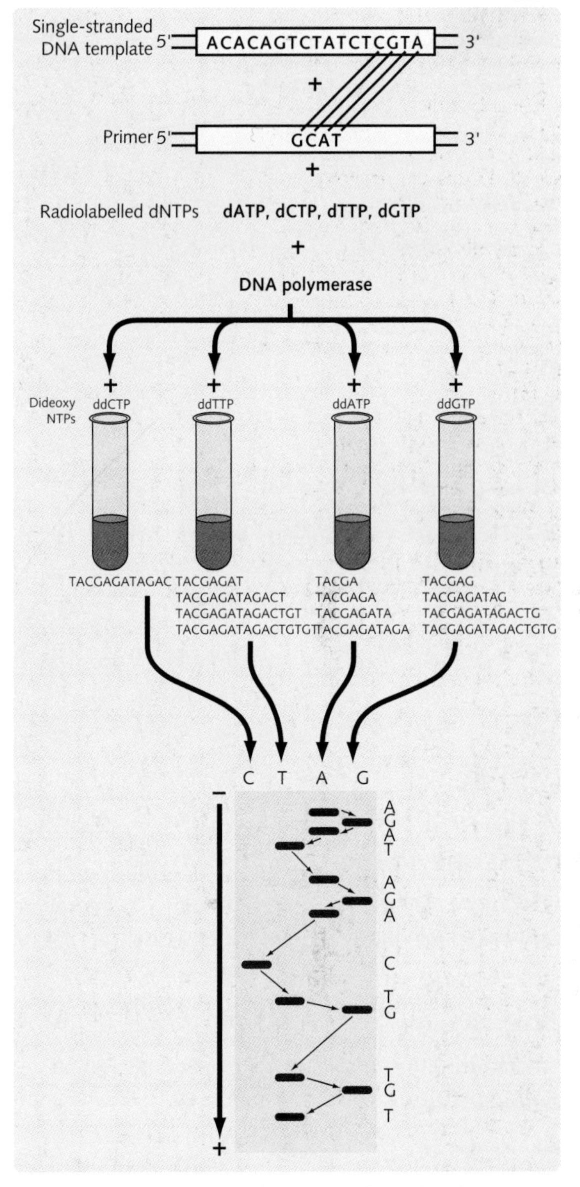

Fig. 7.11 The Sanger method of determining the nucleotide sequence of a segment of cloned DNA. (Adapted from Mueller and Young, 2001.)

> Although rarely used, the methodology behind four-pot Sanger sequencing is a popular pre-clinical exam question. Like most techniques, understanding the prototype approach is key to understanding modern takes on it.

Today, radioactive Sanger sequencing is seldom used, as it is very labour intensive and time consuming when a very high throughput of samples is required, such as in the human genome project or in diagnosis. Fluorophore labelled chain terminators are utilized instead of using radioactivity, with the advantage that the fluorophores can be differentiated by a laser detection system allowing complete sequencing to be performed in a single reaction (Fig. 7.12).

DNA sequence polymorphism analysis

Single nucleotide polymorphisms

Single nucleotide polymorphisms (SNPs), also known as restriction fragment length polymorphisms (RFLPs), are DNA sequence variations that occur

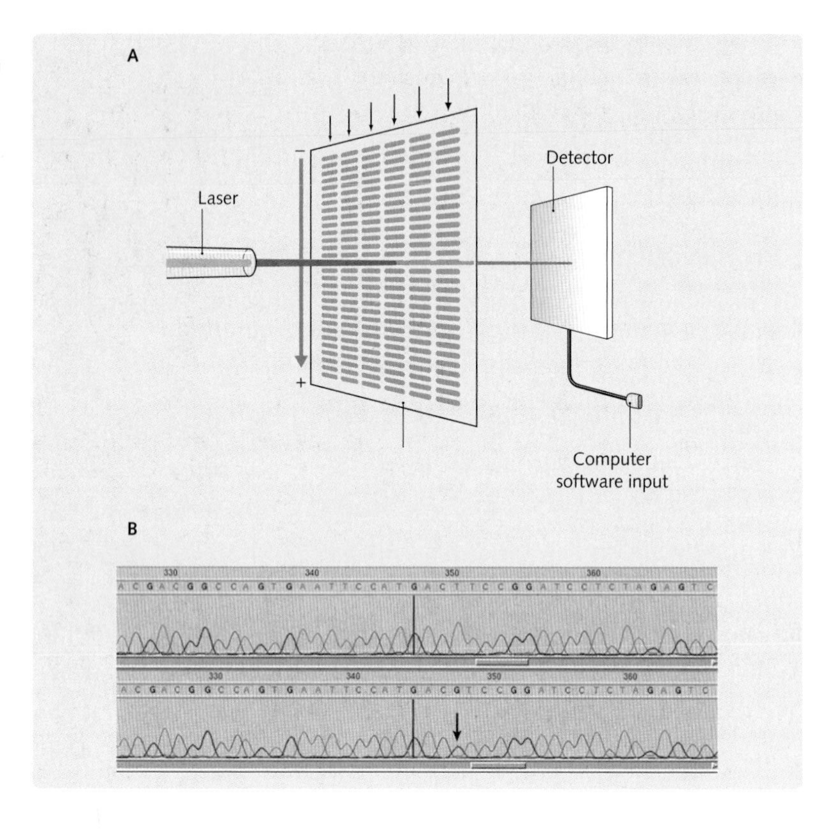

Fig. 7.12 Automated DNA sequencing using fluorescent primers. (A) A laser beam detects the four different fluorochromes that are used for each ddNTP as individual DNA fragments migrate past. (B) An example of a fluorescent DNA sequencing output (only G sequence is shown). The two traces are two different DNA sequences for the same portion of a gene. The arrow marks an allele that differs between the two sequences. In the top trace a T is found in this position, so the top sequence does not include the peak marked in the lower trace, which has a G at this allele. (Reproduced from Mueller and Young, 2001.)

when a single nucleotide in the genome sequence is changed. They occur approximately once in every 200–500 nucleotides, most of which lie in non-coding regions of the genome, and are thought to be silent. For a variation to be considered a true polymorphism, it must occur in at least 1% of the population.

Traditionally, SNPs have been detected using Southern blotting of restriction enzyme digested DNA, and this was the laborious technique used for the first physical maps of the human genome. This technique has been superceded by a number of high throughput PCR based genotyping and microarray methods.

Variable number tandem repeats

Variable number tandem repeats (VNTRs) are short nucleotide sequences up to 100 bp long that are organized into clusters of 'tandem repeats'.

There are two main families of tandem repeat:

1. Microsatellite DNA
2. Minisatellite DNA.

Microsatellite DNA, found throughout the genome, consist of tandemly repeated sequences characteristically of between 1 and 6 bp. Microsatellite repeats rarely occur within coding sequences, but trinucleotide repeats in or near genes are associated with certain inherited disorders (see pp. 154–155).

Minisatellite DNA consists of generally GC-rich repeats with a repeat unit length of 15–100 bp running over a length of 0.5 to 30 kb. The term 'minisatellite DNA' encompasses:

- telomeric DNA (see pp. 99–100)
- hypervariable minisatellite DNA.

Variable number tandem repeats (VNTRs) (minisatellites) are highly polymorphic, and there are many of them scattered throughout the genome. Hybridization with a probe that binds to these sequences is the basis of 'genetic fingerprinting', which is used in paternity testing and in forensics.

VNTRs are detected by restriction enzyme digestion of DNA and electrophoresis of the resultant fragments (smaller fragments with fewer repeats will travel further), which are then transferred to a nitrocellulose membrane. Probes specific for the given microsatellite region under investigation are then added and the number of repeats determined.

Cytogenetics

Cytogenetics is concerned with the study of chromosomes and their abnormalities. Chromosomes derived from specially treated metaphase cells can be viewed directly by microscopy. Techniques such as G-banding and fluorescence in situ hybridization (FISH) enable individual chromosomes, or specific sequences within them, to be identified, while comparative genomic hybridization (CGH) allows for the detection of the loss or amplification of specific chromosome regions.

The visualization of individual chromosomes is achieved by manipulation of the cell cycle using chemicals that specifically arrest the cell cycle in metaphase, but without inducing DNA damage. Chemicals used include colcimid and colchicine.

G-banding

G-banding is the most commonly used chromosome staining technique and is the mainstay of cytogenetic diagnosis. Chromosomes are subjected to a controlled protein digestion with trypsin and stained with Giemsa. This results in a pattern of dark and light bands that is specific for each chromosome, and which, therefore, allows each chromosome to be identified (Fig. 7.13). It is used in the diagnosis of:

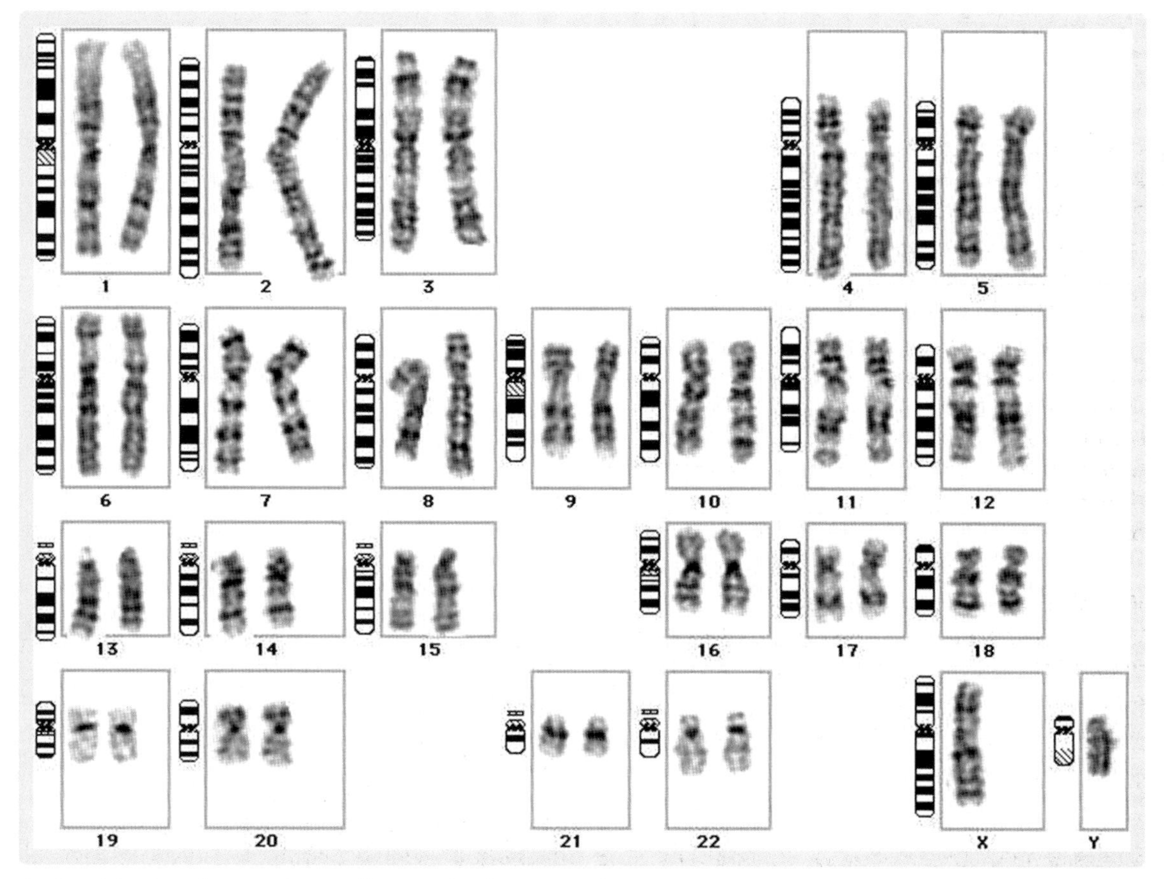

Fig. 7.13 G-banding: a karyotype for normal human male cells. (Courtesy of Dr Linda E. Ritter.)

- monosomies and trisomies
- translocations
- large deletions and insertions.

For many of its applications, G-banding is being superseded by FISH.

Fluorescence *in situ* hybridization

Fluorescence *in situ* hybridization (FISH) is a method of visualizing specific regions of metaphase or interphase chromosomes. The DNA that constitutes chromosomes is fixed to a microscope slide and denatured. It is then hybridized with fluorescently labelled probe DNA that binds specifically to a complementary sequence. The region of the chromosome where hybridization has occurred can be visualized under a fluorescent microscope. FISH probes may bind to a single genomic sequence (Fig. 7.14), stain an entire chromosome (such probes are called 'chromosome paints') or, in the case of spectral karyotyping (SKY) and multiplex fluorescence *in situ* hybridization (M-FISH), allow each individual chromosome to be visualized in a different colour. In addition to being used in research laboratories to assist in gene mapping, FISH is used diagnostically to identify a variety of chromosome abnormalities including:

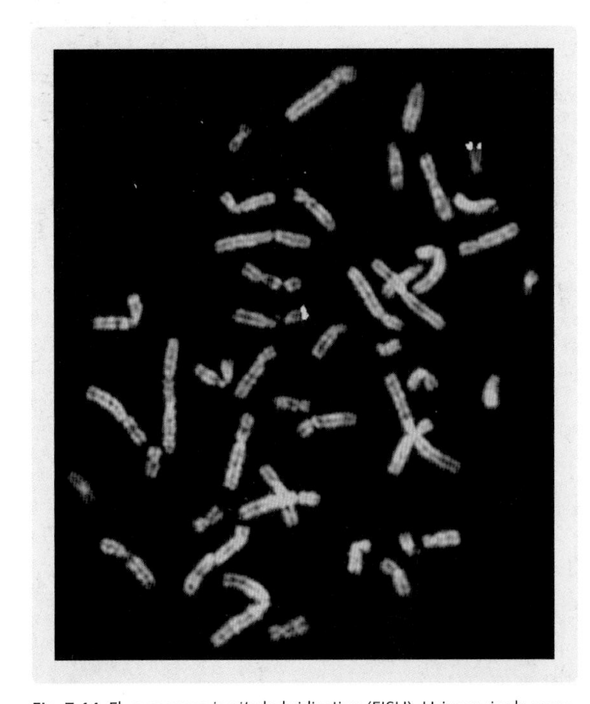

Fig. 7.14 Fluorescence *in situ* hybridization (FISH). Using a single copy probe. Although the probe is said to be single copy, this refers to the haploid genome. On the metaphase spread illustrated, two pairs of dots are visible that correspond to hybridization of the biotin-labelled probe to both chromatids on a pair of homologous chromosomes. (Courtesy of Dr Paul Scriven, GSTT.)

- monosomies and trisomies
- translocations
- microdeletions and insertions.

FISH methodology is more sensitive than G-banding and may be used to identify microdeletions that cannot be detected by this technique.

Comparative genomic hybridization

Comparative genomic hybridization (CGH) is a two colour FISH technique that allows chromosomal losses and duplications to be quantified. Equal amounts of differentially fluorescently labelled test genomic DNA, frequently tumour DNA, and normal reference DNA are mixed together and hybridized to normal metaphase spreads. Areas of chromosomal duplication in the test sample will hybridize, with excess quantities of its label, to the corresponding region of the metaphase spread. Areas of deletion in the test sample will lead to its under-hybridization with the metaphase spread, which will hybridize the labelled normal DNA instead (Fig. 7.15). CGH is used clinically to:

- screen chromosomal copy number changes in tumour genomes within a single experiment
- study tumours that do not yield sufficient metaphases for other forms of cytogenetic analysis
- study archival material allowing correlation of chromosomal aberrations with the clinical course.

A more powerful modification of the CGH technique, known as array CGH, exploits new microarray technology using several thousand probes attached to a glass slide, rather than metaphase chromosomes, this has the advantage of being able to detect much smaller areas of deletion and amplification, in the size region of 5–10 kb.

Multiplex ligation-dependent probe amplification

Multiplex ligation-dependent probe amplification (MLPA) is a new method that is replacing FISH for the detection of many disorders. It detects copy number variation in genomic sequences at high resolution (e.g. single exon deletions) and, unlike FISH, is easily amenable to multiplexing. For each locus of interest, two oligonucleotide probes are synthesized that anneal end to end to the target sequence.

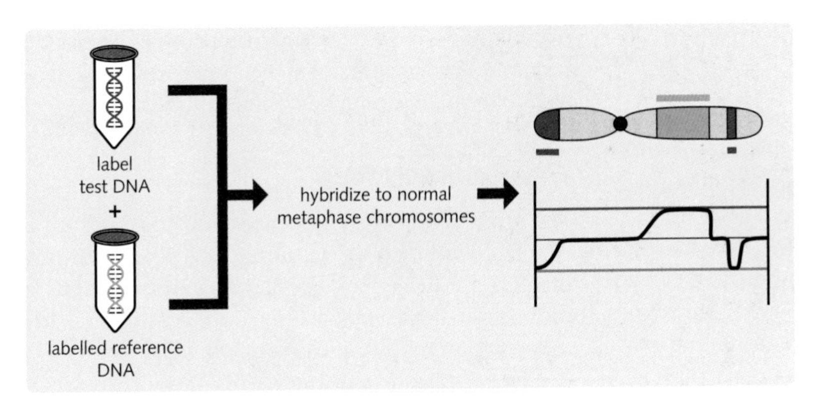

Fig. 7.15 Comparative genomic hybridization (CGH). Fluorescently labelled test sample DNA (FITC stained; green) and reference DNA (rhodamine stained; red) are hybridized to normal metaphase chromosomes. Equal co-hybridization of the green and red probe indicates a stable copy number at that point on the chromosome. DNA losses in the test sample see only reference DNA binding, leading to a predominance of the red fluorophore, while DNA gains lead to excess test sample DNA binding, out competing the reference DNA, and a predominance of the green fluorophore.

The probes are joined by a ligase and then PCR is used to amplify the probe ligation product. Variation in the length of the original probes allows testing of up to 40 loci in one reaction (Fig. 7.16). The relative amount of product is proportional to the copy number and, thus, can be used to detect deletions or trisomies. Allele specific probes can be used to detect single base changes as the mismatch between the mutant genomic sequence and the wild type probe introduced by the SNP prevent ligation (Fig. 7.16).

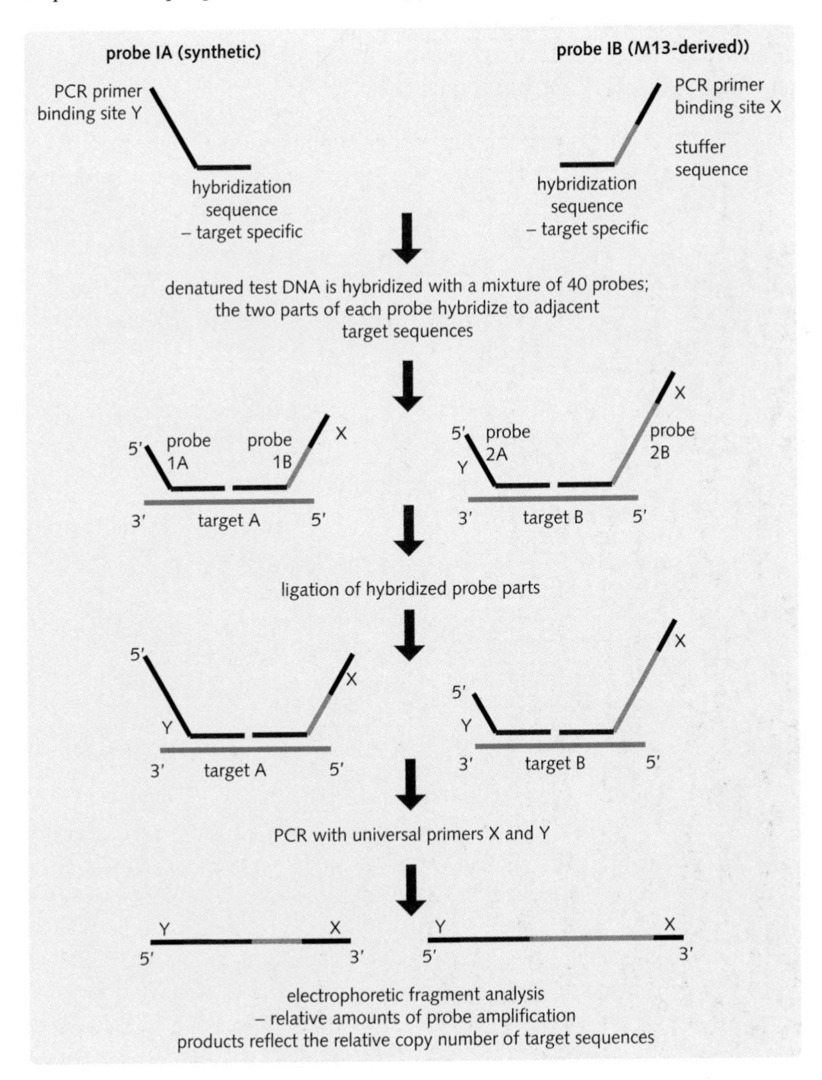

Fig. 7.16 Multiplex ligation-dependent probe amplification (MLPA). MLPA probes consist of two oligonucleotides: a short and synthetically derived oligonucleotide and a longer oligonucleotide derived from phage M13. Each M13 derived oligonucleotide has a different stuffer sequence. Following hybridization, ligation is achieved using a heat stable ligase. Only perfectly matched probes will ligate. As the amplification product of each probe has a different size, varying from around 130 to 480 base pairs, the relative amounts of each probe amplification product reflects the relative copy number of that target sequence.

Quantitative fluorescence polymerase chain reaction

Quantitative fluorescence polymerase chain reaction (QF-PCR) is another molecular technique in which DNA markers (e.g. microsatellites) are amplified using fluorescent primers. It can be applied to diagnose aneuploidy, to further characterize chromosomal abnormalities identified by G-banding and to determine the parent of origin for certain abnormal chromosomes. QF-PCR is quicker and less labour-intensive that FISH.

MAPPING AND CHARACTERIZING HUMAN DISEASE GENES

Cloning and vectors

Historically, the identification of a set of clones spanning the region under investigation is the first step to obtaining a genetic or physical map. Although gene cloning has largely been superceded by technologies resulting from the human genome project, it is a technology that was fundamental to its success.

The need for clones

The human genome is too large and complex to be characterized directly. Cloning vectors are used to break it up into small, overlapping chunks that are more amenable to characterization techniques, such as restriction mapping and sequencing.

The process of cloning involves the transfer of a specific sequence of DNA into a single cell of a microorganism. Once cloned, the sequence of DNA is replicated alongside the host's own complement of DNA. Large quantities of microorganisms can easily be cultured in the laboratory, so a plentiful supply of DNA for molecular analysis can be readily generated. Vectors vary with respect to the maximum amount of DNA they are able to accommodate (Fig. 7.17).

Fig. 7.17 Commonly used vectors

Vector	Origin	Features	Size of insert accommodated
Plasmid	Circular double stranded DNA in the cytoplasm of bacteria; undergoes replication with the bacterial genome and is passed on through generations; can be regarded as bacterial parasites	Origin of replication; antibiotic resistance genes; restriction enzyme site, which can break open plasmid and allow DNA to be inserted	Less than 10 kb
Phage	Viruses that infect bacteria	Phage particles can be assembled *in vitro*; DNA of interest is fragmented and ligated to phage *cos* sites; it is mixed with packaging extract, which contains all the proteins needed for phage assembly; phage heads are filled with DNA between two *cos* sites and a phage tail is attached; the assembled phage infects the host cell	16 kb
Cosmid	A genetically engineered hybrid of a plasmid and a phage	Contains plasmid origin of replication, selectable marker and phage *cos* site	45 kb
BAC	Bacterial artificial chromosomes are based on the F-factor plasmid	The vector includes the F-factor origin of replication, a chloramphenicol resistance gene and a marker that allows positive selection of recombinants on media containing sucrose. BACs are introduced into *E.coli* cells by electroporation	100–300 kb
YAC	'Yeast artificial chromosomes' are genetically engineered units that can be replicated in yeast cells	Contain the three DNA sequences essential for yeast chromosome function: telomeres (*TEL*), origin of replication (*ARS*), centromere (*CEN*)	100–1000 kb

Fig. 7.17 The most commonly used vectors in DNA cloning.

Once cloned in a vector, the tools of molecular genetics may be used to introduce specific modifications into the cloned sequence.

Basic cloning techniques

In the simplest case, both insert and vector DNA are cut with the same restriction enzyme to produce DNA molecules that have compatible sticky ends (Fig. 7.18). The vector is treated (phosphorylated) so that its free ends are unable to re-ligate together. The vector and insert sequence are mixed together and incubated with DNA ligase, which catalyses the formation of phosphodiester bonds between molecules of double-stranded DNA, joining the DNA fragments together.

The products of the ligation reaction are incubated with 'competent' host cells, some of which will take up DNA (transformed cells) and are plated onto a selective media, on which only cells that contain vector sequence can grow (Fig. 7.18). Special techniques (e.g. 'blue/white selection') are used to identify clones containing vector sequence that includes insert DNA. In addition to the desired recombinant molecule, ligation may produce other events (e.g. two vector molecules ligated together). In order to identify the desired clones, several individual colonies are grown up to produce DNA for molecular analysis.

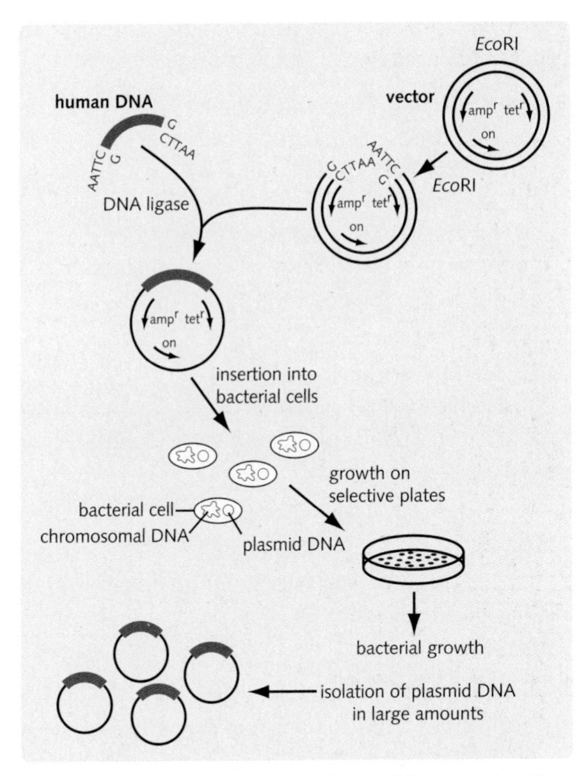

Fig. 7.18 The cloning process. Insert and vector DNA are produced by restriction enzyme digestion with enzymes that produce compatible ends. The plasmid contains an origin of replication (ori) and antibiotic resistance genes (amp^r and tet^r) that enable selection of transformants (bacteria that have taken up the plasmid). Vector and insert are ligated and used to transform competent cells. Transformed cells are plated onto media containing ampicillin or tetracycline, so that only competent cells that have taken up the vector sequence are able to grow. Many different clones are produced by this process, which each contain a different fragment of DNA. (Adapted from Nussbaum, McInnes and Willard, 2001.)

Cloning by PCR

In some cases, PCR is used to generate insert DNA. PCR primers are designed to include the appropriate restriction enzyme sites. After amplification, the product is digested with the restriction enzyme and then added to the ligation reaction as described above.

As well as being useful for characterization and manipulation of DNA fragments, cloning has been exploited for medical applications, such as the large-scale production of therapeutically useful proteins (e.g. insulin) and the generation of transgenic animal models for important human diseases.

Maps

Genetic maps

The genetic map of the human genome is based on the probability of recombination between paternally derived and maternally derived chromosomes at meiosis (see pp. 120–121):

- loci are assigned to linkage groups
- distances are quoted in recombination units (centimorgans).

One centimorgan (1 cM) is equivalent to a 1% chance of recombination. (A Morgan is defined as a length of chromosome segment that undergoes one exchange with its homologue per meiosis.)

Recombination is more frequent in female meiosis, so the male and female maps are different, with markers appearing further apart on the female version. Some areas of the genome are more prone to recombination than others, so there is not a perfect correlation between genetic and physical maps.

Genetic maps are fundamental to positional cloning (see pp. 144–145). The most recent high-resolution genetic maps have polymorphic markers spaced at intervals of less than 1 cM.

Genetic and physical maps are not perfectly correlated, but on average 1 cM is equivalent to 1 Mb.

Physical maps

Physical maps locate genes at cytogenetically defined locations, and they give distances in bp, kb or Mb. Physical maps were constructed using restriction mapping, sequencing and FISH to analyse overlapping clones that together span almost the whole human genome. These arrays of clones are anchored to specific named chromosomes. The ultimate physical map is the full human genome sequence.

The data generated by the human genome project have greatly simplified positional cloning. When the approach was first suggested in the 1980s, individual research groups had to identify polymorphic markers themselves with which to establish linkage. Once linkage was found, they then had to construct their own arrays of overlapping clones in order to construct physical maps before they could even begin to look for the disease-causing gene. Now, all this information is freely available thanks to the efforts of the human genome project.

Identifying disease genes

The successful identification of the genes that are mutated in monogenic disorders has a number of possible consequences:

- definitive diagnosis
- more accurate assessments of risk or prognosis
- presymptomatic diagnosis
- prenatal diagnosis.

Furthermore, an increased understanding of the molecular basis of the disease may lead to the identification of new drug targets, enable rational drug design and facilitate the production of animal models of the disease on which such therapies can be tested.

Positional cloning

Positional cloning (previously called reverse genetics) is a method of identifying a disease gene when nothing is known about the nature of the corresponding protein. This approach has been used to identify the genes responsible for many diseases, including cystic fibrosis and Huntington's disease.

In positional cloning, the gene responsible for a condition is identified from knowledge of its position in the human genome. This knowledge comes from linkage analysis of polymorphic markers in families in which the disease is segregating (see 'genetic linkage analysis', below).

Polymorphic markers

A marker is any Mendelian characteristic used to follow the transmission of a segment of chromosome through a pedigree. DNA polymorphisms that have a known position on the genetic map of the human genome are used as markers in linkage studies:

- DNA polymorphisms are inherited differences in DNA between normal, healthy people.
- They frequently arise in non-coding regions of the genome.
- Examples include restriction fragment length polymorphisms (RFLPs), microsatellites and single nucleotide polymorphisms (SNPs) (see pp. 137–139).

Microsatellite markers

Microsatellites are an extremely useful type of marker that are well suited to genetic linkage analysis for several reasons:

- They can be detected by PCR.
- They are extremely polymorphic.
- Tens of thousands of microsatellite polymorphic loci have been identified scattered throughout the genome.

An example of the results obtained after PCR with a pair of primers that spans a microsatellite is shown in Fig. 7.19.

Each allele of a microsatellite may give rise to many bands on a polyacrylamide gel, which makes interpreting a single lane difficult. To identify which bands are parts of the same allele, you must compare the bands across all the lanes (see Fig. 7.19).

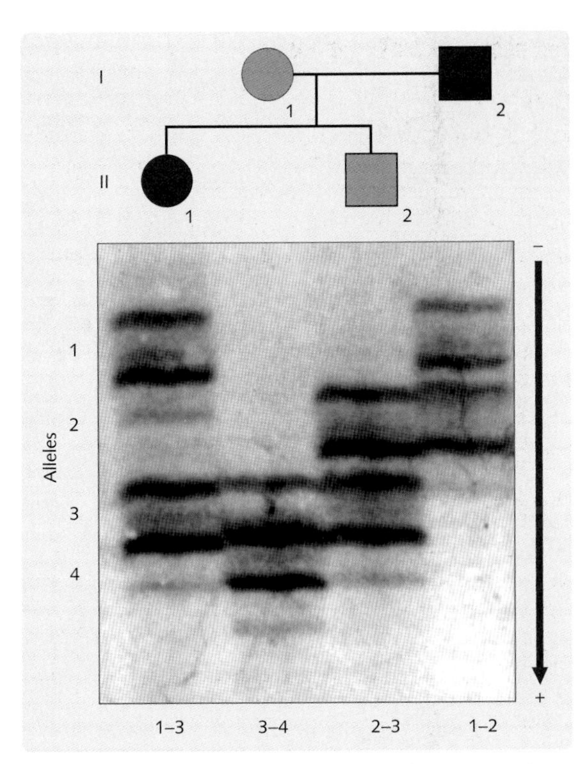

Fig. 7.19 An autoradiograph showing the results of a PCR using primers that span a microsatellite repeat. The microsatellite is closely linked to a gene that causes a dominant disorder and allele 1 is segregating with the condition. Note that each allele gives rise to multiple bands, which is due to an artefact of PCR. (Reproduced from Mueller and Young, 2001.)

Genetic linkage analysis

Linkage

Linkage analysis uses pedigree data to determine whether loci are linked and to estimate the recombination fraction (see below). At meiosis, homologous chromosomes exchange segments before separating into two daughter cells.

Linked markers segregate together in meiosis more frequently than expected by chance because they lie close together on the same chromosome (Fig. 7.20). Crossovers are statistically unlikely to form between markers that are close together. However, if markers are sufficiently far apart on a chromosome it is likely that a crossover will form between them and recombination will occur.

Recombination

Recombination occurs as a result of the formation of chiasmata (crossovers) during meiosis. Recombinants are children who inherit a different combination of alleles at two loci compared with the combination found in the gametes that made the parent.

The recombination fraction (RF) is the proportion of recombinants. For unlinked loci it is 0.5; for linked loci it lies between 0 and 0.5. Recombination fractions are used to produce genetic maps (Fig. 7.21).

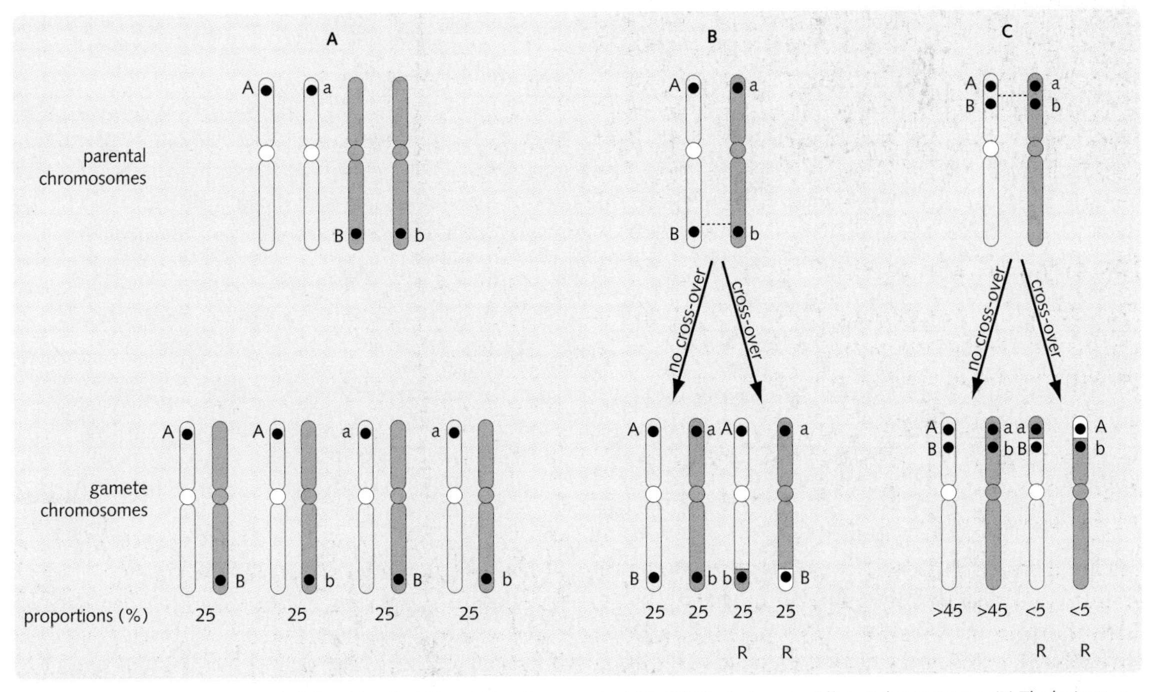

Fig. 7.20 Segregation at meiosis of alleles at two loci. Recombinants are marked (R). (A) The loci are on different chromosomes. (B) The loci are on the same chromosome, but widely separated (>1 M apart). (C) The loci are closely adjacent and so segregate together most of the time (i.e. they are linked). (Adapted from Mueller and Young, 2001.)

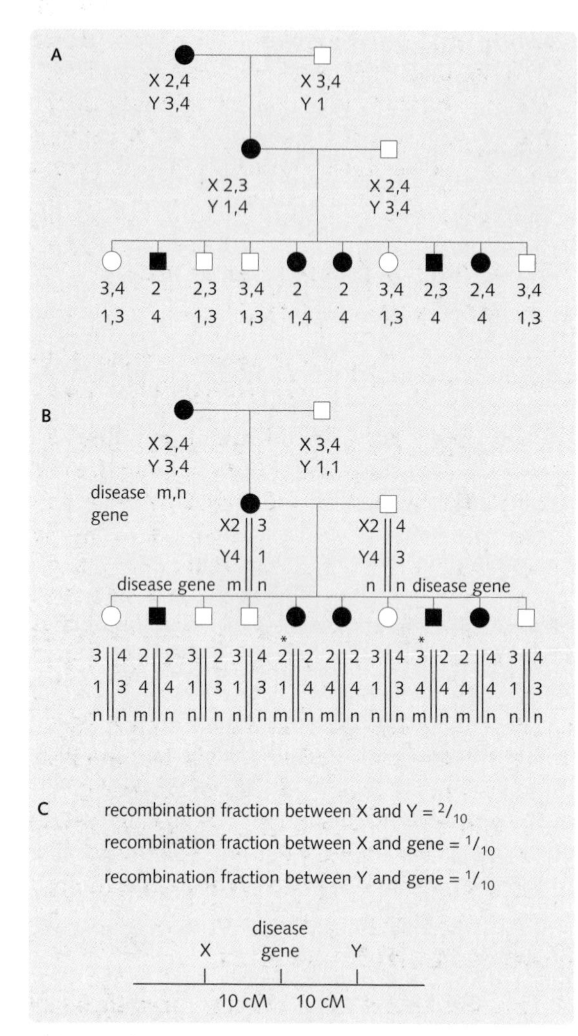

Fig. 7.21 Genetic mapping of two microsatellite markers and a dominant disease gene. (A) Family members are genotyped with respect to two microsatellite markers (X and Y). (B) The markers are rearranged according to parental inheritance. (C) The recombination fraction for each pair of markers is calculated separately using the children's chromosomes and the genetic map is constructed (1% recombination fraction is equal to 1 cM). *, recombinant allele; m, mutant disease gene; n, normal disease gene; X and Y, microsatellite markers; 1–4, allele sizes. (Courtesy of Dr Kathy Mann.)

(1 cM is defined as a 1% chance of recombination, which is equivalent to an RF of 0.01.)

The log of odds score

The random assortment of chromosomes at meiosis means that unlinked alleles on different chromosomes will segregate together on average 50% of the time. Thus, even in large families with many meioses, markers may appear to be linked because they have repeatedly segregated together by chance. A log of odds (LOD) score is a statistical estimate of whether two loci are likely to lie near each other on a chromosome, and are, therefore, likely to be inherited together. A score of three or more is generally taken to indicate that the two loci are close.

Family selection and pedigree analysis

The first step in identifying a disease gene is identifying sufficient families with the gene to identify linkage. The role of the clinician is vitally important at this stage, as it is imperative that all affected members of the family are identified and that all the families included in the study have the same disease. The mode of inheritance of the disease (recessive, dominant, etc.) should also be considered, as this information is required for linkage analysis.

Establishing linkage to a chromosome

To establish linkage to a specific chromosome, all members of the family are genotyped with respect to about 400 polymorphic markers:

- These polymorphic markers are nearly always microsatellites.
- They are freely available through the HGP.
- They are scattered throughout the whole genome.
- They have known positions on genetic and physical maps.

The data generated are reviewed to determine whether any one of these markers is segregating with the disease (Fig. 7.22).

Fine genetic mapping

Once linkage to a particular chromosome or chromosomal region has been identified, all the members of the family are genotyped with respect to a further set of more densely packed markers that map to the appropriate chromosomal region. This aims to:

- confirm linkage is genuine
- narrow the region of the genetic map in which the disease gene can be shown to lie.

Even before the disease-causing mutations have been identified, closely linked genetic markers (5 cM or less) can be used to track the inheritance of a disease gene in individual families for diagnostic purposes.

From linked genetic marker to cloned gene

The output of the HGP means that, once the relevant genetic region has been identified, researchers and diagnosticians can go straight to the HGP sequence maps. From here, candidate genes can be identified and investigated as a cause of the disorder under

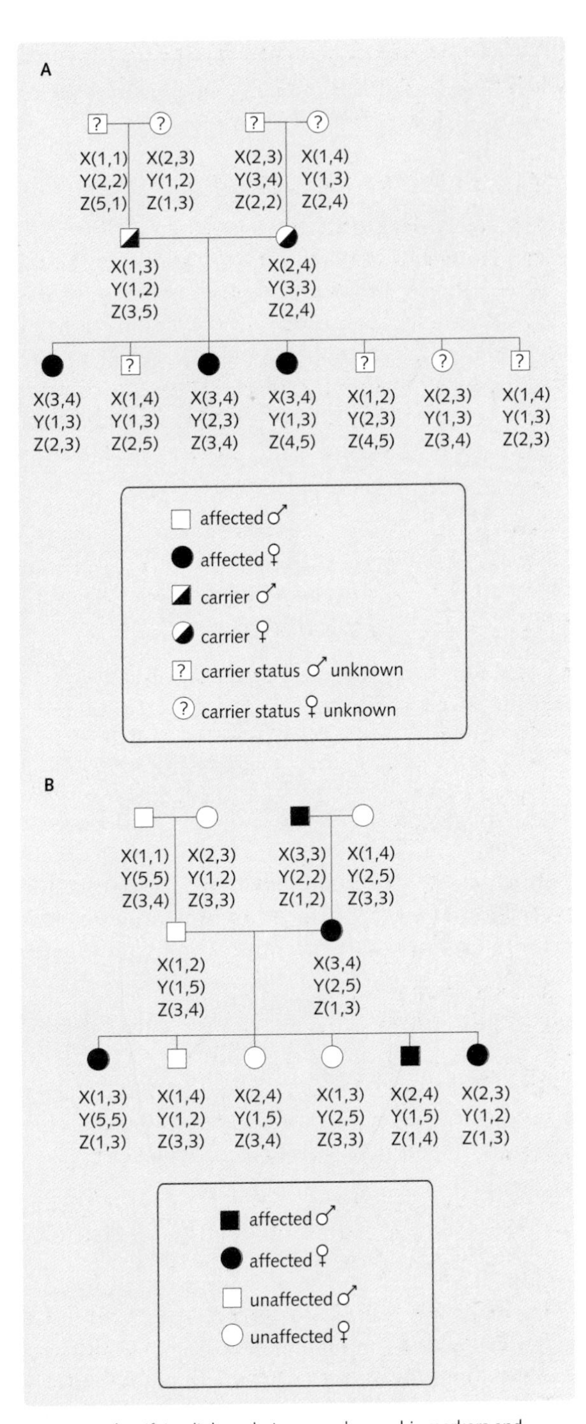

Fig. 7.22 Identifying linkage between polymorphic markers and disease genes. In these examples, three polymorphic markers are used (X, Y and Z) that map to three different chromosomes (2, 3 and 5, respectively). (A) In this pedigree showing autosomal recessive inheritance, the disease is segregating with the '3' and '4' alleles of marker X, suggesting that the disease maps to chromosome 2. (B) In this pedigree showing autosomal dominant inheritance the disease is segregating with the '1' allele of marker Z, suggesting that the disease maps to chromosome 5.

investigation. There should be biological evidence that would make it plausible that mutations in the candidate gene might result in the specific disease, for example knowledge of the appropriate biochemical pathway, the gene is expressed in the tissues affected by the disorder, or there is an animal model.

Other methods of identifying human disease genes

Historically, other strategies have been employed to identify the genes that are mutated in genetic diseases:

- Functional cloning – the identification of a gene responsible for a disease based on knowledge of the underlying molecular defect. If the nature of the deficient protein is known, it is often possible to isolate the appropriate mRNAs and use them (or cDNAs derived from them) as probes for the gene. This approach was used to define the genes responsible for phenylketonuria and sickle-cell anaemia.
- Candidate gene approach – the use of previously isolated genes as candidates for causing a specific genetic disease on the basis of being known to have a role in the physiology of the diseased tissue. This approach was used to identify rhodopsin mutations associated with retinitis pigmentosa.

Multifactorial diseases

Linkage mapping has successfully identified the genetic basis of several monogenic diseases, but does not have adequate power to identify multiple gene effects responsible for common multifactorial diseases (see p. 167). However, very recent advances in high-throughput SNP genotyping technology have allowed large-scale population- (rather than family) based studies which have proved to be powerful in the identification of genes associated with diseases such as diabetes and inflammatory bowel disease.

Mutation analysis

If mutation within a particular gene is responsible for a specific genetic disease, then patients with the disorder should have mutations in the gene that are not found in unaffected individuals.

Where the normal structure and base sequence of the gene have not already been established by the HGP, these are determined by molecular analysis of the cDNA and clones isolated from genomic libraries.

After the normal base sequence has been established, patient DNA is screened for mutations by methods such as:

- single strand conformation polymorphism (SSCP) analysis
- heteroduplex analysis
- direct sequencing.

DNA diagnostics

Once a disease-causing mutation has been identified, molecular methods are developed that enable its identification in patient DNA:

- PCR-based techniques are used to detect base changes wherever possible.
- Cytogenetic techniques are used to detect microdeletions, insertions and translocations.
- Southern blotting is used to detect some triplet repeat expansions.

THE HUMAN GENOME PROJECT

The human genome project (HGP) is an international cooperative research effort, the purpose of which is to investigate the human genome in its entirety. The six main aims of the HGP are:

1. Mapping human genes and markers
2. Sequencing the human genome
3. Functional analysis and post-genomic genetics
4. Comparing the human genome with the genomes of model organisms (*E. coli, S. cerevisiae, C. elegans, etc.*).
5. Developing new DNA technologies (e.g. automated sequencing).
6. Developing bioinformatics (systems for collecting, storing, and disseminating the information generated by the project).

The HGP has overseen the development of several types of map of the human genome and of the genomes of model organisms.

Sequencing the human genome

The base sequence of the genome is the most detailed type of physical map. The ultimate aims of the HGP are the production of a single, continuous sequence of bases for each of the human chromosomes, and the delineation of the position of all the human genes. The sequence was determined by sequential steps, as described in Fig. 7.23.

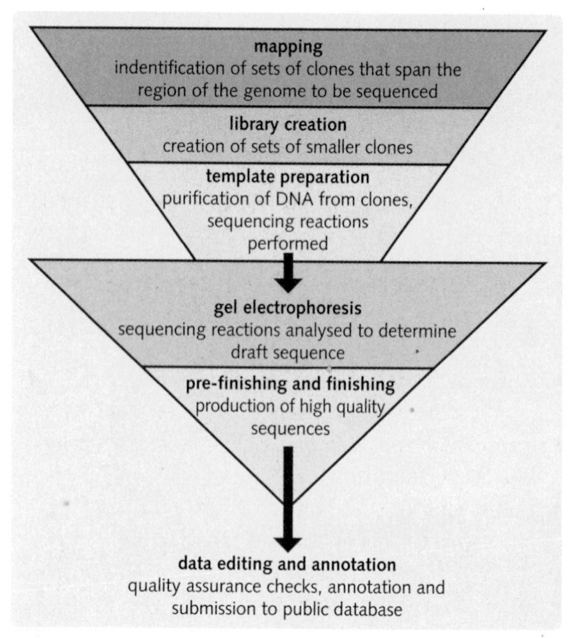

Fig. 7.23 Stages involved in the human genome project. The processes involved in mapping the human genome can be thought of like two large sieves. The first produces progressively smaller, accurate clones; while the second is concerned with increasing sequencing specificity.

A 'rough draft' of the human genome was completed in 2003, and final HGP papers were published in May 2006. It is envisaged that the correction of minor errors (currently estimated at 1 error in every 10 000 nucleotides) will continue for the foreseeable future.

Sequencing of the human genome has also enabled the construction of high-density SNP maps. Such maps will allow researchers and clinicians to:

- search for and isolate specific disease causing genes – single gene disorders
- compare SNP patterns from populations with a given multifactoral disease (i.e. cancer, heart disease, type two diabetes) with those from unaffected individuals, allowing associations to be made
- guide the development of new drugs, identifying individuals in whom a drug might not be as efficacious.

In silico techniques: bioinformatics

Bioinformatics is an integration of mathematical, statistical and computer algorithm methods to analyse molecular data. With the advent of high throughput, automated techniques, the storage of raw data in databanks allows for comprehensive study of:

- normal biological processes
- abnormal biological processes and relationship to disease
- modelling of biological systems
- improved and tailored drug discovery.

Methods used within the field of bioinformatics include:

- DNA informatics, including sequence analysis, open reading frame (ORF) analysis and gene scanning
- genomics, the application of DNA informatics to analyse the entire genome of an organism
- protein informatics, using protein sequence to model structure and function
- proteomics, the application of protein informatics to analysis of the protein complement of the genome
- metabolomics, defined as '*the quantitative measurement of all low molecular weight metabolites in an organism's cells at a specified time under specific environmental conditions*', looks not only at all the end products of gene expression, but does so in a particular physiological background or developmental state.

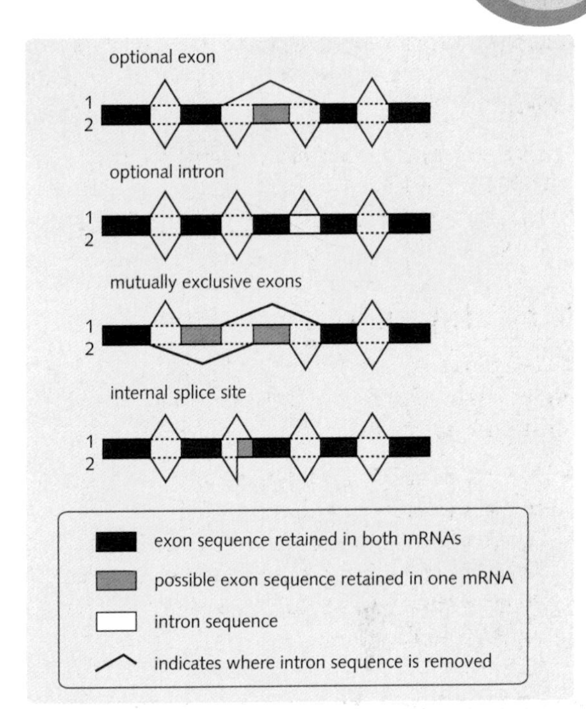

Fig. 7.24 Alternative splicing. Non-consecutive exons are spliced together in some, but not all, of the transcripts produced from the gene. With each of the four mechanisms, a single type of RNA transcript is spliced in two alternative ways to produce two distinct mRNAs (1 and 2).

The human proteome

The proteome is the complete set of proteins encoded by the genome. Compared with invertebrates, vertebrate proteins have a larger repertoire of motifs and domains available, and they include more complex domain architectures and multidomain proteins with multiple functions.

A surprise revealed by the HGP is that humans appear to require only 20 000–25 000 protein-coding genes (although this may yet prove to be an underestimate), which is only double the number that flies and worms have. However, human genes are more complex, and it is believed that processes such as alternative splicing greatly increase the number of proteins produced (Fig. 7.24). This, in association with the greater complexity of protein domains, may be sufficient to account for the increased phenotypic complexity of vertebrates.

GENE THERAPY

Our increased understanding of the molecular basis of disease has led to the proposal that it may be possible to treat some disease by gene therapy (GT). GT can be defined as the treatment of a disease by addition, insertion or replacement of a normal gene or genes. Many diseases have been identified, both genetic and acquired, that could potentially be treated in this way (Fig. 7.25).

Gene therapy trials

Several countries have established regulatory bodies whose remit is to oversee the technical, therapeutic and safety aspects of GT trials. There are two possible strategies for GT:

1. Germ-line GT – genetic changes would be introduced into every cell type, including the germ line.
2. Somatic-cell GT – the genetic modifications are targeted specifically to the diseased tissue.

There is unanimous agreement between all regulatory bodies that only somatic cell GT strategies should be allowed. Germ-line strategies are considered unethical on the grounds that genetic changes would be transmitted to future generations. Before a GT trial is commenced, the following requirements should be fulfilled:

Fig. 7.25 Diseases that can potentially be treated by gene therapy. (Adapted from Mueller and Young, 2001.)

Fig. 7.25 Diseases which can potentially be treated by gene therapy

Disorder	Defect
Immune deficiency	Adenosine deaminase deficiency Purine nucleoside phosphorylase deficiency Chronic granulomatous disease
Hypercholesterolaemia	Low density lipoprotein receptor abnormalities
Haemophilia	Factor VIII deficiency (A) Factor IX deficiency (B)
Gaucher's disease	Glucocerebrosidase deficiency
Mucopolysaccharidosis VII	β-Glucuronidase deficiency
Emphysema	α_1-Antitrypsin deficiency
Cystic fibrosis	*CFTR* mutations
Phenylketonuria	Phenylalanine hydroxylase deficiency
Urea cycle abnormalities Hyperammonaemia Citrullinaemia	 Ornithine transcarbamylase deficiency Argininosuccinate synthetase deficiency
Muscular dystrophy	Dystrophin mutations
Thalassaemia/sickle-cell disease	α- and β-globin mutations
Cancer Malignant melanoma Ovarian cancer Brain tumours Neuroblastoma Renal cancer Lung cancer	
Acquired immune deficiency syndrome (AIDS)	
Cardiovascular diseases	
Rheumatoid arthritis	

- The gene involved should be cloned and characterized.
- The specific tissue to be targeted should be accessible and identified.
- A safe and efficient vector system for the gene should be defined.
- The scientific rationale for the GT approach should be sound and the perceived risks commensurate with the potential benefits.

Vector systems

The vector system is the means by which DNA is delivered to the target cells. Achieving efficient delivery of DNA is one of the major difficulties that must be overcome. The ideal vector for GT would have the following properties:

- It would be easily produced.
- It would give sustained and regulated expression of the delivered gene.
- Immunologically, it would be inert.
- It would deliver the gene only to the required tissue.
- It would be able to deliver the largest of genes and the controlling elements required for their expression.
- It would integrate precisely into the genome in a site-specific manner, or else would be maintained as a stable episome.
- It would be able to infect both dividing and non-dividing cells.

Fig. 7.26 *CFTR* delivery vectors: advantages and disadvantages

Vector	Advantages	Disadvantages
Recombinant adenovirus	Targets the respiratory tract epithelium; infects non-replicating cells; in preliminary experiments 40% of respiratory epithelial cells took up the vector	Expression in target cells transient; can create an inflammatory response in recipient; not known whether re-administration is safe (possibility of secondary immune response)
Liposome	Can be delivered directly to the lung (in an aerosol or by direct irrigation or intravenous infusion); re-administration is unlikely to cause an immune response	Low uptake of liposomes: only 5% of epithelial cells transfected in preliminary experiments, which is not enough to have a therapeutic effect

Fig. 7.26 Advantages and disadvantages of adenovirus and liposome vectors for cystic fibrosis gene therapy.

There are two main types of vector system:

1. Physical (non-viral) vector systems (e.g. liposomes).
2. Viral vectors, which may integrate into the genome (retroviral, lentiviral and adeno-associated vectors) or be maintained as an episome (adenovirus).

Viral vectors are derived from viruses by replacing the genetic components essential for further propagation with the therapeutic gene. Liposomes are artificial lipid vesicles that fuse with the cell membrane to deliver their contents (DNA in the case of GT).

Both types of vector system, each with its own inherent advantages and disadvantages, have been used in clinical trials, for example, the treatment of cystic fibrosis by GT (Fig. 7.26).

Objectives

By the end of this chapter you should be able to:

- Understand the common processes that lead to mutagenesis, and appreciate how different classes of mutation yield different effects on protein structure and function.
- Draw out example pedigrees representing autosomal dominant, autosomal recessive, X-linked dominant, X-linked recessive, holandric and mitochondrial inheritance.
- Understand the concept of consanguinity and its effect on the incidence of autosomal recessive conditions.
- Appreciate how skewed Lyonization can result in X-linked recessive disorders appearing in female patients.
- Define anticipation and understand its relevance in trinucleotide repeat diseases.
- Appreciate the concept of heritability in the context of complex diseases and describe the threshold model of multifactorial disorders.
- Understand the importance of heterozygous cancer predisposition and describe features suggestive of inherited cancer susceptibility.
- Define aneuploidy, triploidy, trisomy and monosomy, with examples of resultant diseases.

Genetic disease is a term that encompasses not only single-gene disorders (see pp. 156–162) and chromosomal defects (see pp. 173–180), but also complex, multifactoral disease (see pp. 164–167), where the impact of genetics may not be immediately apparent. Although each genetic disorder may be rare, combined together genetic disease is common, can affect any body system and have a major impact on both morbidity and mortality. An understanding of genetics is important, not only for the diagnosis and management of such disorders, but also for the identification of genetic disease 'carriers' for genetic counselling referral (see pp. 187–190).

MUTATION

Mutations are permanent changes in the amount or structure of genetic material. They can be inherited or occur spontaneously and can be subdivided into germline or somatic mutations. Mutation plays a key role in the pathogenesis of genetic disorder, by altering gene sequences and ultimately their protein products. Mutations occurring in the DNA of gametes (sperm and egg) are heritable and are passed to subsequent offspring.

Mechanisms of mutation

At the single-gene level mutations may result from:

- substitution (point mutation)
- deletion
- insertion
- inversion
- triplet repeat expansion (unstable expansion).

Substitution

Substitution is the replacement of a single nucleotide by another with no net gain or loss of chromosomal material. Point mutations may arise as a result of mistakes in DNA replication, mistakes in repair following DNA damage or (most commonly) as the result of the spontaneous deamination of methylated cytosine to thymine. Substitutions are classified as:

- transition – purine to purine or pyrimidine to pyrimidine
- transversion – purine to pyrimidine or vice versa (Fig. 8.1).

Point mutations may be silent or deleterious depending upon their type (see p. 155) and site (Fig. 8.2). Rarely, a mutation may be advantageous and favoured by natural selection.

Deletion and insertion

Deletion is loss of DNA involving from one to many thousands of base pairs. Sequences at the ends of deletions are often similar, predisposing to recombination errors (Fig. 8.2A).

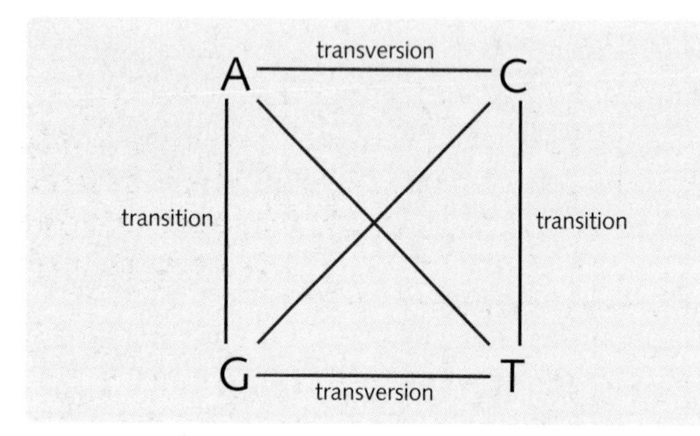

Fig. 8.1 The relationship between transition and transversion substitution mutations. Transitions are an interchange between a purine and a purine, or a pyrimidine and a pyrimidine. Transversions are an interchange between purine and pyrimdine bases.

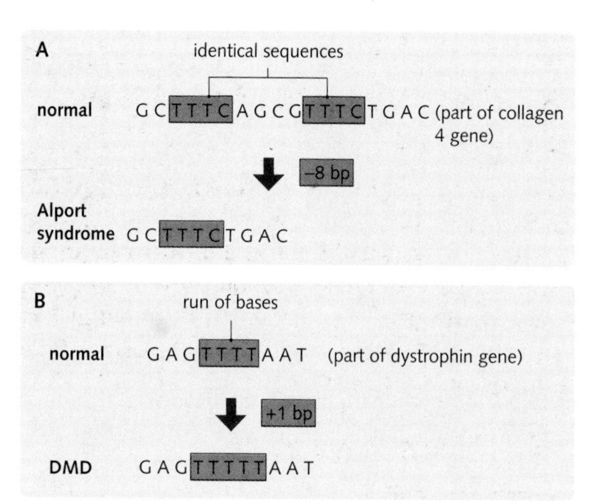

Fig. 8.2 (A) Deletion in Alport's syndrome, a hereditary disease of basement membranes, characterized by sensorineural deafness and renal failure. (B) Duplication in Duchenne muscular dystrophy (DMD). A, adenine; bp, base pairs; C, cytosine; G, guanine; T, thymine.

Insertion is a gain of DNA. Duplication, a type of insertion, occurs when runs of bases and repeated motifs predispose to duplication by replication slippage (Fig. 8.2B).

The effects on the protein of deletion and insertion depend on:

- the amount of material lost
- whether the reading frame is affected (Fig. 8.3).

If the mutation involves the insertion or deletion of nucleotides that are not multiples of three, a frameshift mutation will occur (see p. 155).

Inversion

Inversions may involve anything from two to many thousands of base pairs. They occur in areas of sequence homology (sequences at each end of the inverted segment often resemble each other). In haemophilia A, 40% of mutations result from an inversion of several hundred thousand base pairs within the factor VIII gene.

Triplet repeat expansions

Triplet or trinucleotide repeat expansions are a subset of unstable microsatellite repeats (see

A	wild type	TACAACGTCACCATT	DNA
		AUGUUGCAGUGGUAA	mRNA
		met-leu-gln-trp-STOP	protein
B	silent mutation	TACAA**T**GTCACCATT	DNA
		AUGUUACAGUGGUAA	mRNA
		met-leu-gln-trp-STOP	protein
C	missense mutation	TACAA**G**GTCACCATT	DNA
		AUGUUCCAGUGGUAA	mRNA
		met-phe-gln-trp-STOP	protein
D	nonsense mutation	TACAACGTCAC**T**ATT	DNA
		AUGUUGCAGUGAUAA	mRNA
		met-leu-gln-STOP	protein
E	frameshift mutation	TA**T**CAACGTCACCATT	DNA
		AUAGUUGCAGUGAUAA	mRNA
		ile-val-gly-val-ile-	protein

Fig. 8.3 The structural effects of mutation on protein. (A) Wild type. (B) Silent mutation. (C) Missense mutation. (D) Nonsense mutation. (E) Frameshift mutation.

pp. 138–139), typically involving cytosine/guanine-rich trinucleotides (CGG, CCG, CAG, CTG), that occur throughout all genomic sequences. They can involve a few copies to several thousand repeats. If such repeats are present in the coding region of a gene, their expansion results in a defective gene product, yielding disease (see below). Expansion with each successive generation leads to 'anticipation' (see p. 162). Triplet repeat expansions may be inherited in an autosomally dominant or recessive manner, or be X-linked.

In Friedreich's ataxia the most common abnormality (FRDA1) is an expansion of the trinucleotide sequence GAA within the first intron of the *FXN* gene. While a normal individual has 8 to 30 copies of this trinucleotide, Friedreich's ataxia patients have as many as 1000. The larger the number of GAA copies, the earlier the onset of the disease and the quicker the decline of the patient. This expansion is intronic and is thought to make the DNA 'sticky', interfering with the process of transcription.

Structural effects of mutation on protein

Silent mutations

Silent mutations arise as a result of point mutations and are those which have no effect on the amino-acid sequence of a protein (Figs 8.3A, B). They have long been considered to be 'evolutionary neutral', but recent studies have demonstrated that silent mutations can exert an effect on the control of differential splicing.

Missense mutations

Missense mutations arise where a base change alters a codon leading to the incorporation of a different amino acid into the growing protein chain (Figs 8.3A, C). The effect of the mutation on protein function depends upon its location relevant to the tertiary or quaternary structure of the protein and whether the two amino acids involved are from the same or different groups (i.e. hydrophobic to hydrophilic is more disruptive than hydrophobic to hydrophobic substitutions). Missense mutations that grossly disrupt protein folding have a more marked effect than those that do not compromise this function.

In sickle-cell disease, the substitution of A by T at the 17th nucleotide of the β-globin gene changes the codon for the 6th amino acid from GAG (for glutamic acid) to GTG (which encodes valine). This missense mutation changes the solubility and molecular stability of the resultant protein and the resulting haemoglobin has the physical property of forming polymers under conditions of low oxygen tension, leading to sickling of red blood cells.

Nonsense mutations

Nonsense mutations are point mutations that lead to the conversion of a codon to a stop codon (UAG, UAA, UGA) (Figs 8.3A, D). They lead to a truncated protein product, with those that occur early in a gene sequence having a higher probability of completely inactivating a gene.

Frameshift mutations

Insertions and deletions of nucleotides, if not a multiple of three, lead to 'frameshift' mutations, whereby the open reading frame of the gene and the corresponding amino-acid sequence is altered, usually leading to complete inactivation of the gene (Figs 8.3A, E).

Functional effects of mutation on protein

With the exception of imprinted genes (see pp. 162–163), genes on both the maternal and paternal chromosomes are expressed. If either the maternal or the paternal gene contains a mutation, the cell will express two different protein products. Mutations exert their phenotypic effects by one of two mechanisms: loss of function or gain of function.

Loss of function mutations

Amorphic mutation

Amorphic mutations, also known as 'null mutations', are associated with a complete absence of gene product function.

Hypomorphic mutation

Hypomorphic mutations, also known as 'leaky mutations', lead to a partial loss of function. They usually result from:

- an altered amino acid that makes the polypeptide less active
- a reduction in transcription that results in less normal transcript.

They are usually recessive to wild type.

Haploinsufficiency

The majority of heterozygous states are haplosufficient; that is one functional copy of a gene is adequate for the manifestation of a wild type phenotype. Haploinsufficiency describes a situation whereby a reduction of 50% of gene function results in an abnormal phenotype and, as such, is not sufficient to permit the cell to function normally. Such mutations are invariably dominant.

Gain of function mutations

These mutations result in either:

- increased activity of the gene product (hypermorphic)

or

- the gain of a novel function or a novel pattern of gene expression of the gene product (neomorphic).

They are usually dominant.

Trinucleotide repeat expansions also represent gain of function mutations, usually a toxic gain of protein function, which predisposes to protein misfolding and protein aggregation and leads to neurodegeneration.

Dominant negative mutations

Dominant negative mutations are also known as antimorphic mutations. They arise when the null allele product of a heterozygote adversely affects the normal gene product, for example by dimerizing with and inactivating it. The classical example is that of an amino-acid change that prevents a polypeptide from functioning in a multimeric protein complex, as seen with fibrillin in Marfan syndrome.

MONOGENIC DISORDERS

Monogenic or single gene disorders are caused by individual mutant genes, and frequently they show obvious and characteristic patterns of inheritance. There are approximately 6000 single gene disorders. Individually they are rare, usually affecting from 1 in 10 000, to 1 in 100 000. However, taken all together single-gene disorders are common, affecting 1% of the population. Certain single-gene disorders depend on an environmental trigger before the phenotype is expressed:

- Lactose intolerance is not seen in the absence of lactose in the diet.
- Severe emphysema in individuals homozygous for α_1-antitrypsin deficiency mutations is largely confined to smokers.

Single-gene disorders generally follow Mendelian patterns of inheritance, and relatives in families in which these disorders are present are at a much higher risk of developing the condition than the general population as a whole.

An introduction to pedigrees

A pedigree is a record of one's ancestors, offspring, siblings and their offspring that can be used to determine the pattern of certain genes or disease inheritance within a family (Fig. 8.4).

Mendelian inheritance of single-gene disorders

Mendelian traits generally occur in predictable proportions among the offspring of parents with that trait. The pattern of inheritance seen depends on the chromosomal location of the gene (sex-linked or autosomal) and whether the phenotype is dominant or recessive. Therefore, there are five patterns of Mendelian single-gene inheritance:

1. Autosomal dominant
2. Autosomal recessive
3. X-linked dominant
4. X-linked recessive
5. Y-linked/holandric.

A sixth pattern of single-gene inheritance, mitochondrial, is non-Mendelian (see pp. 163–164).

In medical genetics, a dominant phenotype is one that is expressed in heterozygotes, whereas a recessive trait is expressed only in homozygotes. If the expression of each allele can be detected in the presence of the other, the two alleles are termed codominant.

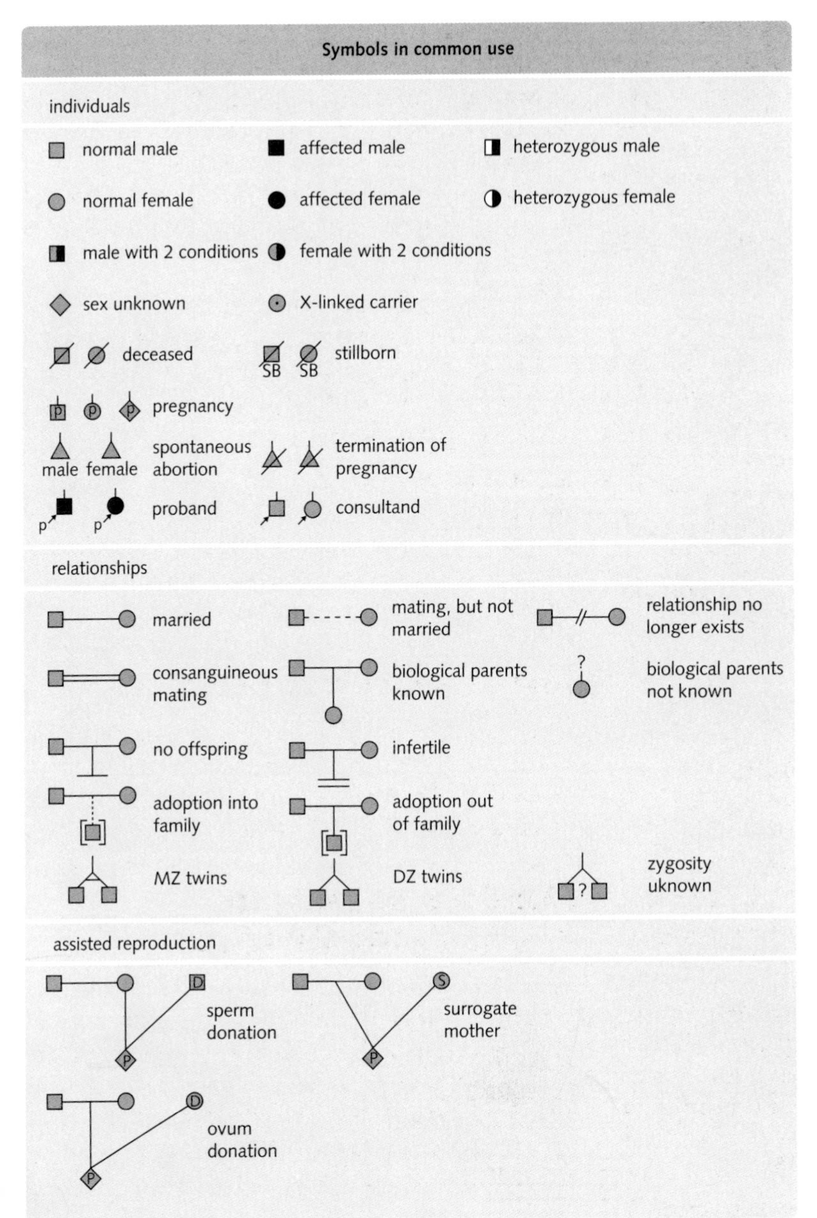

Fig. 8.4 Symbols and configuration of pedigree charts.

Dominant and recessive inheritance is defined according to clinical phenotypes, which may not always reflect the behaviour of the allele at the molecular level. Mutations in the retinoblastoma protein are recessive at the cellular level because one allele expresses enough protein for biological function. However, at the phenotypic level the predisposition to cancer is inherited dominantly because random loss of the compensating allele always occurs in at least one cell (see also pp. 170–172).

Autosomal dominant disorders

Autosomal dominant inheritance

A dominant gene is phenotypically expressed in homozygotes and heterozygotes for that gene (Fig. 8.5). In this pattern of inheritance:

- affected parents have affected children – this is sometimes termed 'vertical inheritance', to reflect the pattern seen in genetic pedigrees
- unaffected family members usually have unaffected partners and they produce normal children

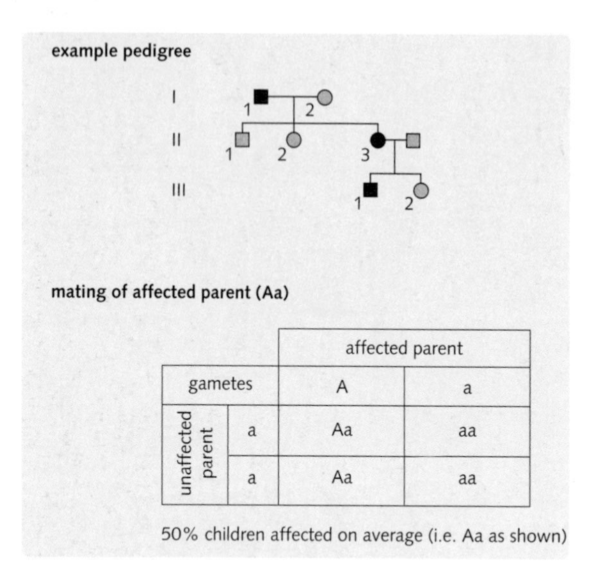

	gametes	affected parent A	affected parent a
unaffected parent	a	Aa	aa
unaffected parent	a	Aa	aa

Fig. 8.5 Example pedigree and typical offspring of mating in autosomal dominant inheritance. A, disease allele.

- affected family members usually have unaffected partners, and they produce a 1:1 ratio of normal and affected children
- usually both sexes are equally affected, and they are equally likely to pass on the disease.

Homozygotes for the trait are rare. In some autosomal dominant (AD) conditions, new mutations account for a substantial proportion of cases (e.g. achondroplasia, familial adenomatous polyposis). AD genes can show:

- sex limitation (e.g. balding, gout, male-limited precocious puberty, familial breast/ovarian cancer)
- reduced penetrance (e.g. retinoblastoma)
- variable expressivity (e.g. tuberous sclerosis)
- imprinting (see p. 162)
- anticipation (see pp. 162–163).

Molecular basis of autosomal dominant inheritance

In AD disorders, the disease occurs despite the presence of the normal gene product expressed from the wild type allele. This may arise as a result of:

- haploinsufficiency – e.g. familial hypercholesterolaemia (Fig. 8.6)
- dominant negative effect – e.g. osteogenesis imperfecta (Fig. 8.7)

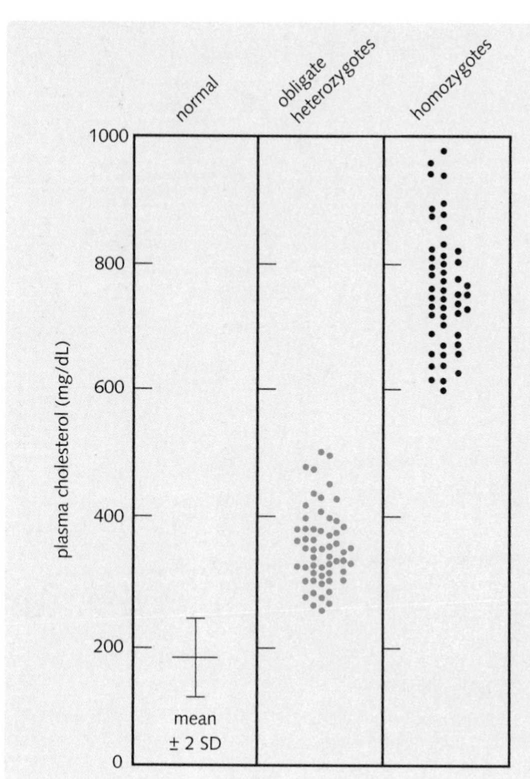

Fig. 8.6 Gene dosage in familial hypercholesterolaemia. This disease results from mutations in the gene that codes for the low-density lipoprotein (LDL) receptor, a cell surface protein that binds extracellular LDL and delivers it to the cell interior. Heterozygotes show levels of plasma cholesterol intermediate between normal and homozygotes for the mutation. However, plasma cholesterol level in heterozygotes is sufficient for the development of premature heart disease, so the condition shows an AD pattern of inheritance. (Adapted from Nussbaum, McInnes and Willard, 2001.)

Familial hypercholesterolaemia is an autosomal dominant disorder, resulting from mutation in the LDL receptor gene on chromosome 19, with an incidence of 1:500. This lack of LDL receptors increases the half-life of LDL cholesterol in the blood from 2.5 days to around 5 days, leading to markedly elevated LDL levels, and normal levels of other cholesterols. The excess LDL cholesterol is taken up by macrophages and foam cells. The sequelae of high LDL levels includes atherosclerotic disease, tendon xanthomata, xanthelasma and premature corneal arcus.

- simple gain of function – from increased expression of the normal protein, e.g. Charcot–Marie–Tooth disease type 1a; or expression of an abnormal protein with novel properties, e.g. Huntington's disease.

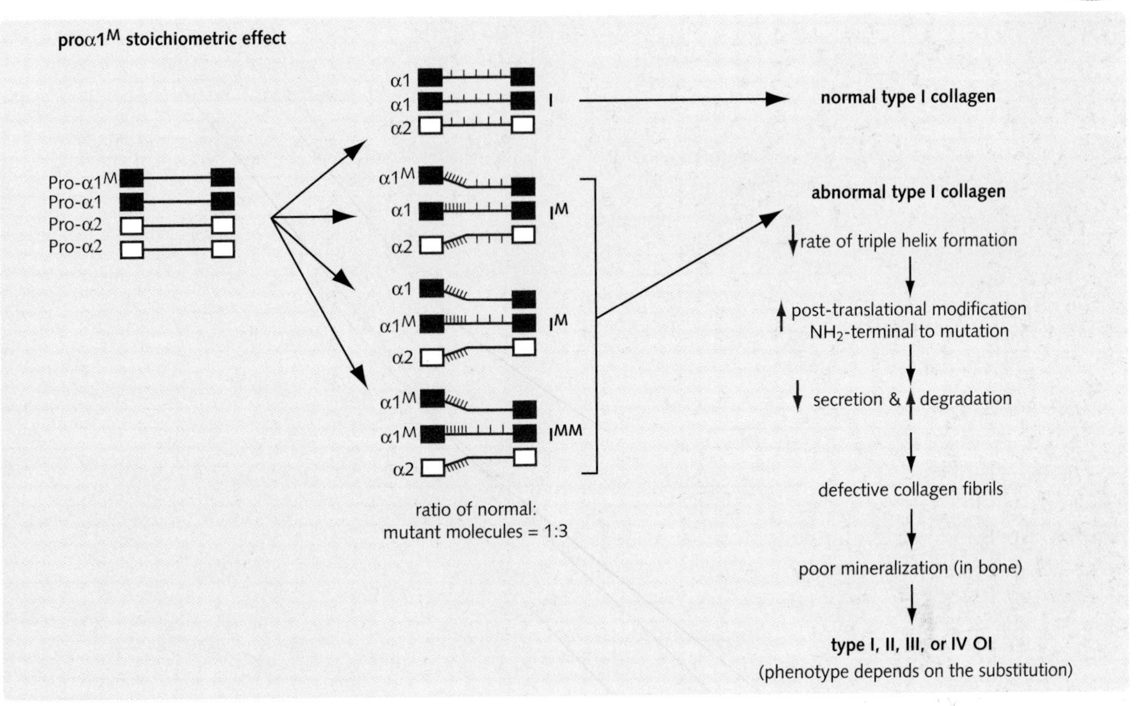

Fig. 8.7 Dominant negative effect. Normal collagen is formed by a triple helix that consists of two molecules of pro-α1 and one molecule of pro-α2. Procollagen containing a missense mutation (pro-α1^{M}) destabilizes the triple helix, resulting in increased degradation and reduced secretion of the mature protein, which ultimately results in osteogenesis imperfecta. (Adapted from Nussbaum, McInnes and Willard, 2001.)

Autosomal recessive disorders

Autosomal recessive inheritance

An autosomal recessive (AR) gene or trait is expressed only in homozygotes for the abnormal gene (Fig. 8.8). In autosomal recessive inheritance:

- affected individuals will usually have phenotypically normal parents – this is sometimes termed 'horizontal inheritance', to reflect the pattern seen in genetic pedigrees

The term horizontal inheritance was coined to describe the pedigree pattern seen in families with a history of an autosomal recessive condition. It does not imply that siblings can spread a condition among themselves.

- affected individuals usually have unaffected partners and all their children will be carriers
- if a carrier has an unaffected partner, there is a 50% chance of the children being carriers

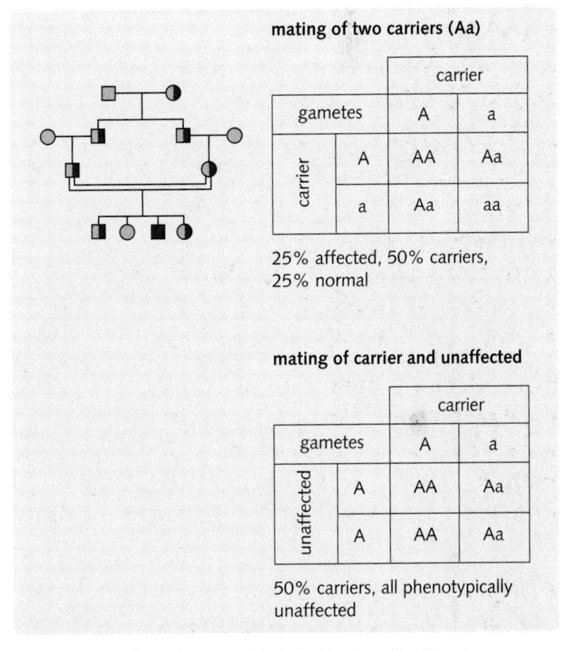

mating of two carriers (Aa)

		carrier	
	gametes	A	a
carrier	A	AA	Aa
	a	Aa	aa

25% affected, 50% carriers, 25% normal

mating of carrier and unaffected

		carrier	
	gametes	A	a
unaffected	A	AA	Aa
	A	AA	Aa

50% carriers, all phenotypically unaffected

Fig. 8.8 Example pedigree and typical offspring of matings in autosomal recessive inheritance. a, disease allele.

- only matings between heterozygotes will produce affected individuals, with an expected frequency of 1 in 4
- both sexes are equally affected
- there is an association with consanguinity due to sharing of genes in families (rare recessive genetic disorders are more likely to arise through consanguinity).

Consanguinity occurs where there is a mating between two people who have a familial relationship closer than that of second cousins.

Complementation

Complementation is the ability of two different genetic defects to correct for one another. In recessive conditions, affected individuals are homozygous for mutations in a particular gene, so it is expected that two parents with the same phenotype will always have affected children. However, occasionally two such parents will have an unaffected child as a result of complementation. Thus, complementation arises if parents who have the same disorder are homozygous for mutations in different genes in the same pathway (Fig. 8.9).

Molecular basis of recessive inheritance

AR disorders include many enzyme defects where expression from the wild-type allele in the heterozygote provides sufficient functional protein to prevent disease. However, homozygotes that express no functional protein develop the associated disorder.

Examples of autosomal recessive conditions include:

- sickle-cell anaemia
- β thalassaemia
- phenylketonuria
- haemochromatosis.

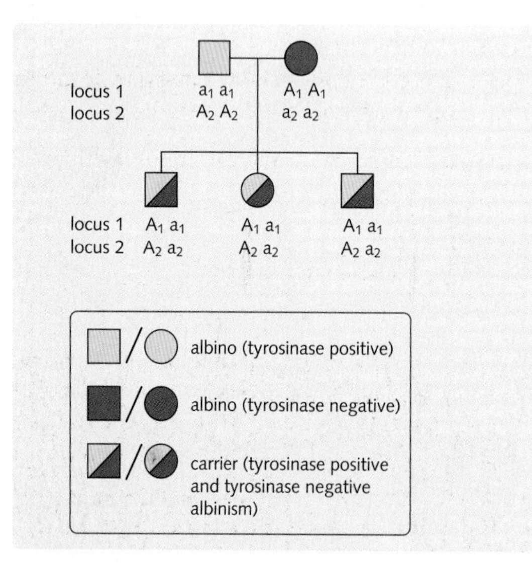

Fig. 8.9 Pedigree showing complementation. Both parents are albino, but all their children are normal. Given that albinism is an autosomal recessive condition, this observation can be explained if the parents are homozygous for mutations in different genes in the same pathway. The children are thus unaffected carriers of both mutations. A_1, wild-type at locus 1, a_1, mutant at locus 1, A_2, wild-type at locus 2, a_2, mutant at locus 2.

Phenylketonuria (PKU) is an autosomal recessive disease, resulting from mutation of phenylalanine hydroxylase encoded on chromosome 12, and the inability to convert phenylalanine into tyrosine leading to the accumulation of phenylalanine and its metabolic products in body fluid. If unrecognized and untreated, PKU leads to irreversible mental disability, microencephaly and fits. Other signs include blue eyes, fair hair, eczema and mousy odour. PKU is controllable by diet. Restricting phenylalanine intake from early life ensures normal intellectual development.

X-linked disorders

X-linked dominant inheritance

The X-linked dominant (XD) inheritance pattern is rare and difficult to distinguish from AD except that affected males have normal sons, but all daughters are affected (Fig. 8.10).

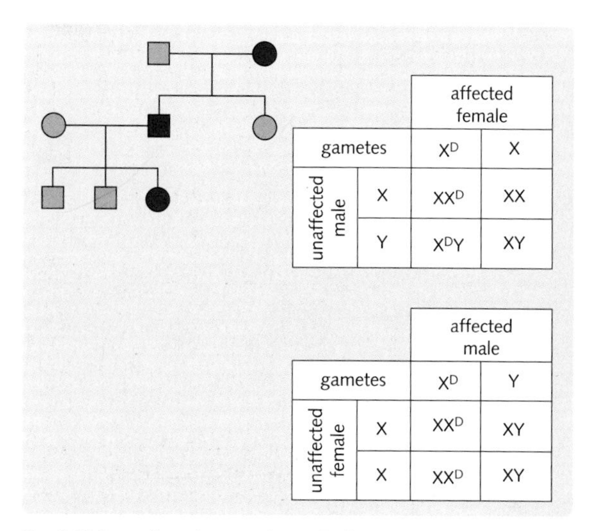

Fig. 8.10 Example pedigree and typical offspring of matings in X-linked dominant inheritance. X^D, disease allele.

Examples of X-linked dominant conditions include:

- Aicardi syndrome
- vitamin D resistant rickets
- Rett's syndrome.

Vitamin D resistant rickets, also known as X-linked hypophosphataemia, is an X-linked dominant condition characterized by hypophosphataemia, normocalcaemia and normal or low levels of calcitriol. It arises from mutations in the *PHEX* gene, leading to decreased reabsorpion of phosphate by the renal tubule and hyperphosphaturia. Classically, presentation includes a short stature, bowing of the lower limbs and rachitic changes in the long bones.

X-linked recessive inheritance

For X-linked recessive (XR) genes the inheritance pattern (Fig. 8.11) is as follows:

- many more males than females show the recessive phenotype
- the disease is transmitted by a carrier female, who is usually asymptomatic
- if a mother is a carrier, her sons have a 50% chance of being affected and her daughters a 50% chance of being carriers

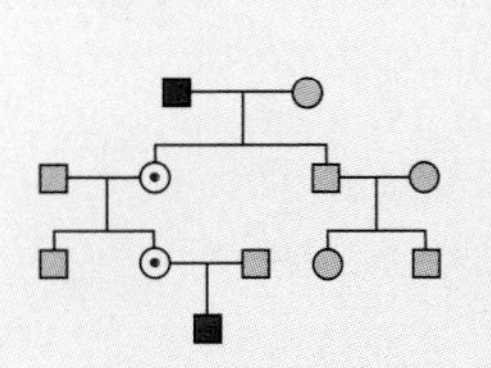

mating of affected male

		affected male	
gametes		X^D	Y
unaffected female	X	X^DX	XY
	X	X^DX	XY

all daughters carriers, all sons normal

mating of carrier female

		unaffected male	
gametes		X	Y
carrier female	X^D	X^DX	X^DY
	X	XX	XY

half of children inherit gene regardless of sex, 50% daughters carriers, 50% sons affected, 50% children normal

Fig. 8.11 Example pedigree and typical offspring of matings in X-linked recessive inheritance. X^D, disease allele.

- an affected male will usually have no affected offspring, but all his daughters will be carriers
- no sons of the affected male will inherit the gene (i.e. there is no male-to-male transmission)
- affected males may have unaffected parents, but they will often have an affected maternal uncle or cousin.

Molecular basis of X-linked recessive inheritance

Males (XY) have only one X chromosome and they are, therefore, said to be hemizygous. Since males receive only one copy of X-linked genes (except for those in the pseudo-autosomal region) they will express any XR traits because they do not have a compensating wild-type allele.

To compensate for the double complement in females, X chromosome inactivation (Lyonization) ensures that X-linked genes are only expressed from one of the two X chromosomes in any given cell. The selection of an X chromosome for inactivation within a specific cell is random, but once established early in development inactivation patterns are transmitted to daughter cells. Females do not tend to show XR disease because many of their cells express the wild-type allele. Women can be affected with X-linked recessive conditions in the following situations:

- if she is the daughter of an affected male and a carrier female
- if there is skewed Lyonization of a 'non-diseased', normal X chromosome
- if there is X chromosome-autosome translocation
- if XO (Turner's syndrome) is present.

Examples of X-linked recessive conditions include:

- G6PDH deficiency
- haemophilia A
- Duchenne muscular dystrophy
- Fabry's disease.

Duchenne muscular dystrophy (DMD) is inherited in an X-linked recessive manner. Mutation in the dystrophin gene, a protein involved in the tethering of muscle fibres to the extracellular matrix, leads to a deficiency of dystrophin in the plasma membrane, uncontrolled entry of calcium into the cell, and the production of high amounts of reactive oxygen species. The main symptoms are rapidly progressive muscle atrophy and weakness, typically of the proximal muscles.

Y-linked inheritance

Since only males have the Y chromosome, only males can pass on the Y to offspring and only male offspring can receive it. The Y chromosome is inherited with no interchromosomal genetic recombination, however, in order to allow it to pair with the X chromosome at cell division; the Y chromosome contains pseudo-autosomal regions.

The Y chromosome contains the genes for determining 'maleness'. These include:

- *SRY* – sex-determining region
- *TDF* – testis determining factor
- *DAZ* – deleted in azoospermia.

Mutation of any of these genes results in azoospermia. Males with a mutation in the *SRY* gene, as well as having fertility problems, have short stature.

Heterogeneity in Mendelian disorders

Genetic heterogeneity

This is the phenomenon by which identical or similar phenotypes arise by different genetic mechanisms. Mechanisms include different allelic mutations in the same gene, or mutations in different loci (genes). Such disorders might (but will not necessarily) show more than one mode of inheritance; for example, Charcot–Marie–Tooth is usually inherited in an autosomal dominant manner, but autosomal recessive and X-linked dominant variants exist.

Allelic heterogeneity

Different mutations in the same gene may result in different phenotypes. This is an important cause of clinical variation, for example, specific mutations within the *CFTR* gene are associated with pancreatic sufficient as opposed to pancreatic insufficient forms of cystic fibrosis.

Locus heterogeneity

This is the situation in which mutations at two or more distinct loci can produce the same or similar phenotype. For example, retinitis pigmentosa may result from mutations in many different genes, including the rhodopsin gene on chromosome 3 and a GTPase regulator gene on the X chromosome.

Non-Mendelian inheritance of single-gene disorders

A number of disorders have been identified that do not follow classic patterns of Mendelian inheritance. The molecular mechanisms underlying these observations are now beginning to be understood.

Mechanisms underlying non-Mendelian inheritance

Anticipation

Anticipation is the occurrence of a hereditary disease with a progressively earlier age of onset in successive generations. The mutations in genes associated with anticipation are trinucleotide repeat expansions, which have a tendency to get larger in successive generations (e.g. Huntington's disease and myotonic dystrophy). Large expansions of the triplet repeat are associated with early age of onset.

Imprinting

This is differential expression of genetic material at chromosomal or allelic level, depending upon which parent (male or female) it has been inherited from. It is thought to result from the selective inactivation of genes (probably through methylation) in different patterns in the course of male and female gametogenesis. Hydatidiform moles illustrate the different roles of paternal and maternal genomes:

- A complete mole (46,XX) has chromosomes that are all paternal in origin (both X chromosomes being of paternal origin, i.e. extra paternal set, but no maternal set), and it results in either no foetus or a normal placenta with severe hyperplasia of the cytotrophoblast.
- A partial mole (69,XXX or 69,XXY) is triploid with an extra set of chromosomes of maternal or paternal origin. An extra paternal set (diandric) results in abundant trophoblast, but poor embryonic development. An extra maternal set (digynic) results in severely retarded embryonic development with a small fibrotic placenta.

Imprinting is important in the aetiology of Prader–Willi (PW) and Angelman syndromes (AS), which both arise from the same microdeletion of chromosome 15q11–13, the so-called 'critical region'. The phenotype varies between PW and AS according to whether the critical region deleted chromosome is paternally or maternally inherited.

AS results from imprinting switched off in the critical region on the paternally derived chromosome and PW from the same region being switched off on the maternally derived chromosome (Fig. 8.12). When a child inherits a chromosome 15 with the microdeletion:

- AS results from the loss of imprinted genomic material within the maternal 15q11–13 locus, thus removing the only active copy of the responsible gene *UBE3A*
- PW syndrome results from the loss of imprinted genomic material within the paternal 15q11–13 locus, removing the active copies of the several genes postulated as having a role in PW.

Fifteen per cent of cases of PW are thought to arise from functional deletion of 15q, as a result of maternal uniparental disomy.

Uniparental disomy

This is caused by duplication of a chromosome from one parent with loss of the corresponding homologue from the other parent (Fig. 8.13). For example, uniparental disomy of maternal chromosome 15 can result in the same phenotype as PW, but with no deletion because there is no paternally contributed chromosome 15. Beckwith–Wiedemann syndrome can be due to paternal duplication of 11p15.

Mitochondrial inheritance

Mitochondrial DNA (MtDNA) is maternally inherited (the sperm contributes no mitochondria to the zygote); therefore, affected males can not transmit the disease to their offspring (Fig. 8.14). Mitochondria are distributed randomly in daughter cells, so these may contain entirely normal mitochondrial DNA or entirely mutant DNA (homoplasmy), or else a mixture of both (heteroplasmy), leading to variable expression of disease depending upon the relative proportion of normal to mutant DNA. Examples of mitochondrial disease include:

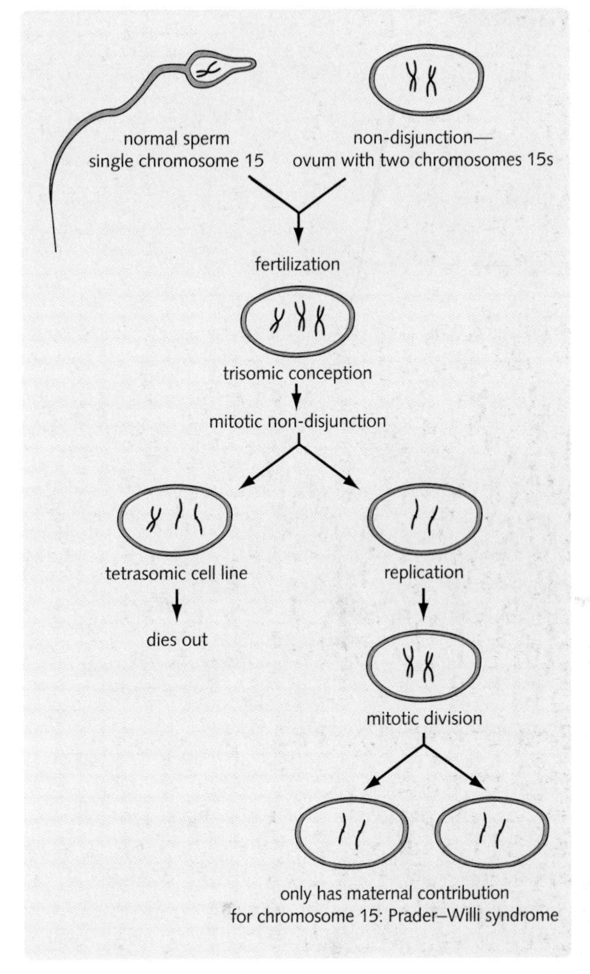

Fig. 8.13 Mechanism of uniparental disomy (here for chromosome 15).

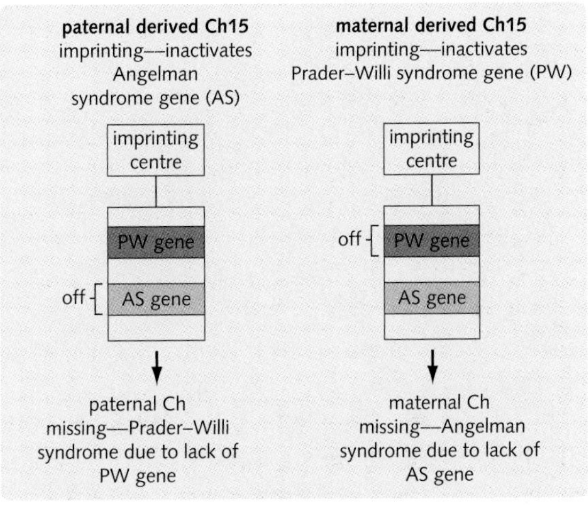

paternal derived Ch15
imprinting—inactivates Angelman syndrome gene (AS)

imprinting centre

PW gene

off AS gene

paternal Ch missing—Prader–Willi syndrome due to lack of PW gene

maternal derived Ch15
imprinting—inactivates Prader–Willi syndrome gene (PW)

imprinting centre

off PW gene

AS gene

maternal Ch missing—Angelman syndrome due to lack of AS gene

Fig. 8.12 How deletion of one parental chromosome 15 with imprinting of the present chromosome 15 can lead to Prader–Willi or Angelman syndrome. Normal development requires both maternally and paternally derived genes to be expressed.

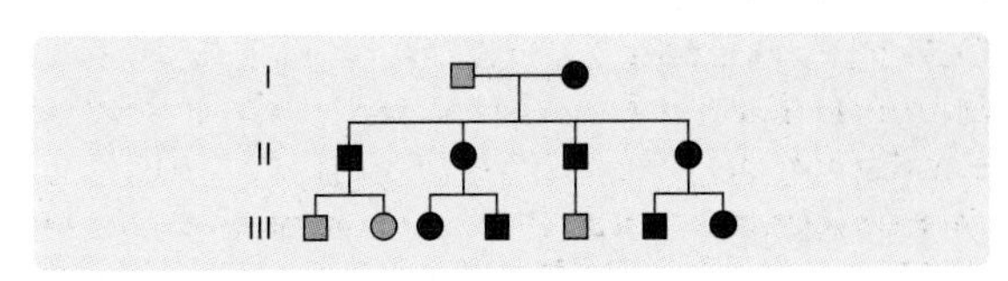

Fig. 8.14 Example pedigree in mitochondrial inheritance. (Adapted with permission from Turnpenny & Ellard Emery's Elements of Medical Genetics 12e, Churchill Livingstone, 2005.)

- Leber hereditary optic neuropathy
- mitochondrial encephalomyopathy, lactic acidosis and stroke-like syndrome (MELAS)
- myoclonus with epilepsy and with ragged red fibres (MERRF).

Mosaicism

A mosaic is an individual with multiple cell lines (which exhibit different genotypes) that arise from a single zygote. Germline mosaicism occurs when an abnormal cell line is confined to the gonads, and it may account for apparently unaffected parents producing more than one child with an AD condition.

Somatic mosaicism is where either the paternal or maternal representative of a chromosome is randomly inactivated in each somatic cell. Cells differ with respect to which chromosome is inactivated, but, once inactivation is established, all the cell's progeny will retain the same pattern of inactivation. It may account for:

- unusually mild symptoms in AD metabolic conditions, if there is disproportional inactivation of the aberrant gene
- expression of X-linked disease alleles in female carriers if there is disproportionate inactivation of the normal gene.

Mosaicism can be an important clinical factor in some instances of chromosomal disorders (e.g. mild cases of Down syndrome (DS) may be mosaics (46,XY/47,X+21 or 46,XX/47,XX+21)) (Fig. 8.15) and in tumour development (see p. 164).

All females are genetic mosaics for the X chromosome. As X chromosomal inactivation (Lyonization) is a random process, in some cells the maternally derived X chromosome will be Lyonized to yield a Barr body, while in others it will be the paternally inherited copy that is 'switched off'.

POLYGENIC INHERITANCE AND MULTIFACTORIAL DISORDERS

Introduction

Multifactorial disorders result typically from mutations in several genes, in combination with environmental factors. Although multifactorial disorders tend to recur in families, they are much more prevalent than single-gene disorders and they do not show Mendelian patterns of inheritance. Multifactorial inheritance is implicated in the aetiology of many common conditions including:

- congenital malformations – e.g. neural tube defects, developmental dysplasia of the hip (DDH), pyloric stenosis, cleft lip and palate and congenital heart disease
- common disorders of adult life – e.g. diabetes mellitus, obesity, epilepsy, hypertension and schizophrenia
- normal human characteristics – e.g. blood pressure, height, dermatoglyphics (finger ridges) and intelligence.

Multifactorial disorders

A continuous normal (Gaussian) distribution curve of the trait within the population as a whole is typical, but not diagnostic, of multifactorial disorders. Abnormalities do not usually have a distinct phenotype, but are extremes of the curve. The number of genes involved may be very few as, even when a single locus is implicated, variation in the environment can ensure normal distribution of the trait.

Twin studies highlight the relative importance of genes and the environment. For example, cleft lip and palate has a population incidence of 1 in 1000, but:

- concordance in monozygotic twins is 40% – if due to genetics alone there would be 100% concordance, so genes are important, but other factors are also involved
- concordance in dizygotic twins is 4% – these twins are not genetically identical (having on average, like all siblings, 50% of their genes in common), but they generally share a similar environment, showing the importance of environmental factors

where the concordance rate = [both affected / (one affected + both affected)] × 100.

DDH is an example of a condition with multifactorial inheritance. Implicated genetic factors include: acetabular dysplasia, familial general joint laxity and transient joint laxity at pregnancy term. Related environmental factors include: breech presentation and response to maternal hormones;

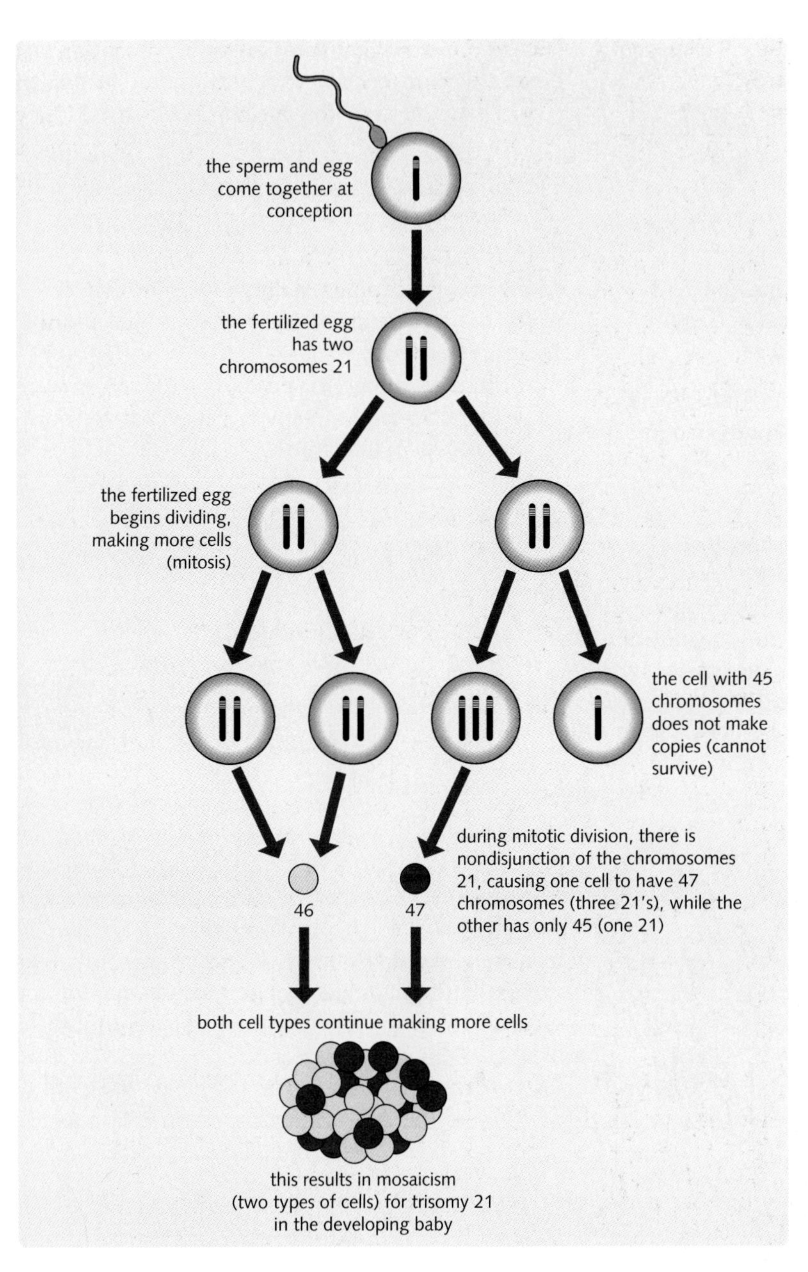

Fig. 8.15 Genetic mosaicism. Mitotic non-disjunction leads to the presence of two distinct populations of cells. Severity of any resultant condition is linked to the proportion of mutated cells to normal cells. The earlier in development the non-disjunction event occurs, the higher the ratio of mutated cells, and the higher the likelihood of expressing a mutant phenotype.

oestrogen in particular, which increases ligamentous laxity, is thought to contribute. The higher incidence of DDH seen in females is attributed to additional oestrogen produced by the female foetus.

Threshold model of multifactorial disorders

Fig. 8.16 shows the threshold model of multifactorial disorders. The liability of a population to a particular disease follows a normal distribution curve, with an

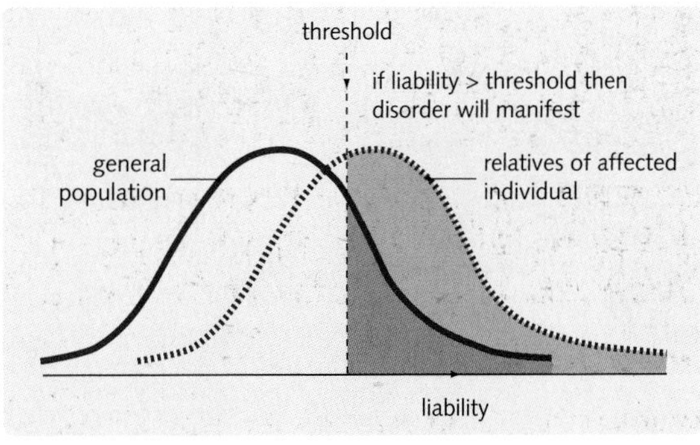

Fig. 8.16 Threshold model of multifactorial inheritance.

individual's genetic susceptibility and environmental factors defining their personal liability:

- The disorder is manifested in an individual when a certain threshold of liability is exceeded.
- The liability of the majority of the population lies below the threshold level.
- Population incidence is equivalent to the proportion whose liability is greater than the threshold.

The liability curve is shifted to the right for relatives of an affected individual, such that more members of the family are likely to be above the threshold required for the disease to be manifested.

Heritability

Heritability is the proportion of the total phenotypic variation that is genetic in origin in a given population. It is expressed as a percentage, with higher values denoting that the genetic contribution is more important (e.g. heritability of schizophrenia is 85% and that of rheumatoid arthritis 60%).

Heritability estimates are population specific since the variation of environmental and genetic effects may not be identical in different geographical areas and ethnic populations.

If heritability is high, there is a high correlation in relatives (e.g. finger ridge correlation in first-degree relatives is 49%). Usually heritability is low, so the incidence in relatives falls off sharply (Fig. 8.17).

Recurrence risks in multifactorial disorders are based upon population and family studies. Risk of recurrence rates are influenced by:

- the severity of the disorder in the affected person (the 'proband') (e.g. severe bilateral cleft lip is more likely to recur in siblings than unilateral cleft lip)
- the number of affected individuals in a family
- the proband being of the less commonly affected sex
- the relationship to the proband – the recurrence risk for first-degree relatives is approximately the square root of the population incidence of the trait.

There may be a sex difference in population incidence. For example, pyloric stenosis is more common in boys, so children of an affected female will be more likely to develop the condition than children of an affected male, as the female needs a high number of risk genes to manifest the condition.

Environmental factors

Multifactorial disorders result when environmental and genetic risk factors combine to bring the susceptibility of an individual to the disease above a threshold value. The liability that can be attributed to genetic factors is fixed. However, the judicious manipulation of environmental factors may enable the reduction in an individual's susceptibility to below the threshold value.

For example, atherosclerosis has a heritability of 65%, so the environmental contribution to the aetiology of the disorder is 45%. Epidemiological studies have identified environmental components, and not smoking, healthy eating and taking regular exercise can significantly reduce an individual's risk of developing heart disease.

Fig. 8.17 Risk to relatives for multifactorial disorders. First-degree relatives are parents, siblings and offspring (share 50% of genome); second-degree relatives are grandparents, aunts and uncles, grandchildren (share 25% of genome); third-degree relatives include cousins and great-grandchildren (share 12.5% of genome).

Fig. 8.17 Risk to relatives for multifactorial disorders

Disorder	Relative risk disorder			
	1° relative	2° relative	3° relative	General population
Cleft lip and palate	4	0.6	0.3	0.1
Neural tube defects	4	1.5	0.6	0.3
Epilepsy	5	2.5	1.5	1.0

Examples of multifactorial disorders

Type 1 diabetes mellitus

There is 50% concordance for type 1 diabetes mellitus (T1DM) in monozygotic twins. The human HLA complex and the insulin gene are considered to be the loci that contribute the most to T1DM susceptibility. Ninety-five per cent of affected individuals have human leukocyte antigens (HLA) DR3 and/or DR4 (see Fig. 6.18), compared with 50% of non-diabetics. Aspartate at the 57th amino-acid residue of the DQ locus is protective – 19% of the general population and 95% of the T1DM population are aspartate negative. The influence of the insulin gene (INS), mapped to 11p15, on T1DM correlates to the variation in the number of tandem repeats of the INS VNTR upstream to the gene. Those with 26–63 repeats have a recessive predisposition to T1DM, while those with 140 plus repeats enjoy a dominant protection. Around a further 20 'minor' contributing loci have also been discovered.

Essential (primary) hypertension

The heritability of essential hypertension is 62%. Epidemiological studies have implicated environmental lifestyle exposures including dietary sodium intake, advancing age, decreased physical activity and body weight. Genes are also important in the response to treatment, and polymorphisms in several genes have been associated with reduced response to diuretics, β-blockers and ACE-inhibitors.

Schizophrenia

The heritability of schizophrenia is 85%. Linkage and association studies have implicated several loci suspected of conferring risk of developing schizophrenia. These include:

- 6p22: *DTNBP1* – Dysbindin
- 8p12–21: *NRG1* – Neuregulin 1
- 13q32–34: *DAOA* – D-amino acid oxidase activator
- 22q11: *COMT* – Catechol-O-methyltransferase.

Asthma

The heritability of asthma is 60%. More than 25 genes, including *DPP10*, *HLA-G* and *ADAM33*, have been associated with asthma pathology and susceptibility. The increasing prevalence of asthma in the developed world has been attributed to a changing environmental allergen exposure profile.

Alzheimer's disease

About 10% of patients inherit Alzheimer's disease as a monogenic AD condition. All of these cases arise from mutations in the amyloid precursor protein gene on chromosome 21 or from presenilin 1 on chromosome 14. Mutations in other genes (e.g. presenilin 2 on chromosome 1, ApoE on chromosome 19 and α-synuclein on chromosome 4) are also associated with an inherited susceptibility to Alzheimer disease.

Rheumatoid arthritis

Rheumatoid arthritis (RA) has an estimated heritability of 60%. Specific polymorphisms in the HLA-DRB1 gene have been demonstrated to be strongly correlated to RA susceptibility. Recent microarray studies comparing disease-discordant monozygotic twins have revealed expression differences in over 800 different genes. The three most strongly altered expression profiles were seen in:

- laeverin, a metallopeptidase enzyme that breaks down certain types of proteins
- 11β-hydroxysteroid dehydrogenase type 2, implicated in steroid pathways linked to inflammation and bone erosion
- cysteine-rich angiogenic inducer 61, involved in the formation of new blood vessels.

Subsequent investigation revealed that all three genes were overexpressed in joint tissue samples taken from patients with RA.

GENETICS OF CANCER

In the majority of cases cancer is a multifactorial disorder in which genetic and environmental factors interact to initiate carcinogenesis. However, in a minority (about 5%) the disease follows a familial pattern of transmission, suggesting a genetic cancer syndrome. Characterization of the mutations segregating in such families has helped elucidate the molecular events that underlie tumour genesis in the more common multifactorial form of the disease.

Cancer is characterized by abnormal growth and proliferation. Normal cell proliferation and survival is controlled by growth promoting proto-oncogenes and growth inhibiting tumour suppressor genes (see Ch. 6), and mutation in these genes may result in cancer (Figs 8.18, 8.19).

Multistage process of carcinogenesis

Carcinogenesis requires the accumulation of many mutations in both oncogenes and tumour suppressor genes. Tumours evolve from benign to malignant as subpopulations of cells acquire further mutations (Fig. 8.20). Given that mutations may arise in any tumour cell, tumours tend to be mosaics. However, when a cell acquires a mutation that is associated with a further loss of growth inhibition this cell type will tend to divide rapidly compared with the other cell types.

Oncogenes

Oncogenes drive cell growth (see Ch. 6); they are usually dominant at the cellular level, with most mutations being gain of function mutations that result in increased expression of the gene. They can be classified according to their position in the normal signal transduction pathway (Fig. 8.21).

Activation of oncogenes may occur by:

- translocation (e.g. Burkitt's lymphoma t(8;14) or t(8;2) activates c-*myc* on 8 q
- amplification (e.g. n-*myc* amplified in neuroblastoma)
- point mutation in oncogene (e.g. Ha-*ras* mutation in bladder cancer).

Tumour suppressor genes

Tumour suppressor genes (Fig. 8.22) inhibit oncogenesis (see Ch. 6). Tumours develop if there is loss of both wild-type alleles (Fig. 8.23). Loss of activity can occur through damage to the genome (e.g. mutation, rearrangement or mitotic recombination), or interaction with cellular or viral proteins (e.g. MDM2, HPV E6 antigen or adenovirus E1b protein).

Tumour suppressor genes are important in:

- inherited predisposition to cancer
- early events in tumourigenesis (they cooperate with dominant transforming genes to cause neoplasia).

The detection of mutations in tumour suppressor genes can be used for presymptomatic diagnosis (e.g. of adenomatous polyposis coli by demonstrating mutations in the *APC* gene).

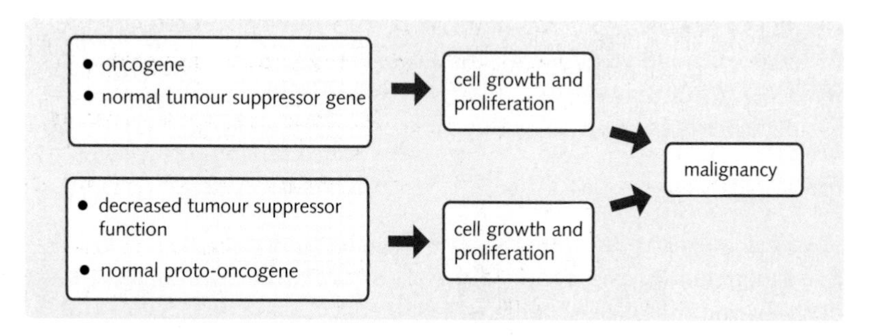

Fig. 8.18 Pathways to malignancy.

Fig. 8.19 Differences between oncogenes and tumour suppressor genes.

Fig. 8.19 Differences between oncogenes and tumour suppressor genes

Oncogene	Tumour suppressor gene
Gene active in tumour	Gene inactive in tumour
Specific translocations/point mutations	Deletions or mutations
Mutations rarely hereditary	Mutations can be inherited
Dominant at cell level	Recessive at cell level
Broad tissue specificity	Considerable tumour specificity
Especially leukaemia and lymphoma	Solid tumours

Fig. 8.20A Carcinogenesis

Phase	Mechanism	Clinical appearance
Initiation	Mutation:	No noticable change
Promotion	Mutated cells stimulated to grow	Usually only detected histologically/biopsy
Conversion	Mutation: Uncontrolled growth & expansion	Benign tumour expansive growth
Progression	Mutation: Complete loss of cellular control	Cancer: Invasion & metastasis

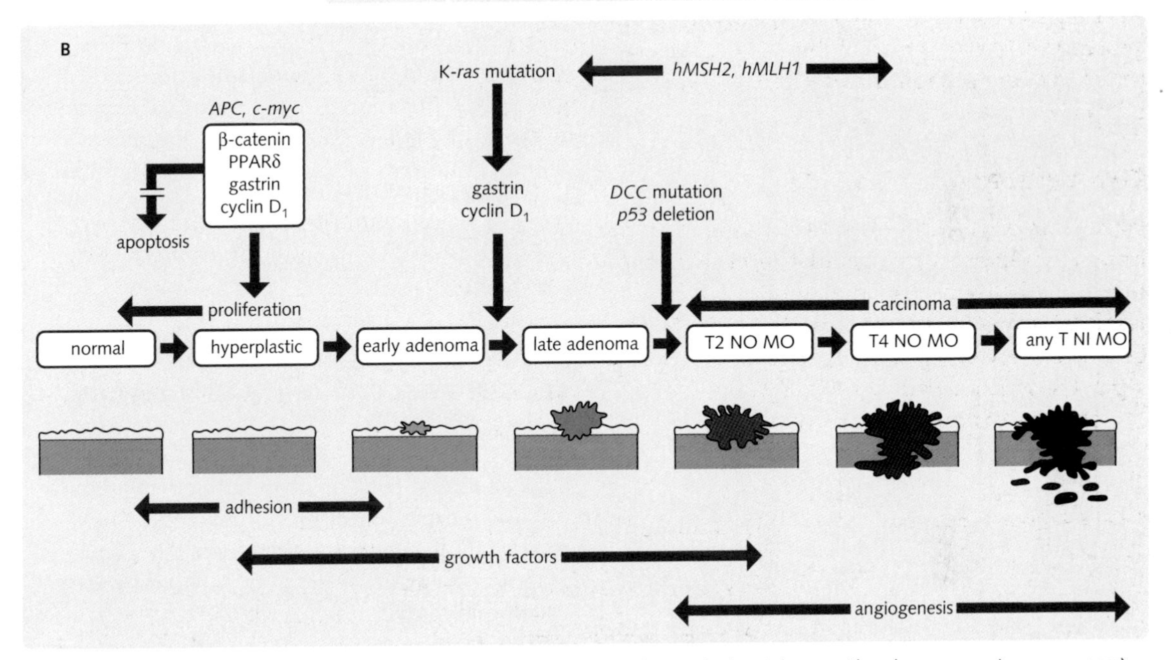

Fig. 8.20 (A) The multistep model of carcinogenesis and (B) the stages in the evolution of colorectal cancer (the adenoma – carcinoma sequence) (Adapted from Kumar and Clark, 2005.)

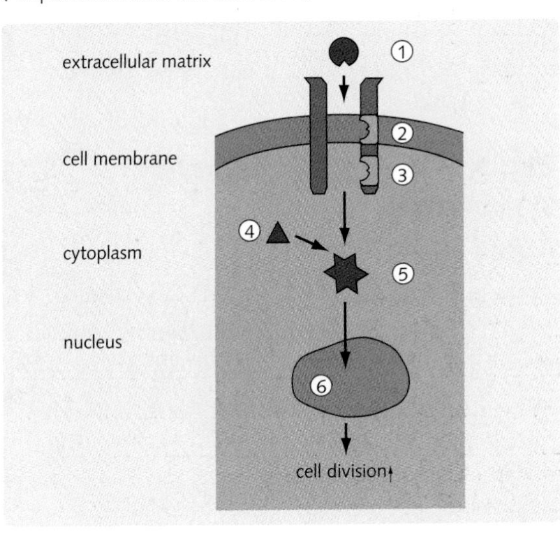

Fig. 8.21 Classes of oncogene by transduction position. (1) Growth factors (e.g. platelet-derived growth factor (PDGF)-*sis*). (2) Growth factor receptors (e.g. EGFR-*Gb*). (3) Post-receptor proteins (e.g. *ras*). (4) Post-receptor tyrosine kinase (e.g. *abl, src*). (5) Cytoplasmic proteins (e.g. *raf*). (6) Nuclear proteins (e.g. *myc*).

Fig. 8.22 Examples of tumour suppressor genes.

Fig. 8.22 Tumour suppressor genes

Gene	Locus	Function	Tumour
Rb	13q14	Substrate of CDK (cell cycle regulation)	Retinoblastoma, osteosarcoma
p53	17p13	Growth arrest and apoptosis	Mutated in 70% of all human tumours
WT-1	11p13	Zinc finger	Wilms' tumour
NF-1	17q	Transcription factor	Neurofibromatosis
APC	5q21	Cell adhesion	Adenomatous polyposis
DCC	18q21	Cell adhesion	Colorectal cancer, pancreatic cancer, oesophageal cancer

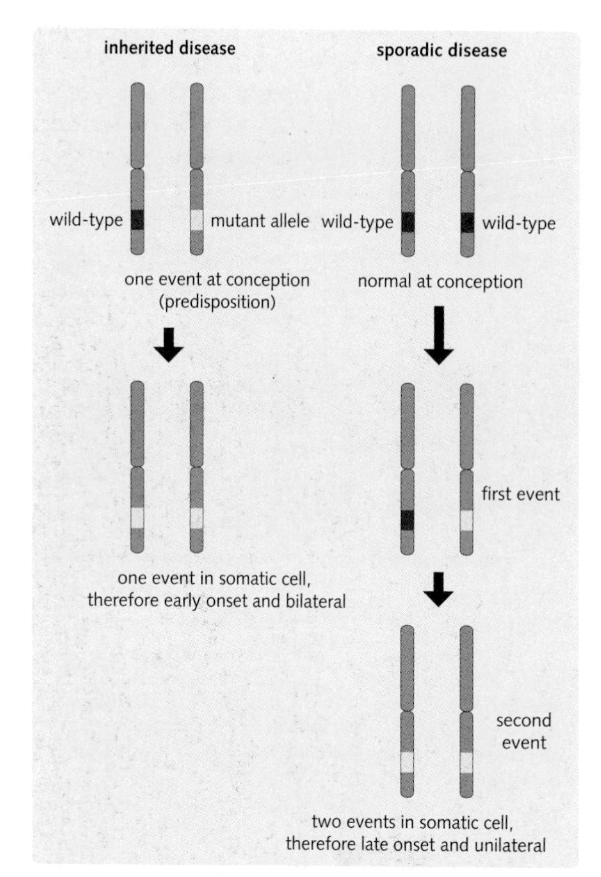

Fig. 8.23 Knudson's 'two-hit' hypothesis. Note that two events must occur to lose tumour repressor function.

Tumour suppressor genes in inherited cancers

Tumour suppressor genes are recessive at the cellular level. Therefore, if an individual inherits a mutation in one allele, every cell in the body will be relying on the product of the wild-type allele to regulate cell growth. Given the number of cells in the body, it is statistically likely that the wild-type allele will undergo somatic mutation in at least one cell. Thus, inherited mutations in tumour suppressor genes, such as in retinoblastoma and Li-Fraumeni syndrome (p. 171) tend to be inherited dominantly. This phenomenon, in which a single functioning allele is lost, is called 'loss of constitutional heterozygosity' (LOCH).

Features suggestive of inherited cancer susceptibility in a family

These are:

- several close (first or second degree) relatives with a common cancer
- several close relatives with genetically associated cancers (e.g. breast and ovary, or bowel and endometrial)
- two family members with the same rare cancer
- an unusually early age of onset
- bilateral tumours in paired organs
- synchronous or successive tumours
- tumours in two different organ systems in one individual.

Examples of genetic cancer syndromes

Ataxia telangiectasia

This is an autosomal recessive disorder that affects 1 in 50 000. The *ATM* gene involved has been mapped to 11q22.3, and its product exerts three critical functions:

- regulation and stimulation of DNA double-strand break repair
- signaling cell-cycle checkpoints
- signaling apoptosis via p53.

Therefore, in the absence of *ATM*, cells:

- fail to repair DNA
- lose the ability to arrest the cell cycle
- lose the ability to trigger programmed cell death in response to DNA damage.

The onset is usually in childhood with progressive cerebellar ataxia, oculocutaneous telangiectasias, immunodeficiency and recurrent infections, radiosensitivity (especially to X-rays) and predisposition to lymphatic leukaemias and other malignancies. Death usually occurs in early or middle adolescence, with a median age at time of death of 20 years, and typically results from bronchopulmonary infection or, less frequently, from malignancy. AT sufferers have a cancer incidence around 100 times greater than the general population.

Familial breast cancer

This accounts for about 5% of breast cancer, and it often presents at an early age. Mutations in the *BRCA1* gene at 17q21 and *BRCA2* gene at 3q12.3 account for over 50% of these cases. Both gene products are involved in homologous recombination repair (see p. 126) and their mutations both have highly penetrant autosomal dominant inheritance and are associated with specific cancers as follows:

- *BRCA1* mutations are associated with breast and ovarian cancer, and also an excess of prostate and colon cancer.
- *BRCA2* mutations are associated with breast, endometrial, renal, pancreatic, ovarian, biliary tract and bladder cancer. They are also associated with laryngeal carcinomas, lymphomas and, occasionally, male breast cancer.

DNA testing is available for large families with breast cancer (e.g. 2% of Ashkenazi Jews have one of three founder mutations: *BRCA1* 185delAG, *BRCA1* 5382insC and *BRCA2* 6174delT).

Prophylactic bilateral mastectomies are sometimes considered following DNA analysis for women and, increasingly, for men with a very strong family history.

Familial adenomatous polyposis coli

This is a rare autosomal dominant condition caused by a deletion in 5q21 resulting in a loss of function of the tumour suppressor gene, *APC*. This leads to the development of multiple benign adenomatous polyps of the large bowel (Fig. 8.24). A somatic mutation can result in the development of adenocarcinoma due to LOCH (see above).

The diagnosis is based on:

- family history
- the presence of multiple intestinal polyps from childhood
- characteristic retinal changes (in 80% of families).

There is a 90% risk of malignancy, and prophylactic whole-colon resection is often considered.

Li-Fraumeni syndrome

This is a rare autosomal dominant trait and is often due to germline mutation in the *p53* gene at 17p13.1. Unlike other inherited cancer syndromes, which are predominantly characterized by site-specific cancers, Li-Fraumeni syndrome presents with a variety of tumour types. It is commonly characterized by childhood cancers such as:

Fig. 8.24 Polyposis coli in large bowel. The multiple adenomatous polyps are clearly visible. (Courtesy of Dr A Stevens and Professor J Lowe.)

- soft-tissue sarcomas
- adrenal carcinomas
- brain tumours.

Later there is a high frequency of very early-onset breast cancer, astrocytoma and lung cancer. Direct mutation analysis is often possible, but of little benefit. The prognosis depends upon the number and sites of tumours.

Hereditary nonpolyposis colon cancer

Also known as Lynch syndrome, this is an autosomal dominant trait with high penetrance in which there is familial clustering of early-onset colon cancer (70% proximal) together with an increased risk of endometrial, ovarian and renal tract tumours. The condition is associated with germ-line mutations in the DNA mismatch repair family of genes. Mutations in two genes, *MSH2* and *MLH1*, account for up to 90% of all cases, with mutations in four other genes (*PMS1, PMS2, MLH3* and *MSH6*) accounting for the remainder. Those with a strong family history (one first degree relative diagnosed before the age of 45 or two first degree relatives diagnosed at any age) are recommended to have yearly colonoscopy from 20–25 years of age.

Multiple endocrine neoplasia syndrome

Multiple endocrine neoplasia syndrome (MEN) represents a group of rare autosomal dominant conditions resulting in hyperplasia, adenomas and carcinomas of multiple endocrine organs. MEN is classified into three categories:

- MEN I has been associated with the loss of a tumour suppressor gene mapped to 11q13.
- MEN IIa and IIb are caused by mutations in the *RET* oncogene on chromosome 10q11.2.

The clinical features are as follows (Fig. 8.25):

- MEN I is characterized by pituitary adenoma, hyperparathyroidism and pancreatic adenomas (often gastrinomas, accounting for 50% of patients with Zollinger–Ellison syndrome).
- MEN IIa is characterized by phaeochromocytoma, medullary thyroid carcinoma and hyperparathyroidism.
- MEN IIb is characterized by neuromas of the mucous membranes, phaeochromocytoma, megacolon and medullary thyroid carcinoma.

Prenatal diagnosis is now possible by direct mutation analysis and MEN IIa and IIb are monitored by yearly

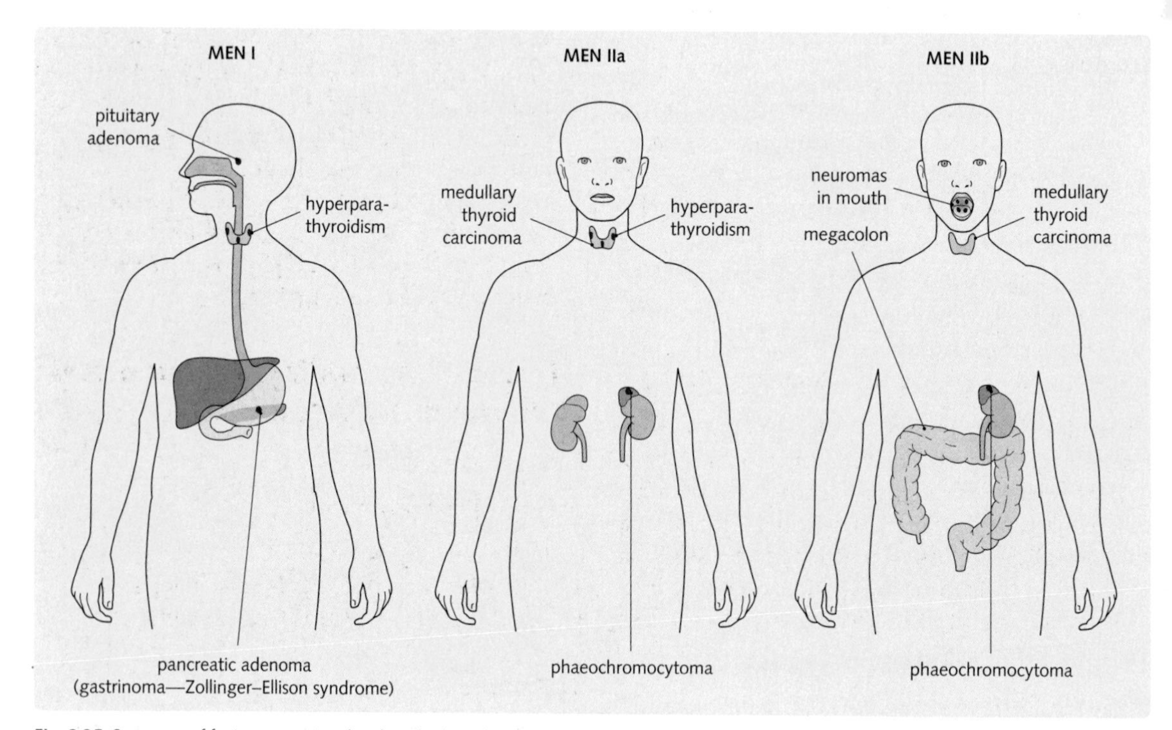

Fig. 8.25 Summary of features associated with multiple endocrine neoplasia (MEN).

calcitonin assays from 5 to 30 years of age for early detection of medullary thyroid carcinoma.

Xeroderma pigmentosum

This has an incidence of 1 in 70 000. There are at least nine subgroups, and these are characterized by:

- defective DNA excision repair
- extreme ultraviolet light sensitivity.

Clinical features of xeroderma pigmentosum are:

- progressive corneal and skin scarring on exposure to sunlight
- multiple skin cancers by 20 years of age.

Gorlin syndrome

This is also known as naevoid basal cell carcinoma syndrome. This is an autosomal dominant condition causing a predisposition to basal cell carcinoma of the skin, medulloblastoma and ovarian fibromas. There are associated congenital malformations including dental malformations, cleft palate and bifid ribs. The gene is *PTCH*, on chromosome 9q22.3.

CHROMOSOMAL DISORDERS

Introduction

Although chromosomal disorders occur in over 7.5% of conceptions, live-birth incidence is only 6 in 1000 since most end in spontaneous abortion. Such chromosomal disorders may be numerical or structural. Numerical disorders concern:

- extra single chromosomes (e.g. trisomy)
- missing single chromosomes (e.g. monosomy – lethal except for X0)
- extra haploid sets (e.g. tetraploids or triploids).

These disorders result in gene-dosage effects. In the case of polyploidy and trisomy, disease results from over expression of the chromosomal genes (simple gain of function). In monosomies disease results from haploinsufficiency (loss of function) of the genes that are expressed from the missing chromosome.

Structural disorders include conditions resulting from:

- translocation
- inversion
- isochromosome (see p. 178)
- duplication and deletion of chromosomal segments involving many genes
- ring chromosomes (see p. 179).

Disease arising from these disorders may result from gene dosage effects or misexpression of critical genes due to disruption of the regulatory regions.

Nomenclature used for chromosome disorders

All chromosomal disorders are individually rare, with no clear patterns of inheritance and minimal risk to relatives. International Standard Chromosome Nomenclature (ISCN) is as follows:

- Numerical disorders are described as follows: number of chromosomes, sex chromosomes, + or – chromosome number. For example, a boy with trisomy 21 is [47,XY,+21], Turner's syndrome is [45,XO].
- Structural disorders are described as follows: number of chromosomes, sex chromosomes, mutation (chromosomes involved); break points, margins or region; p – short arm,q – long arm.

Examples of ISCN for structural disorders are as follows:

- translocation (t) – [46,XY,t(14;21)(q11;p10)]
- inversion (inv) – [46,XY,inv(9)(p12,q14)] pericentric inversion
- isochromosome (I) – [46,X,I(Xq)] long chromosome arm of X duplicated
- duplication (dup) – [46,XY,dup(5)(q20–q30)]
- deletion (del) and ring chromosome (r) – [46,XY,del(15)(q11–q13)] in Prader–Willi syndrome; [46,XX,r(X)(p12,q14)].

Mechanisms leading to numerical chromosomal disorders

Polyploidy

Polyploids arise as a result of:

- fertilization by two sperm
- a diploid sperm due to failure in meiosis
- a diploid ovum due to a failure in meiosis.

Trisomies

Trisomies may result from the failure of separation ('non-disjunction') of homologous chromosomes

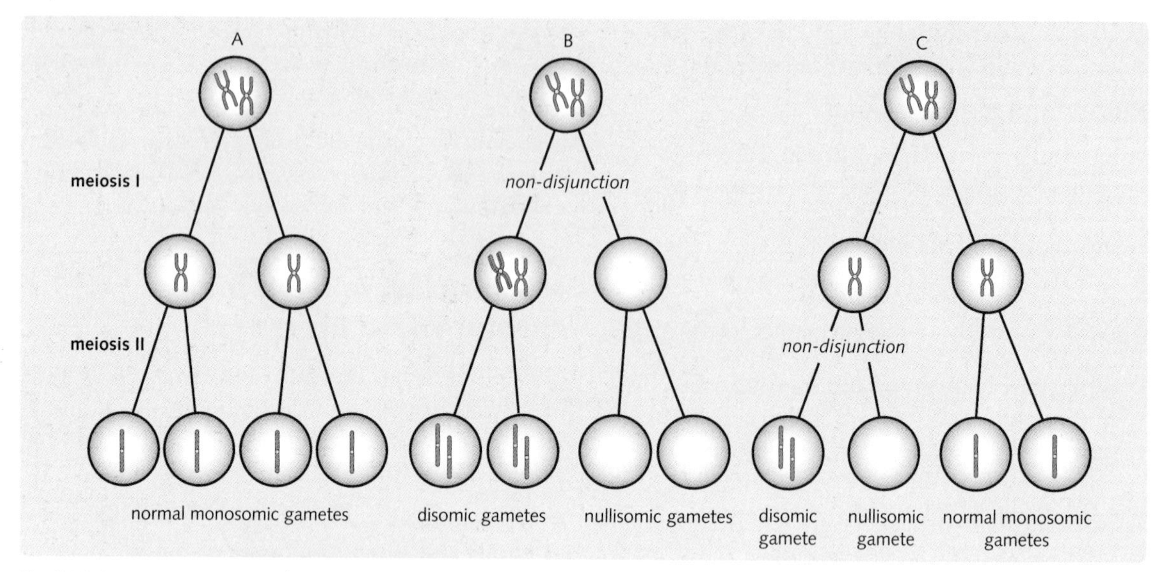

Fig. 8.26 Segregation at meiosis of a single pair of chromosomes. (A) Normal meiosis. (B) Non-disjunction in meiosis I. (C) Non-disjunction in meiosis II. Note that non-disjunction results in disomic and nullisomic gametes, which following fertilization would result in trisomy and monosomy respectively. (Adapted from Mueller and Young, 2001.)

at meiosis I or from the failure of separation of chromatids in meiosis II (Fig. 8.26).

Most cases of autosomal trisomy result from non-disjunction at meiosis I in the female germ line.

Monosomies

Monosomies may result from non-disjunction (Fig. 8.26) or from 'anaphase lag'. Anaphase lag occurs when there is a delay in the movement of one chromosome from the metaphase plate during anaphase. This may result in the loss of a chromosome if it fails to reach the pole of the cell before the nuclear membrane reforms.

Examples of numerical chromosomal disorders

Autosomal disorders

There are only three well-defined autosomal trisomies that are compatible with postnatal survival:

1. Trisomy 21 – Down syndrome (DS)
2. Trisomy 18 – Edwards syndrome (ES)
3. Trisomy 13 – Patau syndrome (PS).

Trisomy 21 – Down syndrome

DS typically arises as a result of trisomy 21 (47,XX+21 or 47,XY+21). In 5% of cases it arises as a result of a Robertsonian translocation (see p. 178) or as a result of mosaicism (see pp. 164–165). The overall incidence of DS, adjusted for the impact of antenatal screening (see Ch. 9), is 1 in 700 live births. The incidence rises with increasing maternal age as a result of increased likelihood of maternal non-disjunction.

Advancing maternal age is a significant risk factor in the aetiology of Down syndrome (DS). However, the absolute numbers of mothers having children with DS is greater in the younger age groups as the total number of mothers in these groups is much higher.

Features associated with DS include:

- typical features, as shown in Fig. 8.27
- hypotonia, often noted at birth

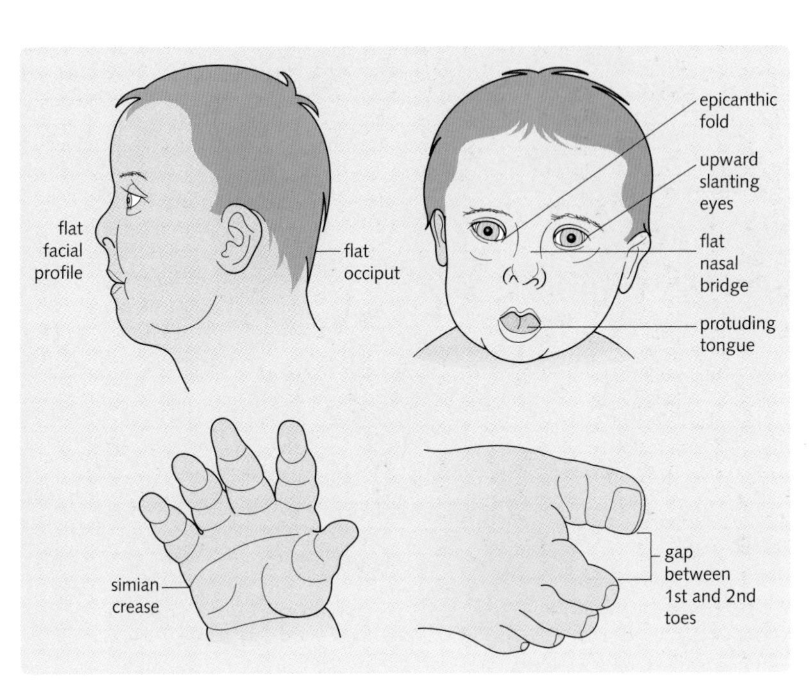

Fig. 8.27 Characteristic features of Down syndrome.

A genetic disorder is one that is determined by genes, whereas a congenital disorder is one that is present at birth and it may or may not have a genetic basis.

- short small middle phalanx of the fifth finger leading to 'fifth finger clinodactyly'
- congenital heart defects – atrial and ventricular septal defect, common atrioventricular canal and patent ductus arteriosus
- learning difficulties and low IQ
- increased risk of leukaemia
- premature senescence leading to early cataract formation and Alzheimer disease.

Life expectancy is generally less than 50 years, and death is usually associated with cardiac disease or acute leukaemia.

Trisomy 18 – Edwards syndrome

ES arises from trisomy 18 – 47,XX+18 or 47,XY+18. The condition has an incidence of 1 in 8000 live births, adjusted for the impact of prenatal screening. Ninety-five per cent of conceptuses with trisomy 18 die *in utero*, and of those born alive, fewer than 10% survive the first year of life. The high mortality rate is attributed to the presence of cardiac and renal malformations and a background of apnoea. Those who survive infancy do so with severe psychomotor and growth retardation. Characteristic clinical features of ES can be found in Fig. 8.28.

Trisomy 13 – Patau syndrome

PS typically arises from trisomy 13 – 47,XX+13 or 47,XY+13. As chromosome 13 is acrocentric, PS can also arise as a result of a Robertsonian translocation (see p. 178), and is thought to do so in 20% of cases. PS has an incidence of 1 in 10 000 live births, adjusted for the impact of prenatal screening, a figure that increases with increasing maternal age. Clinical features of PS include:

- multiple dysmorphic features (Fig. 8.29)
- scalp skin defects – aplasia cutis
- incomplete cleavage of the embryonic forebrain – holoprosencephaly
- congenital heart disease.

The median survival age for children with PS is 2.5 days, with only 1 child in 20 surviving longer than 6 months. Those who do survive usually do so with profound mental and physical disabilities.

Fig. 8.28 Characteristic features of Edwards syndrome.

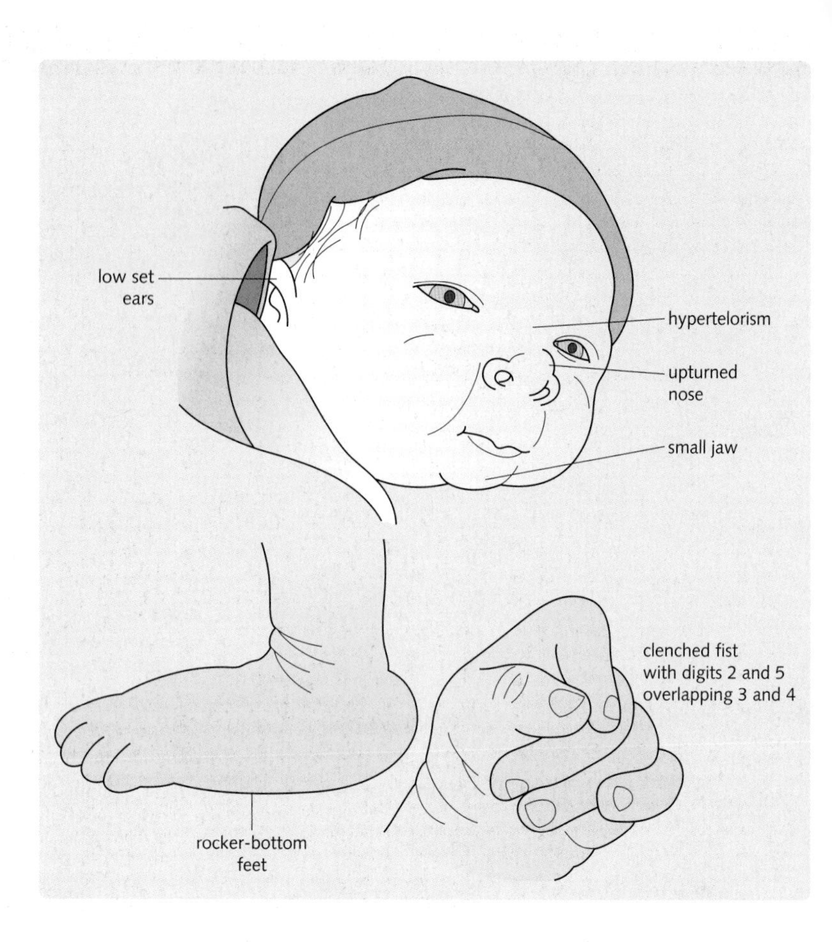

Fig. 8.29 Characteristic features of Patau syndrome.

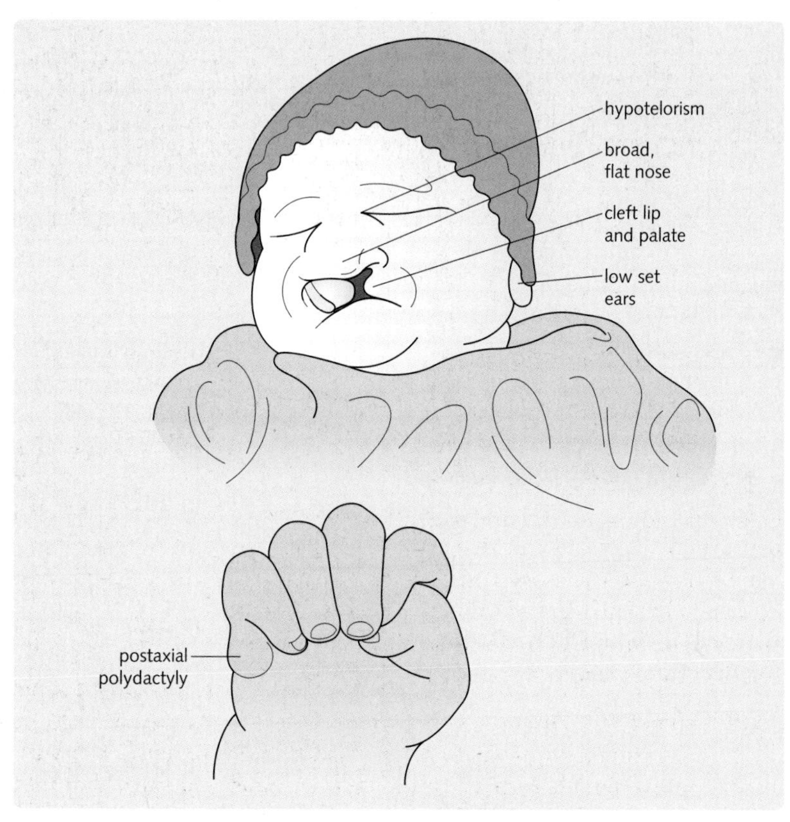

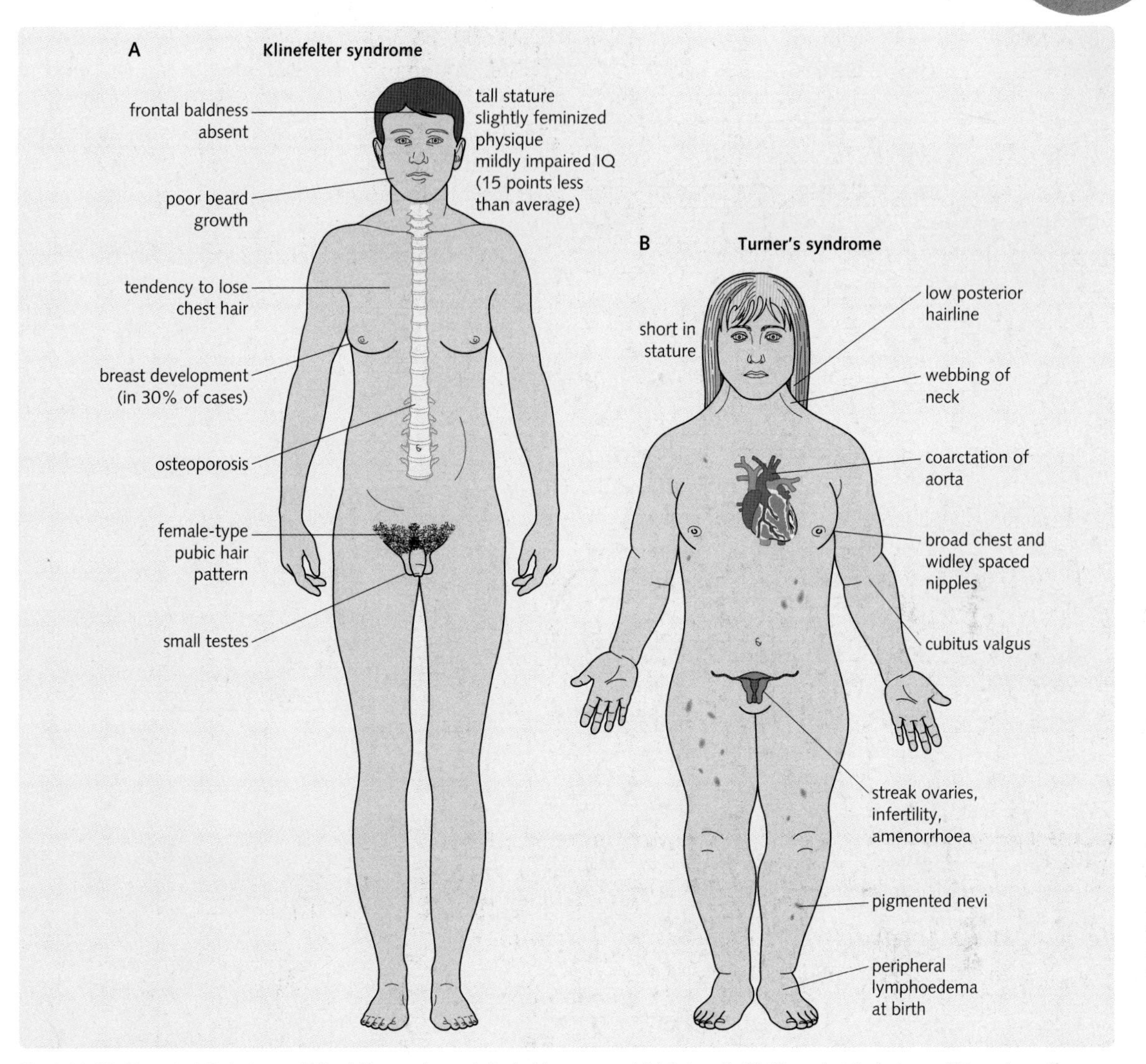

Fig. 8.30 (A) Characteristic features of Klinefelter syndrome (adapted from www.childclinic.net). (B) Characteristic features of Turner's syndrome. (Adapted from Kumar, Abbas and Fausto, 2004.)

Sex chromosome disorders

Examples of sex chromosome abnormalities are shown below.

47,XXY – Klinefelter syndrome

Klinefelter syndrome, the most common karyotypes being 47,XXY, has an incidence of 1 in 1000 live male births. It is the most common cause of primary hypogonadism in males, with an otherwise minimally abnormal phenotype. The classic phenotype associated with Klinefelter syndrome is shown in Fig. 8.30A.

45,X0 – Turner's syndrome

45,X0, or Turner's syndrome, has an incidence of 1 in 2500 live female births. Turner's syndrome is the only viable monosomy in humans, possibly explained by the fact that the normal situation in a cell is to have just one functional X chromosome, as one of the X chromosomes undergoes Lyonization to yield a single X chromosome and one Barr body per cell.

The classic phenotype associated with a single X chromosome is shown in Fig. 8.30B.

47,XXX – Trisomy X syndrome

47,XXX, Trisomy X syndrome, has an incidence of 1 in 1500 live female births. Trisomy X individuals are usually tall, with a lower than average body mass index. Twenty-five per cent are infertile and delays in speech and language development are frequent, but respond well to speech and language therapy; otherwise there are no distinguishing features.

47,XYY

47,XYY has an incidence of 1 in 1000 live male births. XYY males are often tall, but have normal body proportions. Although often asymptomatic, there may be subtle motor incoordination and behaviour problems, with aggression in childhood.

XX male – de la Chapelle syndrome

The XX male phenotype typically arises as a result of unequal crossing over between X and Y chromosomes during meiosis. It has an incidence of 1 in 20 000 live male births. The majority of XX male individuals have the Y chromosome gene SRY attached to one of their X chromosomes. They have a similar appearance to males with Klinefelter syndrome, but:

- are sterile as they lack Y chromosome genes crucial for sperm manufacture
- are usually shorter than the average male, again due to a lack of specific Y chromosomally encoded genes
- have a normal IQ.

XX males usually present at the infertility clinic or when a prenatally predicted female appears to be male. Confirmation is by banding studies, DNA analysis and in-situ hybridization, which identify Y-specific sequences.

Mechanisms leading to structural chromosomal disorders

All structural disorders result from chromosomal breakage. Chromosomal damage is increased by some environmental conditions (e.g. mutagenic chemicals, radiation) and by genetic chromosome instability disorders (e.g. ataxia telangiectasia and Fanconi anaemia).

Translocation

This is the exchange of chromosome segments that generally involves dissimilar chromosomes. Translocations may be:

- reciprocal (non-Robertsonian, balanced)
- Robertsonian (non-balanced).

In balanced translocations there is usually no loss of genetic material, so the individual is phenotypically normal. However, depending on the segregation of chromosomes in meiosis, gametes may result that do not contain a single complete copy of the genome (Fig. 8.31). For this reason, prenatal diagnosis may be offered to known balanced carriers.

A Robertsonian translocation occurs when the long arms of two acrocentric chromosomes (13, 14, 15, 21

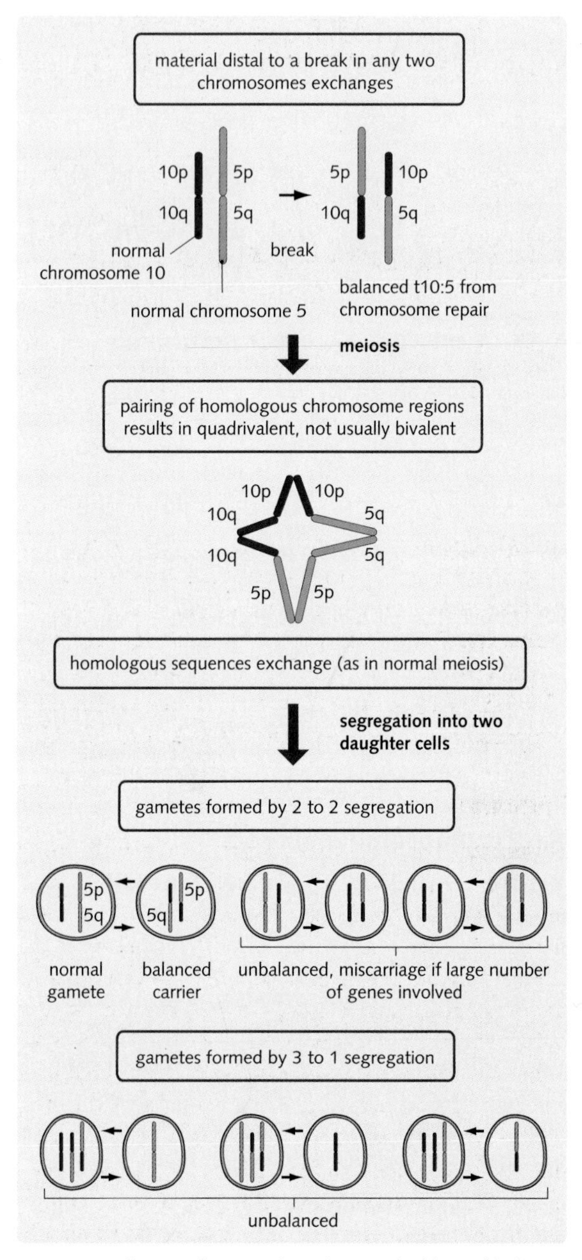

Fig. 8.31 Mechanism of reciprocal translocation (highly simplified).

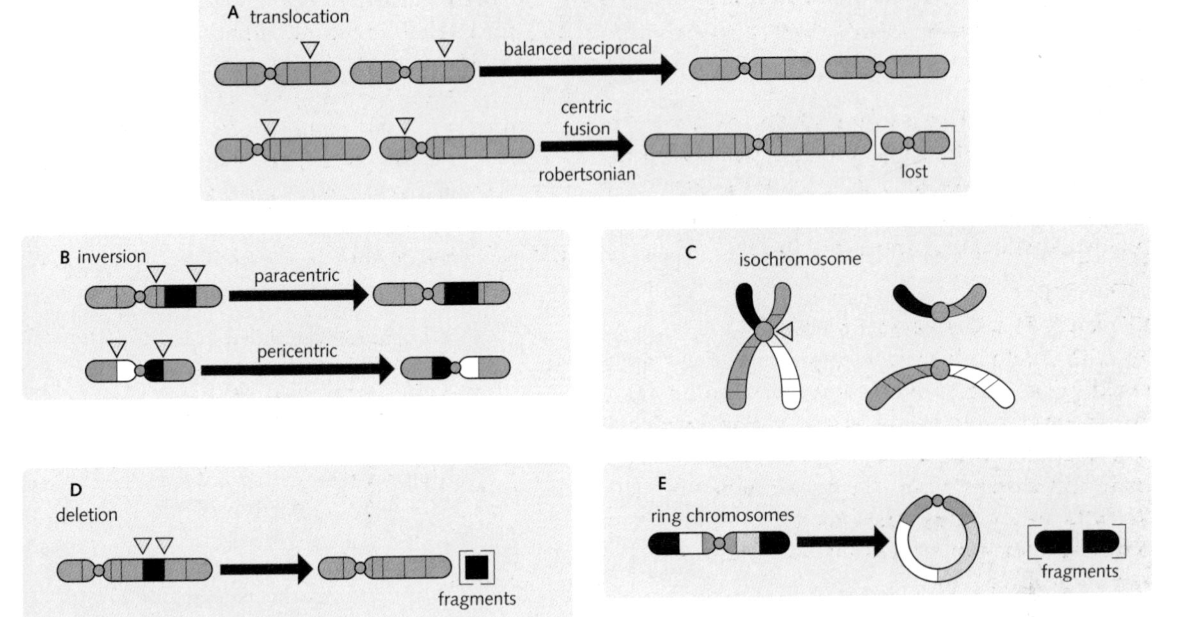

Fig. 8.32 (A) Outcomes of reciprocal and Robertsonian chromosomal translocation. (B) Results pericentric and paracentric chromosomal inversion. (C) Formation of an isochromosome. (D) Outcome of chromosomal deletions. (E) Formation of a ring chromosome. (Adapted from Kumar, Abbas and Fausto, 2004.)

and 22) fuse at a centromere, and the two short arms are lost. As the short arms of these chromosomes are not known to carry essential genetic material, carriers of Robertsonian translocations, although only having 45 chromosomes in each cell, are phenotypically normal. Their offspring, however, may inherit a missing or extra long arm of an acrocentric chromosome (Fig. 8.32A).

Inversion

This arises when two breaks occur in the chromosome and the intervening DNA rotates through 180°. If an inversion includes the centromere it is known as a pericentric inversion (Fig. 8.32B). Those that do not arise from breaks in one arm, and are known as paracentric (Fig. 8.32B). Inversions are usually balanced, that is there is no loss of genetic material. However, in meiosis homologous chromosomes can line up and pair only if a loop is formed in the region of the inversion to bring it in line with its normal homolog. This leads to an increased chance of duplication and deletion of chromosomal regions and for the production of unbalanced chromosome rearrangements in the offspring of carriers.

Isochromosome

This chromosome has a duplication of one arm, but lacks the other. It results from breakage of a chromatid with fusion above the centromere or transverse division (Fig. 8.32C). Isochromosomes of most autosomes are lethal, but those of the X chromosome [46,X,I(Xq)] can yield a phenotype similar to that seen in Turner's syndrome, and isochromosome 18q has been reported in infants afflicted with ES.

Duplication

This implies an extra copy of a chromosome region. Causes include inheritance from parents with balanced structural disorders and *de novo* duplication from unequal crossing-over in meiosis, translocation or inversion.

Deletion and ring chromosome

Deletion results in a loss of genetic material. The telomere is important in chromosome function, so interstitial deletions are more common. If a deleted

Fig. 8.33 Features of some structural chromosome abnormalities

Disorder	Aetiology	Features
Prader–Willi syndrome and Angelman syndrome; incidence 1 in 25 000 live births	Both result from deletion in the same region on 15q11–13; syndromes differ due to genomic imprinting, so depend on which parent the deleted gene is inherited from—Prader–Willi syndrome results from inheritance of the deletion from the father (so only have maternal contribution to the critical area), Angelman syndrome results from inheritance of the deletion from the mother; uniparental disomy can result in these syndromes but does not involve deletions	Prader–Willi syndrome—neonatal hypotonia, initial feeding difficulties, obesity of face, trunk, and limbs (after first year of life), prominent forehead, almond-shaped eyes, triangular upper lips, IQ 20–80, short stature, small hands and feet, hypoplasia of external genitalia, tendency to diabetes mellitus. Angelman (happy puppet syndrome)—hypertonia, ataxic gait, characteristic arm posture, prominent jaw, deep set eyes, happy appearance, laughter, absent speech, learning disability
Wolf–Hirschhorn syndrome	Partial deletion of the short arm of chromosome 4 (4p16.1); male to female ratio is 3:4	'Greek helmet' shaped head, cleft lip and palate, abnormal low-set ears, large beaked nose, hypertelorism (widely spaced eyes), epicanthic folds, microcephaly and learning disability, failure to thrive, heart defects, convulsions, hypospadias
Cri du chat syndrome	Deletion of region on 5p15.2 or the whole short arm of chromosome 5	Round face (in adults the face elongates), cat-like cry (*cri du chat*), hypertelorism, epicanthic folds, strabismus, low-set ears, low birth weight, learning disability (variable degree), appear normal at birth

Fig. 8.33 Features of some structural chromosome abnormalities.

fragment has no centromere it will be lost during mitosis (Fig. 8.32D).

Ring chromosomes result from breaks near both telomeres of a chromosome, which aberrantly repair to form a ring, with the regions distal to the breaks being lost. If the ring has a centromere, it will be passed through generations in mitosis (Fig. 8.32E).

Visible structural deletions are large, with the smallest microdeletion being at least 3 million base pairs (3 Mb). Single gene disorders may be caused by deletions, but if the deletions are not visible by light microscopy they are not considered structural. Structural deletions may cause deletion syndromes with more than one single gene disorder occurring concurrently. Subtelomeric microdeletions, frequently involving several genes, account for a proportion of children with multiple developmental and skeletal disorders. Modern molecular techniques such as fluorescence *in situ* hybridization (FISH) or DNA analysis may reveal the underlying lesions (see Ch. 7).

Examples of structural chromosomal disorders

These contiguous gene disorders are diagnosed with the light microscope by high-resolution banding. Deletion must involve at least 3 Mb to be detected by this technique. It is important to find out whether the deletion has arisen *de novo* in the proband or from a balanced translocation in the parents if the risk of recurrence is to be calculated. Recurrence is:

- negligible with *de novo* mutations
- considerable with a balanced translocation.

Fig. 8.33 shows the features of some structural chromosomal disorders.

Principles of medical genetics

9

Objectives

By the end of this chapter you should be able to:

- Define the Hardy–Weinberg equilibrium and list the factors that may disturb the equilibrium.
- List the Wilson and Junger screening criteria.
- Understand the purpose of carrier detection and give examples of its use.
- Appreciate the importance of maternal serum screening and the use of multiples of the median calculations.
- Understand the probability of a child being affected by an autosomal recessive condition if both parents are carriers, and the probability of a child inheriting an autosomal dominant condition if one parent is affected.
- Understand how Bayes' theorem is used in genetic counselling.
- Appreciate the importance of a good family history in the context of the genetic consultation.
- Understand the main aims of the treatment of genetic disease.
- Consider some of the ethical implications in medical and clinical genetics.

POPULATION GENETICS AND SCREENING

Introduction

It is estimated that between 2 and 5% of all newborns have congenital malformations or genetic disorders. Many common diseases, such as coronary heart disease, diabetes and cancer, have a significant genetic component. The role of the clinical geneticist in the management of individuals and families affected by a genetic disease, includes:

- establishing an accurate diagnosis
- providing information about prognosis
- calculating the risks of developing or transmitting the disorder
- exploring with the patient and family ways in which the development or transmission of the disorder may be modified.

Without established therapies, the emphasis is placed on identifying individuals who are at risk of having an affected child so that they can make informed reproductive choices. Such individuals may be identified:

- in specific families where a genetic disorder has already arisen
- in certain groups at an increased risk of genetic disease (e.g. Tay–Sachs disease in Ashkenazi Jews, Down syndrome in pregnant women over the age of 35 years).

Population genetics

Population genetics is the study of the genetic composition of populations. Allele frequencies, and the frequency of disease-causing mutations, vary between populations, reflecting their specific geographical and genetic origins. Thus, some diseases may be much more common in some groups than others. Cystic fibrosis is common in Europeans of Celtic and Northern European descent (allele frequency 1/40–1/50), but rare in Finnish, Asian and African populations; while Tay–Sachs disease has a high allele frequency in Ashkenazi Jews (1/60), 100-fold more common than in other populations.

If clinical geneticists are to offer an accurate assessment of risk of a genetic disease, they must know how common the disease-causing mutation is in the relevant population. This is particularly important in recessive disorders, where unaffected heterozygotes carry the disease-causing gene.

The Hardy–Weinberg principle enables the frequency of a disease-causing allele in a population to be calculated from the disease incidence (provided that the population is in Hardy-Weinberg equilibrium).

Information on the diagnosis, management and counselling of specific genetic disorders can be found on http://www.geneclinics.org/

Hardy–Weinberg law and equilibrium

The Hardy–Weinberg law is a mathematical equation that forms the basis of population genetics. It states that allele frequencies within a population tend to remain constant from one generation to the next. It assumes that the organism under investigation is diploid, is sexually reproducing and has discrete generations. By extension, the Hardy–Weinberg equilibrium states that allele frequencies remain constant from one generation to the next, providing there is an absence of selection (constant gene and genotype frequencies). Thus, the necessary conditions required to satisfy such an equation are:

- the population is large, to minimize genetic drift
- mating is random
- the mutation rate remains constant
- alleles are not selected for (i.e. they confer no survival or reproductive advantage)
- there is no migration into or out of the population (gene flow).

The Hardy–Weinberg equation is derived by considering a population carrying an autosomal gene with two alleles A and a. The frequency of the dominant allele A in gametes is represented by p, and the frequency of the recessive allele a in gametes is represented by q (Fig. 9.1). Since there are only two alleles, $p + q = 1$ and $q = 1 - p$.

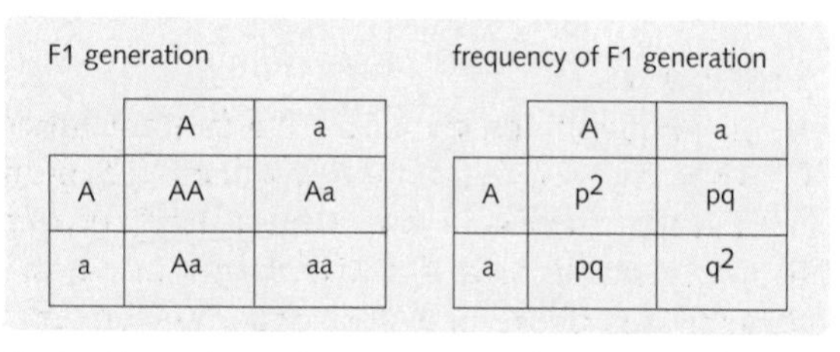

F1 generation

	A	a
A	AA	Aa
a	Aa	aa

frequency of F1 generation

	A	a
A	p^2	pq
a	pq	q^2

Fig. 9.1 The Hardy–Weinberg equation. Consider if two heterozygotes mate, Aa × Aa. The distribution of AA, Aa, and aa genotypes in the population correspond to p^2, $2pq$, and q^2, respectively. If A and a are the only alternative alleles for the same gene locus, then, $p^2 + 2pq + q^2 = 1$. a, recessive allele; A, dominant allele; p, frequency of A; q, frequency of a.

In the combination of alleles that forms the next generation:

- the chance that both male and female gametes will carry the A allele is $p \times p = p^2$
- the chance that the gametes will produce a heterozygote is $(p \times q) + (q \times p) = 2pq$
- the chance that both male and female gametes will carry the a allele is $q \times q = q^2$.

These are the only possibilities, therefore:

$$p^2 + 2pq + q^2 = 1$$

This equation can be used to calculate allele frequency in a population if disease occurrence is known, provided that the population is in Hardy–Weinberg equilibrium.

For autosomal recessive (AR) conditions (disease incidence = q^2; gene frequency = q; heterozygote frequency, i.e. frequency of carriers of AR condition = 2pq). For example, an AR disorder occurs with a frequency of 1 in 1600 live born births:

- incidence, $q^2 = 1/1600$
- gene frequency (a), $q = 1/40$; dominant allele A has gene frequency $p = 39/40$
- heterozygote frequency, 2pq is $2 \times 39/40 \times 1/40$, which is approximately 1/20.

For autosomal dominant (AD) conditions:

- nearly all affected are heterozygotes, so q^2 is approximately 0
- if the condition is rare, p^2 is approximately 1
- disease gene frequency (A) is approximately 2pq, which is approximately 2q.

Factors which disturb Hardy–Weinberg equilibrium

A population is not in Hardy–Weinberg equilibrium if the genotype frequencies do not arise in the proportions predicted by the Hardy–Weinberg law (Fig. 9.2). This may arise as a result of a number of mechanisms, discussed below.

Fig. 9.2A Allele frequencies determined from genotype frequencies

	Genotype			
	TT	Tt	tt	Total
No. of individuals	40	40	20	100
No. of T alleles	80	40	0	120
No. of t alleles	0	40	40	80

Total number of alleles = 120 + 80 = 200
Frequency of T = p = 120/200 = 0.6
Frequency of t = q = 80/200 = 0.4

Fig. 9.2B Using χ^2 test to determine whether a population is in Hardy–Weinberg equilibrium

	Genotype		
	TT	Tt	tt
Observed (O)	40	40	20
Expected (E)	$p^2 \times 100 = 36$	$2pq \times 100 = 48$	$q^2 \times 100 = 16$
O – E	4	–8	4
$(O - E)^2/E$	16/36 = 0.44	64/48 = 1.33	16/16 = 1

$\chi^2 = \Sigma(O–E)^2/E = 0.44 + 1.33 + 1 = 2.77$

Fig. 9.2 Determining whether a population is in Hardy–Weinberg equilibrium given the genotype frequencies. (A) The genotype frequencies are used to determine the allele frequencies. (B) Chi squared (χ^2) tests are used to determine whether the population differs significantly from one in Hardy–Weinberg equilibrium. The observed (O) genotype frequencies are those seen in the population. The expected (E) genotype frequencies are those predicted if the population is in Hardy–Weinberg equilibrium, using the allele frequencies calculated in A. From χ^2 tables (with one degree of freedom and 95% confidence intervals), a population does not differ significantly from one in Hardy–Weinberg equilibrium if χ^2 is less than 3.84.

Non-random mating

Random mating refers to the selection of a partner regardless of that partner's genotype. The tendency of individuals to select partners who share certain characteristics (e.g. height) is called assortative mating, which is a form of non-random mating. Non-random mating may also result from consanguinity.

Mutation

If a locus has a high mutation rate, theoretically there will be an increase in the number of mutant alleles in the population. In practice, this does not occur for populations that are in Hardy–Weinberg equilibrium, because the mutation rate is balanced against a reduction in reproductive fitness in affected individuals.

Selection

The reduced reproductive fitness of affected individuals acts as a negative selection, which leads to a gradual reduction in the frequency of the mutated gene and disturbance of the Hardy–Weinberg equilibrium. In practice, this is balanced by the mutation rate for populations that are in equilibrium.

In some cases, selection of a mutation may increase fitness and, for some autosomal recessive disorders, heterozygotes show an increase in fitness relative to unaffected homozygotes (the 'heterozygous advantage'). For example, carriers of sickle-cell anaemia are comparatively resistant to malaria, which is thought to explain the high incidence of the sickle-cell trait in West Africa.

Genetic drift

In small populations, one allele may be transmitted to a high proportion of offspring by chance, resulting in marked changes in allele frequency between the two generations and a disturbance in Hardy–Weinberg equilibrium. This phenomenon may contribute to the 'founder effect' (see below).

Migration

The introduction into the population of new alleles as a result of migration and subsequent intermarriage will result in a change in the relevant allele frequencies, and thus a disturbance to Hardy–Weinberg equilibrium.

Founder effect

Founder effects arise in small isolated populations that breed amongst themselves. If, by chance, a member of the small group of original founders had a certain disease gene, this will remain over-represented in successive generations as the population expands. For example, several rare autosomal recessive disorders occur at a relatively high frequency amongst the Old Order Amish (an isolated religious group that tend to intermarry).

Population screening and carrier detection

Population screening is the screening of all members of a population regardless of their family history. It is used to identify carriers of recessive traits and to detect pre-symptomatic individuals. Pre-symptomatic testing is sometimes called predictive testing.

The aim of carrier detection is to identify asymptomatic heterozygotes for autosomal recessive (AR) traits, although it is sometimes used to detect carriers of autosomal dominant (AD) disorders that have limited penetrance or late onset. If two partners are heterozygotes for a mutation at the same locus, genetic counselling can be offered before conception and prenatal tests offered as appropriate. Carrier detection tends to be confined to small ethnic populations in which there is an anomalously high incidence of a particular disease due to either the founder effect or heterozygous advantage.

Prenatal diagnosis concerns the use of tests in pregnancy to determine whether an unborn child is affected with a particular disorder (see pp. 186–187). Fig. 9.3 describes some of the tests available in prenatal diagnosis and their relative risk to the pregnancy.

Fig. 9.3 Tests used for prenatal diagnosis

Test	Gestation	Procedure	Abnormalities detected	Risk of procedure
Amniocentesis	16–18 weeks (routinely) 14–32 weeks is possible	Liquor is removed via a long needle inserted transabdominally. Cells are cultured for 2–3 weeks	Foetal sexing, karyotyping, and enzyme assay	Miscarriage rate estimated at 1% above the normal rate at 16 weeks
Chorionic villus sampling	8–10 weeks	Biopsy usually taken transabdominally or transvaginally	Foetal sexing, foetal karyotyping, biochemical studies, DNA analysis (cell culture not necessary)	Miscarriage rate estimated at 2–3% above the average at 10 weeks; rhesus isoimmunization if mother is rhesus negative
Cordocentesis	18 weeks +	Foetal blood sample obtained by inserting a fine needle into foetal umbilical cord	Suspected foetal infection or mosaicism, unexplained hydrops, single gene disorders, fragile X	Procedure-related loss approximately 1%; risk of rhesus isoimmunization
Ultrasound	Routine scan at 16–18 weeks in all pregnancies	Visualization of foetus by transabdominal ultrasound probe	Over 280 congenital malformations	Non-invasive test with low associated risk
Maternal serum screening: infectious screen (TORCH screen), quadruple test (serum AFP, HCG, uE3 and DIA)	13–26 weeks	Sample of maternal blood collected	Detection of infections that may cause congenital malformations (see Chapter 8); quadruple test estimates relative risk of NTDs and Down syndrome based on serum levels of HCG, AFP, uE3 and DIA	Very low

Fig. 9.3 Some of the tests available in prenatal diagnosis. (AFP, α-fetoprotein; HCG, human chorionic gonadotrophin; uE3, uncongugated oestradiol; DIA, dimeric inhibin-A; NTDs, neural tube defects; TORCH, toxoplasmosis, other agents, rubella, cytomegalovirus, herpes simplex.) Associated risk refers to the risk that the test may damage the foetus.

Criteria for a screening programme

Population wide screening programmes for genetic disorder must meet the screening criteria as laid down by the UK National Screening Committee (NSC). They are an extension of the Wilson–Jungner criteria (WHO 1968):

- The condition being screened for should be an important problem.
- There should be a latent or early symptomatic stage.
- The natural history of the condition should be understood.
- There should be a definitive test or examination for the condition.
- The test or examination should be acceptable to the population.
- Case finding should be a continuing process rather than a 'one-off' project.
- There should be an effective treatment for the disease.
- Facilities for diagnosis and treatment should be available.
- There should be an agreed policy on whom to treat as patients.
- The cost of screening should be balanced in relation to overall health expenditure.

Methods used for carrier detection and pre-symptomatic diagnosis

Direct mutation detection

If the mutation(s) that cause a specific disease has been defined, carriers or pre-symptomatic individuals can be identified by direct mutation analysis using DNA techniques, such as PCR or fluorescence *in situ* hybridization (FISH) (see Ch. 7).

Linkage to polymorphic marker

Genetic linkage is the tendency for alleles close together on the same chromosome to be transmitted together through meiosis (see Ch. 6). It allows disease genes for which the causal mutation is unknown to be followed through generations, by using a linked polymorphic marker that lies sufficiently near the disease gene for recombination to be unlikely (Fig. 9.4).

Biochemical tests

Biochemical tests form an important part of prenatal and pre-symptomatic diagnosis. Over 100 inborn errors of metabolism can be detected by enzyme assays using cultured amniocytes or chorionic villus samples. Newborn babies can be screened for errors of metabolism by simple biochemical tests on a blood sample.

> Phenylketonuria (PKU) and congenital hypothyroidism (CHT) are routinely screened for by way of the blood spot test. It identifies around 250 PKU or CHT affected babies a year, allowing treatment to commence before irreversible damage is done. Blood spot testing sees a sample of blood taken from a heel prick from all newborn babies, placed on a Guthrie card and sent to the laboratory for analysis. As part of the Newborn Screening Programme, it is expected that by 2007 all parents will be offered the opportunity to have their newborns screened not only for PKU and CHT, but also for sickle-cell anaemia and cystic fibrosis.

Screening in 'at risk' populations

Haemoglobinopathies

As part of the NHS Sickle Cell & Thalassaemia Screening Programme, all newborn babies will be screened for sickle cell and thalassaemia. Traditionally, screening has been aimed towards at risk populations. Sickle cell testing has been offered to those within a geographical area of high prevalence, and to those deemed as 'at risk' in areas of low prevalence. Adult carrier detection is by the Sickledex test, which demonstrates sickling of red blood cells *in vitro* at very low oxygen concentrations.

Traditionally, screening for thalassaemia has been targeted to those of Mediterranean or South East Asian descent, and achieved by measurement of red cell indices (to detect hypochromic microcytosis) and electrophoresis for abnormal haemoglobins (to detect an increase in haemoglobin A_2 concentration).

Tay–Sachs disease

Carriers are detected by hexosaminidase A (HEXA) enzymatic activity in plasma or white blood cells. Alternatively, if the likely mutation is known, direct mutation detection may be offered. These tests are targeted at the Ashkenazi Jewish population.

Cystic fibrosis

Cystic fibrosis (CF) is the most common serious AR condition to affect Caucasians. Carrier detection is usually limited to affected families, although in some

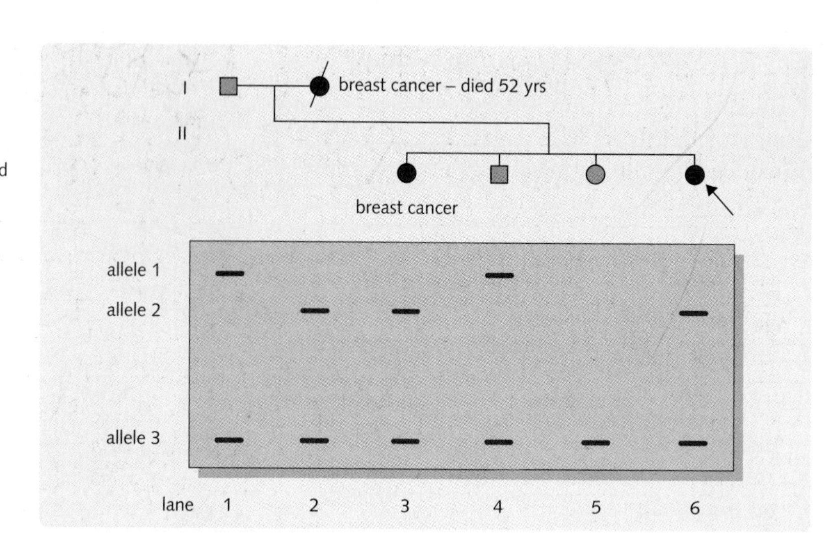

Fig. 9.4 Detection of inherited breast cancer using microsatellite linkage. Using a PCR microsatellite site known to be linked to the disease gene locus, three alleles (each corresponding to a different repeat element) are detected. The affected mother has handed the disease gene and allele 2 to the affected daughter (lane 3) and her normal gene and allele 3 to each unaffected child (lanes 4 and 5). This could be a scenario where the mother and daughter have already both developed breast cancer and the test shows that the youngest daughter (lane 6) also has the disease gene so is at high risk of developing cancer. She should, therefore, have genetic counselling and regular mammograms for early detection (some patients undergo elective double mastectomy in this situation).

areas of the UK all newborn babies are screened. Three main screening modalities are employed:

- Antenatal screening – via amniocentesis or chorionic villus sampling and offered to parents both known to be carriers.
- Newborn testing – offered as routine in some areas of England as part of the 'blood spot' protocol, followed by confirmation by sweat test.
- Mutation detection – approximately 85% of CF carriers have a 'common' mutation, identifiable using DNA sequencing technologies.

Antenatal screening tests

These are usually quick and simple tests that can be used to assess a woman's relative risk of having an affected child. Maternal serum is used in many of these tests. At 13–26 weeks' gestation, women can be tested for maternal serum α-fetoprotein (MSAFP), human gonadotrophin (HCG), uncongugated oestradiol (uE3) and dimeric inhibin-A (DIA) (Fig. 9.5). Measurements are corrected for gestational age and a multiple of the mean value calculated to ascertain the relative risk of a number of foetal conditions including neural tube defects (spina bifida) and trisomy 21 (Down syndrome) (Fig. 9.6):

- MSAFP is usually low in trisomies 18 and 21, and normal in trisomy 13.
- MSAFP is usually high in neural tube defects.
- uE3 is usually low in trisomies 13, 18 and 21.
- HCG is usually high in trisomy 21 and low in trisomies 18 and 13.
- DIA is usually high in trisomy 21.

'Cascade' screening

Cascade screening is the identification within a family of carriers for an autosomal recessive disorder or persons with an autosomal dominant gene following ascertainment of an index case. Once an individual has been diagnosed with a condition, family members at high risk are offered testing. If more people within the family are identified with the gene, the results may be used to identify further family members at risk, who are then offered the test. These results, in turn, redefine risk status for other members of the family – and so on.

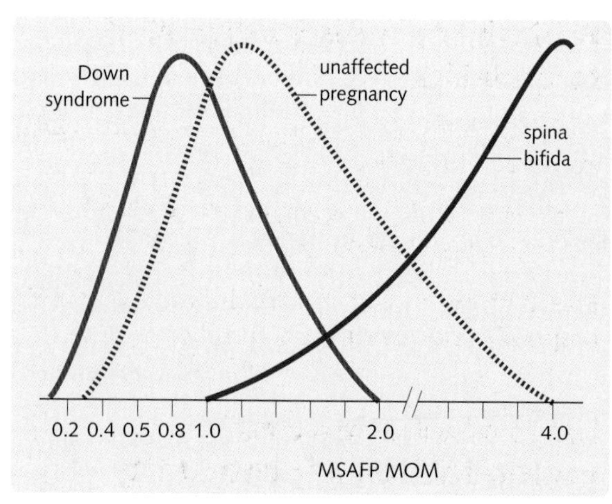

Fig. 9.5 Screening for trisomy 21. Maternal serum α-fetoprotein (MSAFP) levels are used to generate likelihood ratios, which are derived from overlapping distribution of affected and unaffected pregnancies. Likelihood ratio and maternal age are used together to generate a combined risk. If there is a high combined risk, refer for amniocentesis. If there is a low combined risk, provide reassurance. MOM, multiples of the median.

Fig. 9.6 Detection rates using different Down syndrome screening strategies

Screening modalities	% of all pregnancies tested	% of Down syndrome cases detected
Age alone		
40 years and over	1.5	15
35 years and over	7	35
Age + AFP	5	34
Age + AFP, uE3 + HCG	5	61
Age + AFP, uE3, HCG + DIA	5	75
NT alone	5	61
NT + age	5	69
HCG, AFP + age	5	73
NT + AFP, HCG + age	5	86

Fig. 9.6 Detection rates using different Down syndrome screening strategies. AFP, α-fetoprotein; HCG, human chorionic gonadotrophin; uE3, uncongugated oestradiol; DIA, dimeric inhibin-A; NT, nuchal translucency. (Reproduced from Turnpenny and Ellard, 2004.)

RISK ASSESSMENT AND GENETIC COUNSELLING

For families in which a disease is showing a recognizable pattern of Mendelian inheritance, the risk to other family members of developing or transmitting the disorder can be calculated by simple probability theory.

Probability theory as applied to genetics

The probability of a single event occurring can be expressed as a fraction:

p = number of ways events can happen/total number of possibilities

Probability is a useful way of demonstrating the risk of specific genetic event(s) occurring.

Laws of addition and multiplication of probability

The law of addition

If two events could not happen at the same time, they are said to be mutually exclusive. The probability that either event will occur is equal to the sum of their probabilities. The law of addition is:

$$p = p1 + p2$$

The law of addition is sometimes called the 'or' law because it is the probability that one event or another event will occur.

The law of multiplication

An independent event is one that has no effect on subsequent events. The outcome of the first event has no effect on subsequent events. The probability that all the events will occur is equal to the product of the individual probabilities. The law of multiplication is:

$$p = p1 \times p2$$

The law of multiplication is sometimes called the 'and' law because it is the probability that one event and another event will occur.

To determine the probability of two independent events both occurring, you should multiply the probabilities of the individual events together.

Calculating risks from pedigree information

Autosomal recessive conditions

AR disorders are only manifested if two mutant copies of the gene are inherited (see Fig. 8.8). Therefore, for a child to have a recessive disorder both parents must be carriers.

Read questions on autosomal recessive (AR) disease carefully. For example, if you are asked to define the probability that a healthy child sired by two carrier parents is a carrier for cystic fibrosis, the answer is 2/3, because you know the child does not have the condition. However, if you are asked to calculate the probability that an unborn child will be a carrier the answer is 1/2 because they might also be affected.

Autosomal dominant conditions

AD conditions are manifested if one mutant gene is inherited (see Fig. 8.5). Therefore, for a child to have a dominant disorder, either the mother OR the father must have the disease gene.

X-linked recessive conditions

The situation is more complicated with X-linked recessive disorders because the situation varies according to whether the child is a girl or a boy (see Fig. 8.11).

As we learn more about the development of complex disease and the interplay between our genes and our environment, questions relating to social and environmental factors are becoming even more crucially important in mapping disease progression.

Bayes' theorem

Additional information may be used to modify the risk calculated from the pedigree data in order to obtain a more accurate value. For example, given that Huntington's disease generally manifests in middle age, a suspected carrier who has not yet developed the disease at age 30 years may still have the mutation. However, if they have not developed the disease by age 60 years, given that the condition would have been expected to manifest by this age, it is less likely that they carry the mutation. Bayesian analysis is a method for taking such considerations into account by determining the relative probabilities of two alternative outcomes:

- The 'prior' probability is based on classical Mendelian inheritance.
- The 'conditional' probability is based on observations that modify the prior probability, such as existing unaffected offspring, results of screening tests or the age of the offspring.

A 'joint' probability is then calculated as the product of the prior and the conditional probability. A final 'relative' probability is the proportional risk of one alternative with respect to the other (Figs 9.7, 9.8).

Aspects of genetic counselling

Establishing the diagnosis

Genetic counselling is the provision of information to affected individuals or family members at risk of a disorder that may be genetic. The consultands are informed of:

- the consequences of the disorder
- the probability of developing or transmitting it
- ways in which it may be prevented or ameliorated.

An accurate diagnosis is essential to genetic counselling so that the correct advice can be given. A medical history of all affected family individuals is needed and a pedigree constructed. Miscarriages, unexplained learning difficulties, congenital malformations and parental consanguinity should be asked about specifically.

Investigations may involve chromosomal or DNA analysis or specific biochemical tests related to the disease being screened for.

Presenting the risks in context

Once the diagnosis and mode of inheritance have been established, carrier risk and recurrence risk for the consultands can be estimated, based on Mendelian rules and Bayes' theorem.

X-linked recessive disorder—haemophilia A

A

B relative risk calculation using a Bayes' table

II_2	P (carrier)	P (non-carrier)	
prior	$^1/_2$	$^1/_2$	(Mendelian calculation)
conditional	$^1/_2 \times {}^1/_2 \times {}^1/_2 = {}^1/_8$	1	(based on existing information—three unaffected sons)
joint	$^1/_2 \times {}^1/_8 = {}^1/_{16}$	$^1/_2$	(product of prior and conditional)
relative	$^1/_{16} : {}^1/_2 = 1{:}8$	$^1/_2 : {}^1/_{16} = 1{:}8$	(proportional risk, based on joint probabilities)

Fig. 9.7 Estimation of carrier risks using Bayes' theorem. (A) Mother (I_2) is an obligate carrier as she has an affected brother and son. Daughter (II_2) has already had three unaffected sons. What is the risk that she is a carrier? (B) The relative risk calculated can be displayed on a Bayes' table.

isolated X-linked recessive case — Duchenne muscular dystrophy (DMD)

A

B relative risk calculation using a Bayes' table

III_2	P (carrier)	P (non-carrier)
prior	$^1/_3$	$^2/_3$
conditional	$^1/_3$	1 *
joint	$^1/_3 \times {}^1/_3 = {}^1/_9$	$^2/_3 \times 1 = {}^2/_3$
relative	$^1/_9 : {}^2/_3 = 1{:}7$	$^2/_3 : {}^1/_9 = 7{:}1$

* all non-carriers have normal CK test

Fig. 9.8 Estimation of carrier risks using Bayes' theorem II. (A) Isolated case of DMD in III_1; assess carrier risk in his sister, III_2. Thirty-five per cent of cases arise from new mutations – neither mother nor sister would undergo mutation. If mutation was inherited from the mother, the carrier risk for the sister would be half. If the mutation was from the grandmother, the mother would be an obligate carrier (carrier risk) 1 and the daughter's carrier risk would be half. Based on this information, prior risk for the grandmother is one-third, the mother is two-thirds, and the daughter is one-third. The daughter has a normal creatine kinase (CK) test. In general, two-thirds of carriers have raised CK and one-third have normal CK. (B) The relative risk calculated can be displayed on a Bayes' table.

Discussing options, communication and support

The counsellor must aim to be non-judgemental and non-directive towards the consultands when discussing their future options. At best, the counsellor can give the consultand reassurance that the recurrence risk is no greater than the population risk. If the recurrence risk is high, the counsellor must explore the feelings of the consultand towards the disease in terms of the emotional, physical and financial implications. With these in mind the counsellor can provide information about the options open to them such as:

- no further pregnancies – advice about reliable forms of contraception should be given
- prenatal diagnosis – selective termination of affected foetuses
- artificial insemination of donor sperm (AID) – if the male has an AD condition or both partners are carriers for an AR condition
- *in vitro* fertilization (IVF) with pre-implantation genetic diagnosis
- ignoring the risk and coping if an affected child is born.

Follow-up sessions should be offered and, with informed consent, the consultands should be kept on a register so they can be recalled if new prenatal or carrier tests are developed.

Ethical considerations in genetic counselling

Consanguinity and incest

Incest is the mating of first degree relatives. In the UK, double first cousins are the closest relatives allowed to marry (i.e. the two sets of parents are both full siblings). The risk of disease or serious congenital malformation in a child born to first cousins is 1 in 20. This rises to 1 in 11 in a highly inbred family.

Therefore, a detailed anomalies scan is indicated during the pregnancy as well as careful monitoring during and after the pregnancy.

Disputed paternity

Paternity testing uses modern DNA fingerprinting techniques, which use VNTRs (Ch. 7), to generate individual 'barcodes'. The banding pattern that results is unique for each individual and the probability that two unrelated people have the same fragment pattern is less than 3×10^{-11}.

Confidentiality and conflicts of interest

The laws of patient confidentiality must be observed. This can sometimes be a problem if a family has a disease with a very clear pattern of inheritance and only some members of the family want to know their risks of being affected (see pp. 194–195).

GENETIC CONSULTATION AND HISTORY TAKING

An accurate diagnosis is essential if relevant genetic advice is to be given. Some disorders that are superficially similar may have totally different aetiologies and modes of inheritance (e.g. dwarfism). Incorrect assumptions about the underlying diagnosis may, therefore, lead to misleading risk calculations. It is important that information is given in a way that the consultands can understand. It is useful to ask them what they understand about genetics before attempting to explain the nature of the disease.

Outline of the interview

A genetic history is much the same as for any medical condition, and good history taking is paramount. There are certain questions and circumstances that are crucial to explore. For example, particular attention should be given to relatives with relevant disorders. A family tree should be constructed to show how the condition has been passed on through the family. A standardized set of symbols is used, as described in Fig. 8.4. An accurate family tree should reveal the mode of inheritance of the genetic disease and its penetrance and possibly where it originated from in the first place. In this way, realistic risk calculations can be made for the family with regard to future pregnancies.

Enquire specifically about:

- infant deaths, stillbirths and abortions as this may alter recurrence risks (e.g. spina bifida is associated with an increased risk of neural tube defect in subsequent children)
- consanguinity
- non-paternity (discreetly!), as this may explain unexpected disease incidences
- ethnicity and country of descent, as those born in the UK may have a 'hidden' relevant geographical ancestry
- obstetric history (e.g. maternal health, teratogen exposure and viral infections).

COMMON PRESENTATIONS OF GENETIC DISEASE

Some common presentations of genetic disease are given in Fig. 9.9.

TREATMENT OF GENETIC DISEASE

Following genetic screening, testing and appropriate ongoing counselling and support, the final aspect of the management of genetic disease is treatment. Currently, nearly all genetic disorders are incurable, but many are controllable. One of the great hopes of the post-genomic era is that the elusive cures become possible and commonplace.

Current strategies

Current strategies for the management of genetic disorder include:

- conventional supportive treatment
- surgical correction of gross phenotype
- substrate limitation
- environmental modification
- enzyme replacement.

Conventional supportive treatment

Conventional treatment of genetic disorder aims to relieve symptoms and reduce complications. It includes several modalities from pharmacological and surgical treatments to physiotherapy, occupational therapy and speech and language therapy. For example, management of Marfan syndrome requires

Fig. 9.9 Common presentations and associations of genetic disease

Area/system	Example	Association
Development	Social skills	Gregarious personality in Williams syndrome. Inappropriate laughter and absent speech in Angelman syndrome
	Failure to thrive	Pancreatic insufficiency of cystic fibrosis. Metabolic disorders (e.g. mucopolysaccharidoses)
Oral and facial deformities	Hypertelorism	Basal cell nevus syndrome, DiGeorge syndrome and Loeys–Dietz syndrome
	Cleft lip and palate	Chondrodysplasia punctata, trisomy 13 and Pierre Robin syndrome
	Craniosynostosis	Crouzon syndrome
	Low set ears	Patau syndrome, Edwards syndrome, Turner's syndrome
Skin	Ichthyoses	Chondrodysplasia punctate, Conradi's syndrome
	Blistering	Epidermolysis bullosa
Bone and connective tissue	Polydactyly	Patau syndrome; Ellis–van Creveld syndrome
	Brachydactyly	Turner's syndrome
	Rockerbottom feet	Edwards syndrome
Congenital heart disease	Pulmonary stenosis and atrial septal defects	Noonan's syndrome
	Aortic stenosis and pulmonary stenosis	Williams syndrome
	Coarctation of the aorta	Turner's syndrome
	Truncus arteriosus and pulmonary atresia	DiGeorge syndrome
Gastrointestinal disorders	Exomphalos and gastroschisis-exomphalos	Beckwith's syndrome
	Meconium ileus	Cystic fibrosis
	Duodenal atresia	Down syndrome
	Imperforate anus	VATER syndrome
Liver disease	Cirrhosis	Wilson's disease, haemochromatosis
	Cirrhosis and emphysema	α_1-Antitrypsin deficiency
	Isolated jaundice	Gilbert's syndrome
Genitourinary tract disorders	Polycystic kidney disease	Patau syndrome, Edwards syndrome, tuberous sclerosis, Meckel syndrome
	Renal tumours	Wilms' tumour, von Hippel-Lindau syndrome
	Hypogonadism	Klinefelter syndrome, Turner's syndrome and Prader–Willi
Blood disorders	Pancytopenia	Fanconi anaemia
	Disordered coagulation	Haemophilia A and B, von Willebrand's disease
	Immunodeficiency	X-linked Bruton's agammaglobulinaemia, severe combined immunodeficiency

Fig. 9.9 Common presentations and associations of genetic disease.

regular ophthalmology and cardiology review, and may include the prescription of β-blockers to reduce the rate of dilatation of the aortic root and, thus, aortic dissection; annual echocardiogram surveillance to monitor the size and function of the heart and aorta; physiotherapy; mobility aids and home modification; and the prescription of glasses and contact lenses. Surgery is also common, both for prophylactic replacement of the aortic root and for the skeletal complications that arise.

Surgical correction of gross phenotype

Features of certain syndromes can be corrected surgically for cosmetic and functional reasons, such as the correction of cleft lip and palate, and for functional reasons alone in the case of pyloric stenosis and congenital heart defects.

Environmental modification

The severity and effects of some genetic disorders may be reduced by avoiding key environmental compounds. For example, patients with α_1-antitrypsin deficiency may delay the rate and severity of emphysema by avoiding exposure to tobacco smoke. Patients with glucose-6-phosphate dehydrogenase (G6PD) deficiency are advised to avoid certain drugs, such as aspirin and dapsone, in order to prevent intravascular haemolysis. Those with the Mediterranean variant of G6PD should also eliminate broad (Fava) beans from their diet.

Substrate limitation

By limiting exposure to certain key substrates, the effects of genetic disease may be alleviated.

As phenylketonuria patients do not possess phenylalanine hydroxylase, avoiding dietary exposure to phenylalanine from the first month of life ensures normal intellectual development. The effects of Wilson's disease are mediated through deposition of copper in key structures including the basal ganglia and liver. Treatment with long-term penicillamine chelates copper and increases urinary excretion.

Enzyme replacement

Some enzyme-deficient conditions are treatable by enzyme-replacement therapy:

- Type 1 diabetes mellitus is therapeutically managed by replacing the body's inability to produce sufficient natural insulin with exogenous recombinant human insulin.
- The bleeding tendency caused by the congenital deficiency of factor VIII in haemophilia A is corrected by intravenous infusion of a factor VIII concentrate.
- Type 1 Gaucher disease can be treated with a mannose-6-phosphate modified β-glucosidase, allowing the replacement enzyme product to be targeted to macrophage lysosomes, both alleviating symptoms and reducing the severity of organomegaly.

However, enzyme replacement therapy is not the magic bullet once envisaged. In the absence of an efficient enzyme delivery system, enzyme replacement therapy has poor results. For example, types 2 and 3 Gaucher disease are not treatable by enzyme replacement therapy due to lack of a suitable CNS delivery mechanism capable of crossing the blood–brain barrier.

Forward look

Modalities of treatment currently under investigation include:

- gene therapy
- gene-blocking therapies
- stem-cell therapy.

Gene therapy

Gene therapy refers to the genetic alteration of the cells of individuals with genetic diseases to correct the relevant genetic abnormality. It may be achieved by:

- the introduction of the functional gene sequence into target cell DNA
- the introduction of transfected genes expressed from vectors or integrated into the host genome.

Currently, two main methods of gene sequence introduction are utilized:

- Viral. Viruses including oncoretroviruses, lentiviruses, adenoviruses, adeno-associated viruses and herpes viruses have been manipulated to use their inherent ability to introduce genetic material into host cells to deliver recombinant DNA constructs.
- Non-viral. Non-viral delivery methods include the introduction of naked DNA directly in to the cell and the use of liposome-mediated DNA transfer.

Although, in theory, target cells may be either somatic or germline in nature, the ethics of germline manipulation remains controversial (see p. 197), and all current gene-therapy protocols concentrate on somatic gene therapy, in which the alteration of genetic information is targeted to specific cells and are not transmissible to future generations.

ADA-SCID is an autosomally inherited severe combined immunodeficiency disease (SCID).
In the USA, the first somatic gene therapy protocol approved by the National Institutes of Health was for the use of retroviral vectors to target the *ADA* gene into patients' *in vitro* stimulated lymphocytes, which were then injected back into the body. However, as lymphocytes have a limited lifespan, such patients had to undergo reinjections every few months. Research is currently underway to retrovirally insert *ADA* into cultured bone marrow cells, which could then be transplanted into the patient and offer the promise of long-term remission.

For gene therapy to become a more powerful potential treatment method, factors that still need to be surmounted include:

- the short half-life of therapeutic DNA constructs
- the 'host' immune response
- problems with viral vectors – including gene targeting and control issues and the fear of viral reversion to wild type.

Gene-blocking therapies

Traditionally, research into the treatment of gain-of-function or dominant negative (see pp. 155–156) conditions has centred around the use of anti-sense DNA to bind to mRNA and, thus, prevent gene expression (Fig. 9.10), with relatively disappointing results. A more recent research tool, RNA interference (RNAi), offers the possibility to inhibit gene expression in a sequence-dependent fashion to selectively turn off the disease gene by using double-stranded RNA (dsRNA). RNAi was developed from the observation that cells of multicellular organisms utilize an enzyme called 'dicer' to digest viral dsRNA into small fragments, which then become templates, along with RNA-induced silencing complex (RISC), to direct the destruction of single-stranded RNA of the same sequence. By synthesizing double-stranded RNA molecules that correspond to a disease-causing DNA sequence, RNAi can be induced to destroy specific disease-related mRNA sequences (Fig. 9.11).

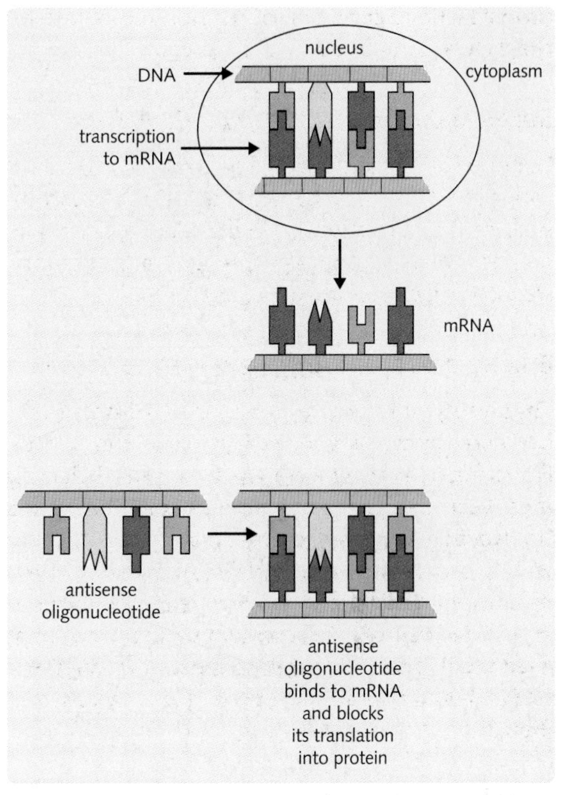

Fig. 9.10 Gene therapy using an antisense technique. Binding of the abnormal mRNA by the antisense molecule prevents it from being translated into an abnormal protein. (Reproduced from Jorde et al., 2006.)

Stem-cell therapy

Rather than replacing defective genes and gene products, stem-cell therapy aims to utilize the pluripotency and multipotency of stem cells to replace defective cells and organ systems. By injecting either embryonic or bone marrow stem cells into the diseased tissue or organ, it is hoped that the stem cells will differentiate into the relevant cell type, replacing the diseased counterparts. It is hoped that stem-cell therapy holds the key to curing, among others, type 1 diabetes mellitus and neurological conditions such as Parkinson's disease. Stem-cell transplants are already common place for some haematological conditions, such as leukaemia.

There is some concern that stem-cell therapy could unwittingly pass viruses and other disease-causing agents to people who receive cell transplants or may, over time, become cancerous.

Stem cells can be harvested from a number of locations. Adult stem-cell transplants can be conducted using multipotent bone marrow or peripheral blood stem cells from either autologous or, more commonly, allogenic sources. Allogenic stem cells bring with them the necessity for a full HLA match and the risk of graft versus host disease (GVHD). Umbilical cord blood stem cells are less differentiated than adult stem cells, are immunogenically immature and are less likely to result in GVHD. The immunogenic immaturity of such cells also means that a perfect HLA-match is not required.

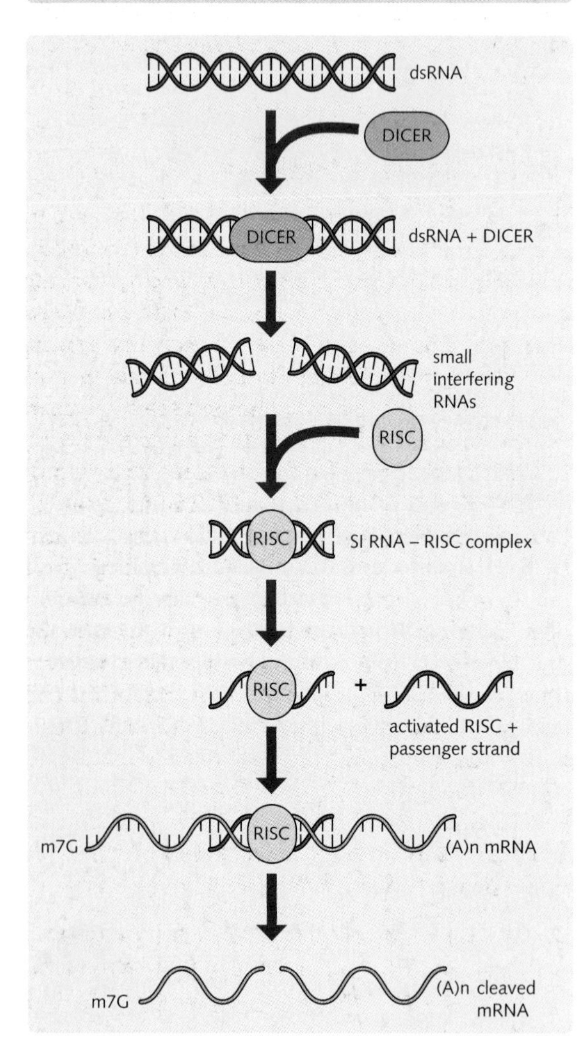

Fig. 9.11 Mechanism of RNA interference. On entering the cell, dsRNAs are processed by dicer (a RNAse III enzyme), producing siRNA. The siRNA is incorporated in to RISC, which cleaves the siRNA and discards the passenger (sense) strand. The anti-sense strand of the siRNA targets RISC to its homologous mRNA sequence, resulting in endonucleolytic cleavage of the targeted mRNA. siRNA, short interfering RNA; RISC, RNA-inducing silencing complex.

ETHICAL ISSUES IN MEDICAL GENETICS

The discipline of genetics attracts ethical debate perhaps unlike any other specialty, as the balance of what could be done is countered by that of what should be done. Advents in modern technology now allows us to potentially cross the border past what is generally taken as morally and ethically acceptable. A full and comprehensive introduction to ethics, ethical theories and practice can be found in *Crash Course, Ethics and Human Sciences.*

Medical ethics and its principles

Ethical principles are the common goals that ethical theory tries to achieve in order to be successful. In medical ethics there are four basic principles:

1. The principle of beneficence – the principle of seeking to do good.
2. The principle of non-maleficence – the principle of seeking, overall, to do no harm.
3. The principle of justice – incorporating fairness and evenhandedness in the context of the resources available, equity of access and opportunity.
4. Respect for patients' autonomy – incorporating respect for the individual and their decisions and highlighting the importance of truth telling, informed consent and confidentiality.

Genethics

Partially as a result of the horrors of World War II and the Nazi drive for genetic supremacy, genetics and ethics have sat somewhat in opposition, as the world shies away from the ghost of eugenics. Ethical dilemmas encountered in the field of medical genetics highlight the importance of genetic counselling (see pp. 187–190) and include issues around:

- presymptomatic and susceptibility testing
- testing of children and adolescents
- preimplantation diagnosis
- prenatal diagnosis and therapeutic abortion
- genetic testing for non-disease traits
- data protection and ownership of genetic data.

Presymptomatic and susceptibility testing

Those individuals known to be at risk of genetic disorder, especially late-onset disorders, may be offered susceptibility testing to determine whether they have inherited relevant mutations before they develop clinical symptoms of the disorder. The usefulness of such tests is affected by the clinical course of the disease and availability of treatment, and the individual's response to the result.

For conditions that do not display full penetrance, one of the most serious limitations of susceptibility testing is interpretation. A positive result only confers that there is a risk, and does not mean that the person will definitely develop the disease for which they carry an associated mutation.

Familial cancers

In the case of familial cancers, a positive result for mutation in causative genes alters the frequency of screening, allowing cancerous changes to be detected earlier, thus improving prognosis. It also allows an individual the chance to consider therapeutic surgical options, such as mastectomy in the case of familial breast cancer.

Huntington's disease

Currently, there is no effective treatment to halt Huntington's disease progression. However, definitive diagnosis may allow an individual to make certain life choices, for instance taking the decision not to have children, or to investigate preimplantation diagnostics (see below) before starting a family. Others simply prefer not to know.

> Mr X Snr has Huntingdon's disease. His granddaughter is thinking of starting a family and wishes to take the test to find out if she is affected before making a decision. Her father, Mr X Snr's son, does not want to be tested or to know his disease status.
> This scenario raises the issue of testing 'by proxy'.
> If Ms X goes ahead and takes the test, and that test comes back positive, it would imply that her father is also positive and is likely to develop the disease before she does. Non-disclosure of such results is difficult within immediate families and Ms X's attendance at counselling or her sudden decision not to have children or even to be sterilized may create severe family tension.

Testing of children and adolescents

Issues surrounding the testing of children are, in part, the same as those for adults, namely the usefulness of the test in guiding treatment options versus the identification of autosomal dominant disease for which there is no known treatment. While the former has very clear benefit to the child, the later constitutes a definite infringement of a child's future autonomy. For example, in the case of children with a mutation in the adenomatous polyposis coli (*APC*) gene, screening for colon cancer by flexible sigmoidoscopy or colonoscopy is indicated from the age of ten and is potentially lifesaving. In the instance of fatal adult-onset autosomal conditions, diagnosis while a child may have a great psychological impact and lead to stigmatization.

Pre-implantation diagnosis

Pre-implantation genetic diagnosis (PGD) is used to diagnose a severe genetic or chromosomal condition in an *in vitro* fertilized embryo to avoid commencing an affected pregnancy. The use of PGD has raised questions about the use of genetic selection, the social power of genetic information and the devaluation of human life based on genes alone. The UK Human Fertilisation and Embryology Authority govern which conditions may and may not be identified by means of PGD (Fig. 9.12). PGD has also has the power to identify an embryo that can serve as a tissue match for a sick child, the so-called 'saviour sibling'. There is no preset list of conditions for which the creation of a saviour sibling is permitted, and each family has to apply individually to be allowed to proceed. This process includes a review of psychological and emotional implications for each individual child and their families.

Prenatal diagnosis and therapeutic abortion

Prenatal diagnosis allows for the detection of both structural and genetic congenital abnormalities. It allows information to be gathered that may influence:

- the management of the rest of the pregnancy, including planning for possible complications with the birth process
- the strategy for dealing with problems that may occur in the newborn infant

Fig. 9.12 Some of the conditions for which the use of PGD has been licensed by HFEA.

Fig. 9.12 Some of the conditions for which the use of PGD has been licensed by HFEA	
Autosomal dominant	
	Charcot–Marie–Tooth
	Crouzon syndrome
	Huntington's disease
	Myotonic dystrophy
	Neurofibromatosis type II
	Osteogenesis imperfecta
Autosomal recessive	
	Autosomal recessive non-syndromic sensorineural deafness
	β Thalassaemia
	Cystic fibrosis
	Plakophilin 1 (PKP1) associated ectodermal dysplasia syndrome
	Sickle-cell anaemia
	Spinal muscular atrophy – types I, II and III
X-linked	
	Adrenoleukodystrophy
	Duchenne muscular dystrophy
	Fabry disease
	Fragile-X syndrome
	Haemophilia A and B
	Hypogammaglobulinaemia
Mitochondrial	
	Mitochondrial myopathies

- the decision whether or not to continue the pregnancy
- the decision whether or not to have more children in the future.

Ethical considerations in prenatal diagnosis include:

- who should be screened and what is the screening threshold that warrants diagnostic testing?
- informed consent – can someone truly understand the implications of each and every condition that may be detected before giving consent?
- the inherent risk of miscarriage associated with diagnostic sample collection by chorionic villus sampling or amniocentesis
- the accuracy, both in terms of specificity and sensitivity, of the test
- when to consider termination, what constitutes a 'serious' defect?

In the UK termination of pregnancy is permitted up to 24 weeks' gestation. This may be extended beyond 24 weeks if the foetus has a lethal condition, if there is a substantial risk that the resulting child would be born with serious handicap or if there is a risk to the mother's life or physical or mental health.

Fig. 9.13 Data protection and ownership of genetic data

Issue	Examples
The misuse of test results	Insurance provision may be limited if a person is shown by genetic test to have a predisposition to chronic ill health and reduced life expectancy. The potential for the creation of a genetic underclass.
Rare mutations with the potential for identifying individuals	Disclosure of data pertaining to very rare mutations and conditions has the potential of breaching a patient's right to confidentiality.
Privacy of genetic information	Who should have access and how widely should genetic data be shared, both within academia and industry?
Ownership of genes and chromosomes	Can you patent life itself? Biomedical research is expensive and commercial companies and their shareholders expect a return on their investment. Some have succeeded in doing so, 'selling' genes to large pharmaceutical companies for hundreds of millions of dollars. Documents such as the Universal Declaration on the Human Genome and Human Rights, although having no basis in law, aim to promote the benefits of altruistic conduct in relation to genetic data – a major principle of the HGP was that the sequence results should be placed in the public domain within 24 hours of their being obtained.

Fig 9.13 Data protection and ownership of genetic data.

- the psychological impact of having to make a 'life or death' decision with regard to your unborn child
- the eradication of disease from the population by 'screening eugenics' and the devaluation of human life.

Genetic testing for non-disease traits

The possibility of screening for lethal and life-limiting conditions, and the completion of the mapping of the human genome, also opens the door for screening for non-disease traits, for instance sex, eye colour, hair colour, athletic endurance and IQ. The automatic assumption often made is that certain phenotypes will be eradicated in a drive for genetic perfection. However, such tests also allow for the selection of mutations associated with disease and disability, such as positive selection for deafness, blindness and dwarfism. Such decisions are usually deeply imbedded in culture and trigger huge debate about the rights of the unborn child against the desires of its parents.

Data protection and ownership of genetic data

Examples of the concerns surrounding the generation, protection and ownership of genetic data can be found in Fig. 9.13.

SELF-ASSESSMENT

Multiple-choice questions (MCQs)

Indicate whether each answer is true or false.

Chapter 1 Cell biology and genetics of prokaryotes

1. **On the prokaryotic cell:**
   a. They lack a membrane bound nucleus.
   b. Cell division is usually through mitosis.
   c. The rate of diffusion is related to cell volume.
   d. Cell surface area expands to the cube of the linear increase of cell diameter.
   e. The larger a prokaryote, the more efficient diffusion of nutrients across the plasma membrane becomes.

2. **On prokaryotic organelles:**
   a. The plasma membrane acts as a selectively permeable membrane.
   b. Mycoplasma possess a thick waxy cell wall.
   c. The large subunit of the bacterial ribosome is '70S'.
   d. Prokaryotic cells do not contain a cytoskeleton.
   e. The flagellum allows bacteria to be motile.

3. **On the transfer of genetic material:**
   a. Transformation is the uptake of DNA from the environment via specific receptors.
   b. Conjugation is the transfer of fragments of DNA from one bacterium to another bacterium by a bacteriophage.
   c. In bacterial conjugation DNA passes into the F^- cell by way of a pilus.
   d. F plasmids always exist as extra-chromosomal bodies.
   e. F′ plasmids are hybrid F plasmids that contain also a segment of the host chromosome.

4. **On prokaryotic DNA replication:**
   a. *Escherichia coli* has an average division time of 40 min at 37°C.
   b. DNA polymerase requires a primer in order to initiate extension of the DNA chain.
   c. The lagging strand is synthesized slowly, but in one continual piece.
   d. DNA chain elongation is achieved via the action of RNA polymerase III.
   e. Okazaki fragments are joined together by DNA ligase.

5. **On prokaryotic transcription:**
   a. Promoters are RNA sequences to which RNA polymerase binds to initiate transcription.
   b. RNAP is a holoenzyme of a core subunit and a σ subunit.
   c. Prokaryotic RNA is synthesized in a 3′–5′ direction.
   d. Intrinsic termination involved the formation of a hairpin at the intrinsic terminator sequence.
   e. Rho-dependent termination is ATP dependent.

6. **On prokaryotic translation:**
   a. tRNA charging is the ATP dependent joining of amino acid and tRNA to form aminoacyl tRNA.
   b. Initiator tRNA is always coupled to the amino acid arginine.
   c. The Shine–Dalgarno sequence is purine rich.
   d. Peptide bond formation is catalysed by peptidyl transferase.
   e. Release factors recognize specific codons, known as termination codons.

7. **On prokaryotic gene regulation:**
   a. Constitutively expressed genes are highly regulated.
   b. An operon is a functionally integrated genetic unit for the control of gene expression in bacteria.
   c. The promoter region serves as the binding site for the repressor protein.
   d. The lactose operon is an example of control by repression.
   e. If tryptophan is present in the environment, the try operon is switched off.

8. **On antimicrobial agents:**
   a. Bactericidal compounds are irreversible in action.
   b. β-lactams inhibit DNA gyrase.
   c. Inhibitors of cell wall synthesis leave cells prone to osmotic lysis.
   d. Ceftazidime is used in the treatment of pseudomonas infection.
   e. Antibiotics with ribosomal toxicity can affect eukaryotic cells.

9. **On viruses:**
   a. The protein viral coat is called the capsid.
   b. The hepatitis B virus is an example of an RNA virus.
   c. Lysogenic infection can result in cancer.
   d. RNA viruses are less prone to mutation than DNA viruses.
   e. Aciclovir is a prodrug, and converted to active substance by a virally encoded enzyme.

Chapter 2 Eukaryotic organelles

10. **The eukaryotic cell:**
    a. Consists of a single cytoplasmic compartment containing all the cellular components.
    b. Divides by mitosis.

c. Receives nutrition by simple diffusion at a rate related to membrane surface area.
d. Has a relatively small surface area to volume ratio.
e. Has a nucleus bound by a single membrane.

11. On the nucleus:

a. The nucleolus contains all of the cell's nuclear DNA.
b. Chromosomes are made up of DNA twisted around carbohydrate molecules.
c. The nuclear membrane contains pores, which allow the passage of mRNA into the cytosol.
d. Nucleolar protein and rRNA complexes are known as chromatin.
e. Most cells have more than one nucleus.

12. On the endomembrane system:

a. The function of the smooth endoplasmic reticulum is to produce mature folded proteins.
b. Ribosomes are fixed to the inner surface of the rough endoplasmic reticulum, so that proteins can be delivered straight into the lumen.
c. The smooth endoplasmic reticulum is involved in the production of steroid hormones.
d. Vesicles travel from the endoplasmic reticulum to fuse to the trans-face of the Golgi apparatus.
e. Secretory cells have a well-defined endomembrane system.

13. On mitochondria:

a. The main function of mitochondria is to detoxify drugs.
b. They have the ability to replicate independently of the cell nucleus.
c. They have a phospholipid bilayer.
d. They are involved in apoptosis.
e. They contain acid hydrolases.

14. On vesicles:

a. Lysosomes carry hormones and neurotransmitters to the cell surface.
b. Macrophages contain many lysosomes.
c. Peroxisomes are involved in the production of bile.
d. Zellweger syndrome results from an accumulation of peroxisomes.
e. Insulin is released from pancreatic islets constitutively.

15. On the cytoskeleton:

a. The cytoskeleton is essential for cell division.
b. Microfilaments are made from tubulin.
c. Centrioles are non-motile extensions of plasma membrane.
d. Primary cilia have a 9+2 arrangement of microtubules.
e. Spermatozoa utilize cilia for propulsion.

16. On cell differentiation:

a. There are six clearly defined tissue types.
b. Similarly specialized cells have the ability to communicate with each other.
c. All cell types are derived from a single cell called the morula.
d. The mature erythrocyte genome is the same as that of a neuron, but different genes are expressed.
e. Differentiated cells have 'memory' and will retain many of its functional characteristics, even if transplanted into a different tissue.

Chapter 3 The cell membrane

17. On the fluid mosaic model:

a. A key concept of the fluid mosaic model is that biological membranes consist of a phospholipid trilayer.
b. Biological membranes are stabilized in part by hydrophobic interactions of the hydrophobic fatty acid molecules in the interior layer.
c. Amphipathic phospholipid molecules are the major component of biological membranes.
d. Integral proteins are only found in the outer layer of the membrane.
e. Biological membranes are static structures, making them very strong.

18. On cell membrane lipids:

a. Lipids are soluble in organic solvent, but only sparingly so in water.
b. Phospholipids are composed of three fatty acid molecules attached to a glycerol backbone.
c. Phosphatidylcholine is not a true phospholipid as it contains a ceramide backbone, rather than a glycerol one.
d. Phosphatidylethanolamine occurs predominantly in the inner monolayer of the plasma membranes.
e. Cholesterol stabilises the plasma membrane by increasing fluidity at physiological pH and decreasing fluidity at low temperatures.

19. On cell membrane proteins:

a. Peripheral proteins possess intracellular and extracellular domains.
b. The cytosolic and extracellular domains of integral proteins are rich in non-polar amino acid residues.
c. Peripheral proteins can be associated with the cytoplasmic leaf of the lipid bilayer.
d. All peripheral proteins are attached to the membrane by covalent linkage.
e. Membrane glycoproteins are conjugated proteins in which the non-protein group is a lipid.

20. On properties of biological membranes:

a. Membranes are more fluid if they have a high proportion of saturated fatty acids.
b. Membranes are more fluid at high temperatures as phospholipid molecules have more kinetic energy.
c. Phospholipid molecules undergo axial rotational movements in the plane of the membrane.
d. Phosphatidylserine is concentrated on the cytoplasmic side facilitating its ability to interact with protein kinase C.
e. Lateral movement of proteins is restricted by interactions with the cytoskeleton.

21. On transport across the cell membrane:

a. Diffusion is the movement of particles from a region of low concentration to a region of higher concentration.
b. The osmotic pressure is the pressure required to prevent the net movement of pure water into an aqueous solution across a semipermeable membrane.
c. One osmol/L of NaCl contains 1 mole of the ion Na^+ and 1 mole of the ion of Cl^-.
d. The range for normal plasma osmolarity is 250–270 mosmol/L
e. Osmolarity is expressed in terms of osmoles of solute per kilogram of solvent.

22. On ionic distributions across the membrane:

a. Distribution of ions across the plasma membrane is controlled, in part, by electrochemical gradients.
b. Polar molecules diffuse freely through the lipid bilayer.
c. Membrane transport proteins are regulated such that only certain molecules can get through at any one time.
d. Pumps are used to maintain energetically unfavourable concentration gradients.
e. There is more sodium inside the cell than out.

23. On transport across the membrane:

a. Passive diffusion does not require energy.
b. Sodium passes through the cell membrane by passive diffusion.
c. Facilitated diffusion by means of carrier proteins shows Michaelis–Menten kinetics.
d. The Na^+/K^+ dependent ATPase pumps two Na^+ ions out of the cell and three K^+ ions in to the cell for every molecule of ATP hydrolysed.
e. Symports transport both ions in the same direction.

24. On transport proteins:

a. Potassium channels are the most common type of ion channel.
b. Carrier proteins transport polar and ionic molecules by active transport and facilitated diffusion.
c. The permeability of a cell is governed by its carrier protein population.
d. The sodium pump is an example of a symport transporter
e. The sodium pump is a heterodimeric protein composed of an α- and a γ-subunit.

25. On membrane potential:

a. Voltage-gated sodium channels play an essential role in the initiation and propagation of action potentials.
b. The Gibbs–Donnan equilibrium can be used to calculate the electrical potential difference required for an ion to be in equilibrium across the membrane.
c. The electrochemical potential difference of chloride is –70 mV.
d. The presence of impermeable ions influence the distribution of permeable ions in an attempt to maintain electroneutrality.
e. The three dominant ions affecting the resting potential of a membrane are Na^+, K^+ and negatively charged proteins.

26. On receptors:

a. Ionotropic receptors are also known as ligand-gated ion channels.
b. The β-adrenergic receptor is an example of an ionotrophic receptor.
c. Enzyme-linked receptors also known as seven pass transmembrane proteins.
d. Activated intracellular receptors bind to hormone responsive elements.
e. Metabotropic receptors activate GTP-binding proteins.

27. On signal transduction:

a. IP_3 is an example of a first messenger.
b. G_s increases cAMP.
c. Second messengers are extracellular molelcules, responsible for activating first messengers.
d. Protein kinase G (PKG) activation is dependent on high levels of cAMP.
e. Second messengers allow signal amplification.

28. The Na^+/K^+ ATPase pump:

a. For every ATP hydrolysed, this transporter pumps two Na^+ ions outward and two K^+ ions inward.
b. The Na^+/K^+ ATPase pump is an example of pure secondary active transport.
c. The pump is a heterodimer consisting of an α-subunit and a glycosylated β-subunit.
d. The binding of sodium to the pump leads to its dephosphorylation.
e. The pump is an important source of energy for co-transport.

29. Receptors and drugs:

a. Agonists bind receptors and stop them transmitting their signal.
b. The effect of competitive antagonists can be reversed by increasing the concentration of agonist.
c. Propranolol is an example of a receptor blocking drug.
d. The binding of a drug to receptor sites other than its intended target is a common cause of side effects.
e. The binding of a drug to its intended target is a common cause of side effects.

Chapter 4 The working cell

30. Concerning microfilaments:

a. They are composed of tubulin dimers.
b. They are a component of the terminal web.
c. They determine axonal diameter.
d. They can undergo 'gel–sol' transitions.
e. They may be bound by α-actinin.

31. Concerning microtubules:

a. They are found in mature erythrocytes.
b. They form the flagella of spermatozoa.
c. They direct the elongation of neuronal axons.

d. They form the nucleoplasmic veil.
e. They extend from organizing centres.

32. Concerning the cytoskeleton:

a. It provides mechanical strength.
b. It maintains erythrocyte shape.
c. Cilia have a 9+2 microtubule arrangement.
d. It has a role in mitosis.
e. It is not involved in phagocyte motility.

33. Concerning lysosomes:

a. They contain neurosecretory products.
b. They are derived from the Golgi.
c. They are found in all cell types.
d. They only digest exogenous material.
e. Their contents are glycosylated in the Golgi.

34. Concerning lysosomal storage diseases:

a. They are multifactoral disorders.
b. They are generally more common in Ashkenazi Jews.
c. They result in a depletion of macromolecules from within the lysosome.
d. Gaucher's disease has an incidence of 1 in 25 000 live births.
e. Tay–Sachs disease results from β-glucosidase deficiency.

35. Concerning cell junctions:

a. Cell–cell contact tight junctions are mediated by the protein occludin.
b. Anchoring junctions are responsible for bladder impermeability.
c. Adherens junctions attach the cytoskeletons of adjacent cells together.
d. Each gap junction is composed of two connexons.
e. Autoantibodies against integrin results in pemphigus.

36. Concerning cadherins:

a. They are calcium dependent.
b. Their C-terminal amino acids are important in ligand binding.
c. P-cadherin is found in nervous tissue and skeletal muscle.
d. Loss of cadherin expression is implicated in cancer.
e. They are associated with catenins.

37. Concerning the immunoglobulin (Ig) superfamily:

a. They have a β-barrel structure.
b. They are Ca^{2+}-independent.
c. They initiate leukocyte–endothelial interactions.
d. ICAM is a member of this family.
e. They are formed by the association of an α and β subunit.

38. Concerning basement membranes:

a. The predominant collagen is type IV collagen.
b. They act as a porous filter in the glomerulus of the kidney.
c. Epithelial cells are anchored to the basement membrane via hemidesmosomes.
d. They line endothelium.
e. They contain integral proteins.

39. Concerning the extracellular matrix:

a. Glycosaminoglycans are hydrophobic.
b. Chondroitin sulphate is found in the skin.
c. Collagen is a double helix.
d. Collagen is secreted by fibroblasts.
e. Elastin provides rigidity to the ECM.

40. Concerning intermediate filaments:

a. Under physiological conditions, IFs dissociate to yield monomers.
b. Neurofilaments are found exclusively in the nucleus.
c. Desmin is found predominantly in muscle cells.
d. Vimentin is found in fibroblasts.
e. Glial fibrillary acidic protein is found in the support cells of neurons.

41. Concerning selectins:

a. Undergo heterophilic binding to carbohydrate ligands.
b. L-selectin is expressed on white blood cells.
c. Sialyated Lewis X is recognized by P-selectin.
d. E-selectin is also known as GMP-140.
e. P-selectin is activated rapidly by histamine.

42. Concerning integrins:

a. Integrins play a crucial role in cell signalling.
b. β-1 integrins are exclusively expressed on leukocytes.
c. The tripeptide sequence RDG is commonly expressed in fibronectin.
d. β-1 integrins are often referred to as VLA molecules.
e. Type 1 leukocyte adhesion deficiency results from the failure of leukocytes to express β-2 integrin.

Chapter 5 Macromolecules

43. On amino acids:

a. All amino acids consist of an amino group (NH_2), a ketone group (CO) and a side group (R).
b. Glycine exists as both D-glycine and L-glycine.
c. Peptide bonds are formed by a condensation reaction between the carboxyl group of amino acid and the amino group of another.
d. Glutamate and serine are examples of non-essential amino acids.
e. Essential amino acids are so called as they cannot be made within the human body and, therefore, must be supplied in the diet.

44. On amino acid structure and properties:

a. Aliphatic amino acids are generally hydrophobic.
b. Large side chains prevent protein bending.
c. Small or hydrophobic amino acids favour β-sheet secondary structure.
d. Disulphide bridges are covalent bonds that form between cysteine residues.
e. In aqueous solution proteins organize themselves so that hydrophobic residues are on the outside.

45. On amino acids as buffers:

a. Zwitterions have a neutral overall electrical charge.
b. The pH at which the net charge on the molecule is neutral is called the 'isoelectric point'.
c. The Henderson–Hasselbalch equation is: $pH = pKa+log[H^+]/[HA]$.
d. In amino acids, only the amino group may act as a buffer.
e. At pH 2.4 the carbonyl group of glycine functions as a buffer.

46. On functions and structures of proteins:

a. Protein conformation is determined by gene sequence.
b. Protein isoforms have almost identical biological activity, but they differ in amino acid sequence.
c. The α-pleated sheet is an example of protein secondary structure.
d. Tertiary structure is determined by interactions between side groups of the constituent amino acid residues.
e. The quaternary structure is the association of two or more polypeptide subunits to form a functional protein.

47. On forces and bonds of protein structure:

a. The peptide bond has a bond energy of 380 kJ/mol.
b. The hydrogen bond is a polar interaction between two electronegative atoms, one acts as a donor and the other as an acceptor.
c. Ionic interactions result from strong interactions between two similarly charged atoms.
d. Small and hydrophobic amino acid residues, such as glycine and proline, favour α-helix formation.
e. Parallel β-sheets are more stable than antiparallel β-sheets.

48. On monosaccharides:

a. Monosaccharides with a ketone carbonyl group are known as ketose sugars.
b. Most naturally occurring monosaccharides are L isomers.
c. Heterocyclic rings with five carbon atoms are called pyranose rings.
d. Monosaccharides are polar molecules.
e. Each cyclic monosaccharide has an α and a β enantomer.

49. On disaccharides:

a. Disaccharides are formed by an addition reaction, yielding a glycosidic bond.
b. α-glycosidic bonds involve C1 of the α anomer.
c. Sucrose is a dimer of glucose and galactose.
d. Maltose is a dimer of two glucose molecules.
e. Cellibiose is a dimer of two glucose molecules.

50. On sugar derivatives:

a. Sugar acids are formed by the oxidation of a sugar to a carboxylic acid.
b. Sugar alcohols are formed by the reduction of the carbonyl group of a sugar to a hydroxyl group.
c. Amino sugars are formed by the substitution of an hydroxyl group for an amino group.
d. Nucleosides are specialized amino sugars.
e. Deoxy sugars are formed from the substitution of one hydroxyl group for a hydrogen atom.

51. On polysaccharides:

a. Have the chemical formula $C_n(H_2O)_n$.
b. Glycogen is a linear storage form of glucose.
c. Cellulose is a polymer of glucose joined by α-1–6 glycosidic bonds.
d. Heteropolysaccharides are polymers of a variety of different sugar monomers.
e. Glycosaminoglycans consist of a repeating disaccharide unit of an amino sugar and a sugar alcohol.

52. On properties of enzymes:

a. They bind specific ligands (substrates) at active sites and catalyse their conversion to products.
b. Isoenzymes are enzymes performing the same function as each other but having a different set of amino acids.
c. The activation energy is the difference in free energy of the reactants and of the products.
d. A catalyst speeds up a chemical reaction but does not itself enter into the reaction.
e. Prosthetic groups are non-protein structures that can be essential to the activity of the protein.

53. On regulation of enzyme activity:

a. In metabolic pathways it is the quickest enzyme that dictates the rate at which the final product is produced.
b. In the 'induced fit' model, substrate binding leads to a conformational change in the active site, which triggers catalysis.
c. Enzymes may be activated and deactivated by covalent modifications, such as phosphorylation.
d. Allosteric regulation involves a molecule other than the substrate binding to the active site, reducing enzyme activity.
e. Enzyme activity may be controlled by proteolytic cleavage.

54. On enzyme kinetics:

a. Increasing enzyme concentration increases the reaction linearly, so follows first-order kinetics.
b. Increasing substrate concentration increases the reaction in a non-linear fashion.
c. For most enzymes, K_m is the substrate concentration at which the reaction rate is equal to V_{max}.
d. Competitive inhibitors do not alter K_m.
e. The plateaux phase of a Michaelis–Menten graph represents first-order kinetics.

Chapter 6 Basic molecular biology and genetics

55. On structures in the nucleus:

a. The nuclear envelope is a lipid monolayer.
b. The nuclear matrix is called the periplasm.

c. Molecules up to 30 kDa pass freely through nuclear pores.
d. A type of intermediate filament called 'lamins' is found in the nuclear matrix.
e. At interphase, actively expressed genes have more histones.

56. On nucleic acids:

a. Nucleotides consist of a nitrogenous base, a sugar molecule and a phosphate molecule.
b. Adenine and thymine are pyrimidines.
c. Uracil is not normally found in DNA.
d. The rate limiting step of pyrimidine biosynthesis is controlled by PRPP synthetase.
e. Carbamoyl phosphate is derived from glutamine and bicarbonate.

57. On DNA packaging:

a. Nucleotide polymers and their associated nucleoproteins are known as 'chromatin'.
b. Euchromatin is inactive, 'resting' chromatin.
c. Nucleosomes consist of 146 bp of DNA wound twice around a heptamer of histone proteins.
d. The solenoid is also commonly known as the 'beads on a string' arrangement.
e. Metacentric chromosomes have centromeres that are off-centre.

58. On gene structure:

a. Promoters are found upstream of a gene.
b. The TATAAT box is the most conserved sequence in all eukaryotic promoters.
c. Exons are never lost to splicing and are always represented in mature mRNA transcripts.
d. Stop codons are found within the 5′ sequence of a gene.
e. Pseudogenes are not transcribed to yield a functional protein.

59. On DNA replication:

a. Replication always proceeds in the 5′to 3′ direction, with new nucleotides being added to the 3′ OH of the growing molecule.
b. Eukaryotic origins of replication commence replication simultaneously during S phase of the cell cycle.
c. Polymerase α has primase activity.
d. PCNA serves as a co-factor for DNA polymerase δ during DNA synthesis.
e. Okazaki fragments are joined together using a DNA endonuclease.

60. On eukaryotic transcription and post-translational modification:

a. Transcription is catalysed by DNA-dependent RNA polymerases.
b. *Trans* acting factors are specific sequences of DNA that lie on the same molecule of DNA as the gene they regulate.
c. All genes that are transcribed and expressed via mRNA are transcribed by the RNA polymerase II complex.
d. Polyadenylation of mRNA regulates its nuclear export.
e. The spliceosome is a complex consisting of RNA and small nuclear ribonuclear particles (snRNPs).

61. On protein synthesis:

a. Eukaryotic translation is monocistronic in nature.
b. The genetic code is degenerate.
c. Eukaryotic translation is almost always initiated at internal ribosome entry sites.
d. Housekeeping genes are so called because they are constitutively expressed.
e. Glycosylation is an example of a post-transcriptional modification.

62. On the cell cycle:

a. The S phase of the cell cycle contains the restriction point, which denotes the start of the cell cycle.
b. Quiescent cells are held in G2, preventing cell division from occurring.
c. The CDK-dependent phosphorylation of Rb protein controls the passage from G_1 to S phase.
d. INK4 is an example of a cyclin-dependent kinase.
e. The main point of p53 action is thought to be the G_2–M DNA integrity checkpoint.

63. On mitosis and meiosis:

a. Mitosis is also known as reduction division.
b. Mitotic chromosomes are maximally visible at anaphase.
c. The exchange of genetic material occurs during diplotene.
d. Meiotic chromosomes separate in anaphase II.
e. In oogenesis, the first polar body contains 46 chromosomes, while the two second polar bodies each contain 23.

64. On DNA damage and repair:

a. The spontaneous deamination of adenine to produce hypoxanthine is an example of an exogenous mutagen.
b. Double-stranded DNA breaks can lead to chromosomal translocations.
c. If a normal cell cannot repair DNA it will advance the cell cycle.
d. HNPCC is an inherited mutation of a *BER* gene.
e. NHEJ is the least error-prone method of repairing single-stranded DNA damage.

65. On eukaryotic RNA molecules:

a. Heteronuclear RNA (hnRNA) is complementary in sequence to the DNA template.
b. Transfer RNA is required for ribosome shape.
c. The final mRNA transcript contains intronic and exonic information.
d. 20% of t-RNA bases undergo post-translational modification.
e. rRNAs are synthesized in the ribosomes.

66. On telomeres and telomerase:

a. Telomeres are tandem repeats of the heptameric sequence 'TATAGGG'.
b. The T-loop is an important part of telomere structure and function.
c. Telomeres are required for chromosome integrity.
d. Telomerase is active in all somatic cells.
e. Telomerase is a reverse transcriptase.

Chapter 7 Tools in molecular medicine

67. On isolation and preparation of nucleic acid:

a. Restriction enzymes always cut to give blunt ends.
b. Restriction enzymes that recognize eight base pair sequences cut less often than those which recognize four base pair sequences.
c. DNA is positively charged, and so moves towards the negative electrode when electrophoresed.
d. Ethidium bromide, which fluoresces under ultraviolet light, is used to visualize DNA products on a gel.
e. PFGE can be used to separate molecules of several Mbs in size.

68. On PCR:

a. Is routinely used to amplify DNA segments of several Mbs in size.
b. Requires a heat stable polymerase.
c. Involves DNA synthesis in a 5′–3′ direction.
d. The optimum working temperature for *Taq* polymerase is 72°C.
e. RT-PCR uses DNA as a template to produce DNA.

69. On nucleic acid and protein hybridization techniques:

a. Blotting allows the transfer of biomolecules from a gel to a membrane.
b. Southern blotting is quicker than PCR.
c. Southern blotting is commonly used clinically to analyse RFLPs.
d. Western blotting describes RNA hybridization.
e. Northern blotting can be used to investigate gene expression levels.

70. On DNA sequencing and polymorphisms:

a. Sanger sequencing relies on dideoxynucleotide triphosphates.
b. The use of fluorophore-labelled dNTPs means only one sequencing reaction needs to be conducted.
c. SNPs are more common in coding-regions of the genome.
d. For a sequence variation to be a true polymorphism, it must occur in 10% of the population.
e. Minisatellites are generally GC rich.

71. On cytogenetics:

a. G-banding is so called as it involved visualizing chromosomes in the G2 stage of the cell cycle.
b. G-banding can be used to detect chromosomal microdeletions.
c. FISH uses fluorescently labelled probe DNA to detect the presence or absence of specific chromosomal sequences.
d. Comparative genomic hybridization techniques detect deletions as small as 1 kb.
e. Multiplex ligation-dependent probe amplification can detect copy number variation in genomic sequences at high resolution.

72. On cloning and cloning techniques:

a. Cloning uses a host microorganism to amplify DNA.
b. BAC vectors can accommodate 1000 kb sections of DNA.
c. Plasmids are viruses that infect bacteria.
d. Blue/white selection is a method for detecting clones containing insert DNA.
e. PCR can be used to generate sequences for cloning.

73. On maps:

a. Physical maps are measured in centimorgans.
b. 1 cM is equivalent to a 10% chance of recombination.
c. Male and female genetic maps are the same.
d. The closer polygenic markers are on a map, the higher the resolution.
e. The full human genome sequence is an example of a physical map.

74. On the human genome project:

a. The 'final' draft of the human genome published in 2006 has an estimated 1 error in every 10 000 nucleotides.
b. One of the aims of the HGP is comparative genomics.
c. The human genome has an estimated 50 000 protein-coding genes.
d. DNA informatics uses protein sequence to model structure and function.
e. The metabolome refers to the total protein complement of a genome.

75. On genetic linkage analysis:

a. The closer two loci are to each other, the more likely they are to segregate to each other
b. Chiasmata form between homologous chromosomes during mitosis.
c. If two loci are linked, their recombination fraction will be less than 0.5.
d. A recombination fraction of 0.01 is equivalent to 1 cM.
e. A LOD score of 10.0 is the minimum required to confirm linkage.

76. On gene therapy:

a. Germ-line gene therapy is universally accepted as more ethically sound than somatic gene therapy.
b. Gene therapy can only be used to target inherited genetic disease.
c. The disease gene under consideration for gene therapy should be cloned and fully characterized.
d. A good vector needs to be immunologically detectable.
e. Liposomes are viruses that have had the genes required for the replication phase of their life-cycle replaced with insert DNA.

Chapter 8 Genetic disease

77. On mechanisms of mutation:

a. Mutations are transient errors within genetic material.
b. Transversion substitution is an interchange between a purine and a purine or a pyrimidine and a pyrimidine.

c. Sequences at the ends of deletion mutations are highly recombinogenic.
d. Insertions or deletions in multiples of three nucleotides lead to frameshift mutations.
e. Cytosine and guanine-rich trinucleotides are most commonly implicated in trinucleotide repeat expansions.

78. On structural and functional effects of mutation:

a. Point mutations always lead to defective gene products.
b. Missense mutations lead to the conversion of a codon to a stop codon.
c. Nonsense mutations lead to a truncated protein product.
d. Leaky mutations may result from an altered amino acid that makes the polypeptide less active.
e. Dominant negative mutations may result from a mutant allele product of a heterozygote inhibiting the wild-type allele product.

79. On dominant monogenic disorders:

a. In autosomal dominant diseases affected individuals generally have affected children
b. Homozygotes for autosomal dominant conditions are very rare.
c. Disorders displaying haploinsufficiency are autosomal dominant.
d. Affected males have normal sons, but all daughters are affected is consistent with X-linked dominant inheritance.
e. Duchenne muscular dystrophy is an example of an X-linked dominant condition.

80. On recessive monogenic disorders:

a. People with autosomal recessive disease are usually homozygotes.
b. If a carrier of an autosomal recessive disease has an unaffected partner, there is a 50% chance of their children being carriers.
c. Females can never be born with X-linked recessive conditions.
d. All children of female carriers of X-linked recessive conditions have a 50% chance of being carriers themselves.
e. There is no male-to-male transmission of X-linked recessive conditions.

81. On non-Mendelian inheritance of single-gene disorders:

a. Anticipation describes the progressively earlier emergence of a condition with each generation.
b. Progressive external ophthalmoplegia is an example of a disease showing anticipation.
c. Uniparental disomy describes the controlled differential expression of an allele, depending on whether it is maternally or paternally derived.
d. Prader–Willi syndrome usually arises as a result of the switching off of the critical region on the paternally derived chromosome.
e. Affected males generally do not transmit mitochondrial mutations.

82. On chromosome disorders:

a. Numerical chromosomal disorders result in gene-dosage effects.
b. Monosomies may result from anaphase lag.
c. Patau syndrome results from 3 copies of chromosome 18.
d. Individuals with trisomy X syndrome are typically of short stature and have a short webbed neck and wide spaced nipples.
e. de la Chapelle syndrome describes an XX male.

83. On multifactoral disorder inheritance:

a. Multifactorial disorders show Mendelian inheritance.
b. Multifactorial traits generally show a Gaussian distribution within a population.
c. Concordance describes the presence of a trait in sets of individuals.
d. The population incidence of a multifactorial condition is equivalent to the proportion whose liability is greater than the threshold.
e. Heritability is the proportion of the total phenotypic variation that is genetic in origin in a given population.

84. On multifactoral disorders:

a. The human HLA complex and the insulin gene are considered to be the loci that contribute the most to T1DM susceptibility.
b. Dietary sodium intake is an important environmental factor in the pathogenesis of primary hypertension.
c. Mutations in the amyloid precursor protein increase your susceptibility to Alzheimer disease by 20%.
d. Key polymorphisms in the *HLA-DRB1* gene have been demonstrated to be strongly correlated to rheumatoid arthritis susceptibility.
e. The heritability of asthma is 10%, with environmental factors accounting for 90% of all cases of the disease.

85. On Down syndrome:

a. Down syndrome (DS) is always caused by trisomy 21.
b. People with DS have an increased risk of developing leukaemia.
c. The link between DS and maternal age has recently been disproved.
d. Aplasia cutis is characteristic of DS.
e. The current acceptable medical term to describe the features of DS is 'mongolism'.

86. On cancer genetics:

a. Most cancers are familial.
b. Carcinogenesis is a multi-stage process.
c. Proto-oncogenes are pro-carcinogenic.
d. Tumour suppressor genes are also known as proto-oncogenes.
e. Oncogenes tend to be dominant at the cellular level.

87. On cancer syndromes:

a. Two family members with the same rare cancer may signify a cancer syndrome.
b. Ataxia telangiectasia results, in part, from an inability to arrest the cell cycle in the presence of DNA damage.

c. *BRCA1* and *BRCA2* may be mutated in familial ovarian cancer.
d. *APC* is an example of a proto-oncogene.
e. Patients with Gorlin syndrome develop squamous cell carcinomas.

88. On chromosomal disorder nomenclature:

a. In the ISCN classification, translocations are denoted by '(t)'.
b. 46,XY,inv(9)(p12,q14) describes an inversion of a segment of chromosome 9 between position 12 of the long arm and position 14 of the short arm.
c. 45,XO describes a monosomy.
d. In the ISCN classification, duplications are denoted by '(d)'.
e. In the ISCN classification '(r') denotes a ring chromosome.

Chapter 9 Principles of medical genetics

89. On the Hardy–Weinberg equilibrium:

a. The Hardy–Weinberg law assumes that an organism is diploid, sexually reproducing and has discrete generations.
b. The Hardy–Weinberg equilibrium states that allele frequencies remain constant from one generation to the next, providing there are active selection forces.
c. The Hardy–Weinberg equilibrium can be represented as $p^2+pq+q^2=1$
d. AR disorder with a frequency of 1 in 1600 live born births has a heterozygote frequency of approximately 1 in 60.
e. For autosomally dominant conditions, disease frequency is approximately equal to the frequency of heterozygotes.

90. On factors that disturb Hardy–Weinberg equilibrium:

a. Assortative mating refers to the random selection of a mate.
b. High mutation rates always lead to an increase in the number of mutant alleles within a population.
c. The heterozygous advantage is an example of positive selection.
d. Migration can lead to the introduction and the removal of alleles to or from a population.
e. The founder effect describes the changes in allele frequencies that occur when a subpopulation moves from a larger one to form a new colony.

91. On screening:

a. Population screening requires a positive history of a disorder.
b. The Wilson–Hardy criteria define the minimum criteria for a successful screening programme.
c. Before a condition is considered for a screening programme, the natural history of the condition should be understood.
d. Multiple of the median calculations are used to assign a risk to a pregnancy based on the level of a specific maternal serum marker.
e. Cascade screening describes the offering of genetic tests to the family of an affected individual.

92. On methods used for detection of carriers and sufferers:

a. Genetic linkage is an example of a definitive test for a genetic condition.
b. The newborn blood spot test is used to diagnose Tay–Sachs disease.
c. Direct mutation detection can be achieved using PCR.
d. Antenatal screening for cystic fibrosis involves amniocentesis or chorionic villus sampling.
e. Sickle-cell anaemia can only be diagnosed using the Sickledex test.

93. On taking a genetic history:

a. Consanguinity has no real effect on risk of genetic disorder.
b. The symbol for an affected male is a filled in circle.
c. The 'index' case of a genetic disease is known as the consultand.
d. The obstetric history of a mother can be significant in a genetic history.
e. Unexpected genetic patterns can always be explained by disputed paternity.

94. On risk assessment:

a. A child with two carrier parents has a 50% risk of being affected.
b. The risk of transmission of an autosomal dominant condition, where one parent is affected, is always 1/2.
c. All girls born to fathers with X-linked recessive disease are affected.
d. The law of addition is sometimes called the 'or' law.
e. It presents risks as absolute values.

95. On Bayes' theorem:

a. It modifies the risk of an individual having a condition with a real time probability of them not having it.
b. It presents risks as absolute values.
c. Careful family history taking is not important, as the theory incorporates the results of screening tests.
d. It is a method for determining risk in non-Mendelian inheritance.
e. A 'joint' probability is calculated as the product of the prior and the conditional probabilities.

96. On genetic counselling and consultation:

a. Genetic counselling should be directive and authoritative.
b. The risk of a serious congenital malformation in a child from a highly inbred family is 1/11.
c. As parents transmit one of their two alleles to their offspring, a child's VNTR banding patterns will be identifiable in the fingerprint of their biological parents.
d. A physical examination is not necessary during a genetic consultation.
e. It is important to enquire about ethnicity and country of descent in genetic consultations.

97. On common presentations of genetic disease:

a. Dysmorphic facies are common in genetic conditions.
b. Hypotelorism refers to an increased interpupillary distance.
c. Cataracts may be seen in individuals with Down syndrome.
d. Congenital sensorineural deafness is always X-linked.
e. Osteopetrosis refers to a decrease in bone mass and density, causing them to break more easily.

98. On current treatment techniques for genetic disease:

a. Supportive therapy is multidisciplinary.
b. Avoiding certain environments is a management strategy for certain genetic disorders.
c. The only indication for surgical correction of gross phenotype is cosmetic.
d. Treatment of phenylketonuria uses the drug penicillamine to chelate excess dietary phenylalanine.
e. The addition of mannose-6-phosphate to β-glucosidase allows it to be targeted to the CNS.

99. On future treatment techniques for genetic disease:

a. Gene therapy is not currently an option for treating any genetic diseases.
b. The short half life of therapeutic DNA constructs limits the potential of gene therapy.
c. Therapeutic gene sequences can be introduced into host cells using wild-type viruses.
d. RNAi can be used to block specific disease related mRNA sequences.
e. Bone marrow transplants are an example of stem-cell therapy.

100. On genetic ethics, its theories and principles:

a. Deontology describes the maxim 'greatest good for the greatest number'.
b. The principle of non-maleficence describes the principle of seeking to do good.
c. Informed consent and confidentiality are integral to the principle of justice.
d. Presymptomatic diagnosis always influences management of the condition.
e. In the UK, an abortion after 24 weeks can only be conducted if the foetus has a lethal condition.

Short-answer questions (SAQs)

1. Mrs H has asthma. She is feeling a little wheezy and so uses her salbutamol inhaler. What type of receptor does it act on, what family does it belong to and how does this family have its effect?

2. What requirements need to be fulfilled before a gene therapy trial can commence?

3. Briefly describe the transfer of genetic material by bacterial cells.

4. Ms S has a tingling sensation on her lip. The last time this happened, it was followed by a cropping of small, fluid filled blisters. She applies an antiviral to the area, which prevents the blister formation. What condition does she have, what antiviral is used, and what is the mode of action?

5. What are peroxisomes and what are there functions?

6. Familial adenomatous polyposis coli is an inherited cancer susceptibility syndrome. What is its familial basis? Give three factors that would make you suspicious of inherited cancer susceptibility within a family.

7. Deficiency in β-glucosidase leads to Gaucher's disease. Describe the classification of Gaucher's disease how each type is defined.

8. Describe the levels of DNA packaging, from DNA strand to chromosome.

9. What are proto-oncogenes and tumour suppressor genes? How are they implicated in cancer?

10. What are the six main aims of the human genome project (HGP)?

11. Mr and Mrs X come to see you in your clinic very distressed. At 38 years old, Mrs X is pregnant with her fifth child. She tells you that her eldest son, who had cystic fibrosis (CF), has just died aged 20 of a respiratory infection. The couple's three other children are not affected with this condition, but Mrs X tells you she has a bad feeling about this pregnancy, and she is worried the foetus may have CF. She asks if you can offer her prenatal diagnosis. To this end you screen her and her husband for the 10 most common CF mutations, but they do not have them. You conclude that they are carriers of rare CF mutations and are relieved to discover that they are both heterozygous for a very closely linked microsatellite marker. You obtain DNA from all the surviving family members and the foetus (CVS sample) and genotype them with respect to this marker. Does the foetus have CF?

12. Compare and contrast prokaryotic and eukaryotic ribosomes.

13. Miss D is 5 years old. Her mother has recently been told by a geneticist that she has Angelman syndrome. What is the molecular basis of this disorder, and name three features commonly associated with the condition.

14. What is the Hardy–Weinberg equilibrium, how is it denoted, and what criteria must be satisfied for it to apply?

15. Draw the common structure of an amino acid.

16. Mr E is recovering from a bout of *S. aureus* food poisoning. Despite his best efforts, he has become fairly dehydrated. What is the effect on his serum osmolality, what hormone is released in response and what is its target?

17. What is situs inversus and which structure has been implicated in its pathogenesis?

18. Draw a diagram of:

    a. A Gram-positive bacterial cell wall

    b. A Gram-negative cell wall.

19. Mrs F and Mr F are first cousins. They are to be married in 6 months time. Ms F has a first-class degree

in biological sciences, and recalls from her genetics lectures that the offspring of first cousins could be at greater risk of genetic diseases. They wisely decide to consult you, a genetic counsellor, prior to marriage to determine if their future children will be at a higher risk of genetic disease. What information do you need and what is the generally accepted risk of disease in consanguineous relationships and what advice might you give this couple?

20. Name four features that, as a physician, would alert you to the possibility that your patient has a lysosomal storage disorder. Name one such disorder.

1. Theme: prokaryotic structure and function. Options:

a. 80S
b. Acid fast
c. Conjugation
d. FtsZ
e. Pili
f. Hfr
g. Transduction
h. Reverse motor protein 1
i. 70S
j. Filament
k. Gram-negative
l. Prong
m. Transformation

1. The receptor mediated uptake of DNA by bacteria.
2. Part of the flagella apparatus.
3. A bacterial cytoskeletal protein.
4. Bacterial ribosomes are described as being this.
5. Lipopolysaccharides are found in cell walls with this staining property.

2. Theme: anti-microbial chemotherapy. Options:

a. RNA dependent DNA polymerase
b. Rifampicin
c. Saquinavir
d. RNA polymerase activity
e. Metronidazole
f. Ciprofloxacin
g. Isoniazid
h. Cefuroxime
i. Aciclovir
j. Teicoplanin
k. Trimethoprim

1. Inhibitors of nucleic acid synthesis can act here.
2. This anti-viral inhibits DNA polymerase.
3. This compound is a glycopeptide.
4. This is a β-lactam antibiotic.
5. This antibiotic works by inhibiting bacterial DNA gyrase.

3. Theme: receptors and signalling. Options:

a. Metabotropic G protein-coupled receptor
b. G_i
c. Ionotropic receptor
d. Apocrine
e. Hormone sensitive lipase
f. G_l
g. Endocrine
h. Tyrosine-kinase associated receptor
i. G_s
j. Hormone response elements
k. Autocrine
l. Receptor guanylyl cyclases
m. G_q

1. This molecule is an example of an enzyme-linked receptor.
2. This receptor directly affects the activity of a cell by opening the ion channel.
3. This receptor activates the inositol lipid pathway.
4. Activated intracellular receptors bind to this.
5. In this method of cell signalling, the secretory cell is also the target cell.

4. Theme: cellular proteins. Options:

a. Globular actin
b. β-1 integrin
c. β-tubulin
d. Keratin
e. Lamins
f. Ankyrin
g. Kinesin
h. Cathepsins
i. β-galactosidase
j. Cadherins
k. β-2 integrin
l. Chondroitin sulphate

1. A glycosaminoglycan.
2. These proteins are only found in the nucleus.
3. A type of intermediate filament.
4. Lysosomal proteases.
5. Expressed exclusively on leukocytes.

5. Theme: mechanisms of genetic disease. Options:

a. Familial hypercholesterolaemia
b. Prader–Willi syndrome
c. Vitamin D resistant rickets
d. Charcot–Marie–Tooth disease type 1a
e. Haemochromatosis
f. Developmental dysplasia of the hip
g. Haemophilia A
h. Fabry's disease
i. Angelman syndrome
j. Leber hereditary optic neuropathy
k. Beckwith–Wiedemann syndrome

1. An autosomal recessive disease.
2. A multifactoral condition.
3. The genes responsible for this condition are mitochondrial.
4. An example of an X-linked dominant condition.
5. This condition results from the loss of imprinted genomic material from the paternal chromosome.

6. Theme: genetic disease. Options:

a. Multiple endocrine neoplasia syndrome
b. Familial adenomatous polyposis coli
c. Cystic fibrosis
d. Li-Fraumeni syndrome
e. Gaucher's disease
f. Ataxia telangiectasia
g. Essential hypertension
h. Hereditary spherocytosis
i. Phenylketonuria
j. Hereditary non-polyposis colon cancer
k. Tay–Sachs disease
l. Friedreich's ataxia

1. Mrs Halphen, an Ashkenazi Jew, has presented to the GP with her 10-month-old son, Daniel. About 2 months ago she noticed he stopped crawling. Mrs Halphen is also worried that Daniel no longer seems to focus on her, saying it is 'as if he can't see me'.
2. Aspene was born in the Philippines 4 years ago, shortly before her parents emigrated to Slough. Her mother has brought her to see their GP as she is worried that Aspene has eczema. Her mum says that she is a very hyperactive child, and that she has no idea where she gets her light coloured hair from. The GP examines her and notes a distinctive mousey odor.
3. Jason is 25 years old. His father, grandfather and three paternal aunts all died of bowel cancer in their late 40s. Blood tests confirmed that he carries a mutation in the gene *MSH2*, and he is attending clinic for a screening colonoscopy.
4. You are the respiratory F1. You have been called to the bereavement office to complete the paperwork for a patient who died last night. Nisha was diagnosed with this disorder at the age of 6, and had been wheelchair bound since the age of 11. Her parents first noticed that she was clumsy and unsteady on her feet, and her decline had been rapid, with loss of ability to walk within a year. In her short life she had many infections, including the bronchopulmonary pneumonia that killed her. She died aged 20. What underlying condition should also be recorded on the death certificate?
5. You are a clinical geneticist. You are reviewing the case of Gareth R, a 3 month old male, whose parents you are due to see in clinic tomorrow. His 'blood spot' test was positive for this condition, and further analysis revealed a mutation (Δ-F508) in the *CFTR* gene.

7. Theme: eukaryotic organelles. Options:

a. Rough (granular) endoplasmic reticulum
b. Mitochondria
c. Peroxisomes
d. Nucleus
e. Microfilaments
f. Smooth (agranular) endoplasmic reticulum
g. Golgi apparatus
h. Nucleoli
i. Microtubules
j. Lysosomes
k. Centrioles
l. Ribosomes

1. An organelle required for the degradation of fatty acids.
2. This part of the cytoskeleton is comprised of tubulin.
3. A major function of this organelle is oxidative phosphorylation.
4. This organelle is involved in the manufacture of steroid hormones.
5. This organelle is the site of rRNA biosynthesis.

8. Theme: chromosomal disease. Options:

a. Down syndrome
b. Trisomy X syndrome
c. (t)
d. (inv)
e. (I)
f. Edwards syndrome
g. Patau syndrome
h. Turner's syndrome

i. Klinefelter syndrome
j. de la Chapelle syndrome

1. The most commonly occurring trisomy.
2. The most common cause of primary hypogonadism in males.
3. The only viable monosomy in humans.
4. Death in people with this condition is usually associated with cardiac disease or acute leukaemia.
5. Males with this condition have two X chromosomes, one of which contains Y-specific sequences.

9. Screening and risk assessment. Options:

a. Non-random mating
b. VNTR
c. Incest
d. A latent or early stage is not required
e. Bayesian analysis
f. The natural history of the condition should be understood.
g. High migration rate
h. Human gonadotrophin
i. FISH
j. Maternal serum α-fetoprotein
k. RNAi
l. Large population size
m. One-off screening is permissible
n. dimeric inhibin-A

1. A requirement for a population to be in Hardy–Weinberg equilibrium.
2. One of the WHO criteria for a screening programme.
3. Used in cases of disputed paternity.
4. This marker is high in cases of spina bifida.
5. Consanguinity is an example of this.

10. Features associated with genetic disease. Options:

a. Mandibular hypoplasia
b. Craniosynostosis
c. Rockerbottom feet
d. Pulmonary stenosis
e. Blue sclerae
f. Coarctation of the aorta
g. Macro-orchidism
h. Squamous cell carcinoma
i. Aortic stenosis
j. Imperforate anus
k. Micro-orchidism
l. Cavernous haemangiomas
m. Basal cell carcinoma

1. Osteogenesis imperfecta.
2. Pierre Robin syndrome.
3. Gorlin syndrome.
4. Fragile X syndrome.
5. VATER syndrome.

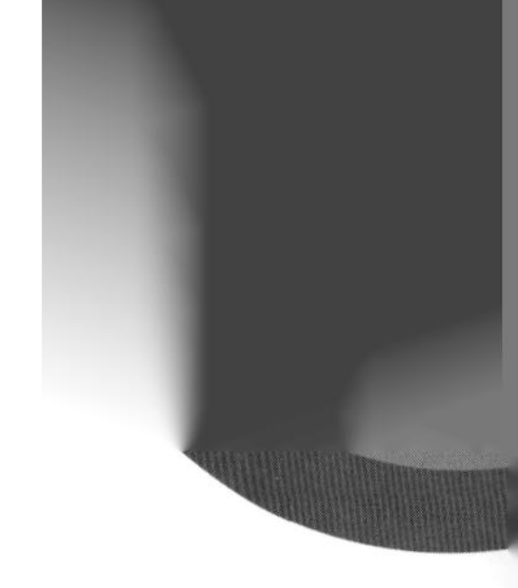

1. On the prokaryotic cell:

- **a.** True they lack membrane bound organelles.
- **b.** False cell division is usually by means of binary fission.
- **c.** False it is related to membrane surface area.
- **d.** False it expands to the square of the linear increase.
- **e.** False smaller cells have a larger surface area to volume ratio, making them more efficient.

2. On prokaryotic organelles:

- **a.** True as per biological membrane.
- **b.** False mycoplasma do not possess a cell wall.
- **c.** False the whole bacterial ribosome is described as being '70S', the large subunit is '50S'.
- **d.** False it is now recognized that they do, and actin- and tubulin-like molecules have been identified.
- **e.** True it enables both positive and negative chemotaxis.

3. On the transfer of genetic material:

- **a.** True if the foreign DNA is compatible it can become incorporated into the bacterial genome.
- **b.** False this describes transduction.
- **c.** True DNA passes from F^+ cells to F^- cells via a sex pilus.
- **d.** False on occasion the F plasmid may integrate into the bacterial genome, resulting in Hfr cells.
- **e.** True they result from integrated F plasmid DNA 'popping' back out of the genome, taking with it host chromosomal DNA.

4. On prokaryotic DNA replication:

- **a.** False at 37°C, *E. coli* divides every 20 min.
- **b.** True the primer is synthesized by RNA polymerase.
- **c.** False the lagging strand is synthesized in short pieces, which are then ligated together.
- **d.** False DNA polymerase III.
- **e.** True the short pieces of DNA are joined together to form one continuous strand.

5. On prokaryotic transcription:

- **a.** False they are DNA sequences.
- **b.** True the core subunit and σ subunit form the initiation complex.
- **c.** False 5′ to 3′.
- **d.** True the intrinsic terminator sequence is a CG rich area, followed by six or more adenines.
- **e.** True Rho is an ATP dependent hexamer.

6. On prokaryotic translation:

- **a.** True also formed is AMP and P_i.
- **b.** False in prokaryotes, the first amino acid is formylmethionine (f–met).
- **c.** True it binds a complementary pyrimidine rich sequence in the small ribosomal subunit unit.
- **d.** True this occurs in the 50S subunit.
- **e.** True these codons are UAA, UAG and UGA.

7. On prokaryotic gene regulation:

- **a.** False constitutive genes are switched on all the time.
- **b.** True genes that encode the enzymes of a metabolic pathway are usually clustered together on the chromosome.
- **c.** False the operator region is the repressor binding site.
- **d.** False it is the lac operon.
- **e.** True tryptophan acts as a co-repressor enabling the tryptophan repressor to bind to the operator.

8. On antimicrobial agents:

- **a.** True bactericidal compounds kill bacteria.
- **b.** False quinolones inhibit DNA gyrase.
- **c.** True bacteria concentrate dissolved nutrients through active transport. Without a strong cell wall, the bacterium bursts from the osmotic pressure of the water flowing into the cell.
- **d.** True pseudomonas is a common cause of infection in the immunocompromised.
- **e.** True although eukaryotic ribosomes are not affected by antibiotics that target prokaryotic ribosomes, the ribosomes of mitochondria can be.

9. On viruses:

- **a.** True the caspid may also be associated with a phospholipid envelope.
- **b.** False HBV is a DNA virus.
- **c.** True one of the fates of lysogenic infection is cell transformation.
- **d.** False RNA viruses do not have the proofreading capability of viral DNA polymerase.
- **e.** True aciclovir is converted to aciclovir monophosphate by α-herpesvirus encoded thymidine kinase.

10. The eukaryotic cell:

- **a.** False The eukaryotic cell is composed of a series of inner membranes that separate the cell into distinct compartments that perform specific functions.

- **b.** True Eukaryotic cells divide by mitosis.
- **c.** False Eukaryotic cells derive nutrition via membrane bound vesicles, that pinch off from the plasma membrane.
- **d.** True As eukaryotic cells are not constrained by simple diffusion, they can have a smaller surface area to volume ratio than an equivalent prokaryotic cell.
- **e.** False the nuclear membrane is a double membrane.

11. On the nucleus:

- **a.** False the nucleus contains the cell's nuclear genome.
- **b.** False chromosomes are composed of DNA wrapped around protein.
- **c.** True the nucleus communicates with the cytoplasm by way of nuclear pores.
- **d.** False DNA and histones make up chromatin.
- **e.** False most cells just have one nucleus.

12. On the endomembrane system:

- **a.** False that is the function of the rough endoplasmic reticulum.
- **b.** False the ribosomes are on the outer surface of the rough endoplasmic reticulum.
- **c.** True the smooth endoplasmic reticulum is involved in the production of steroid hormones and the detoxification of body fluids.
- **d.** False they fuse to the cis face.
- **e.** True secretory cells such as pancreatic acinar cells have well-defined endoplasmic reticulum and Golgi apparatus.

13. On mitochondria:

- **a.** False the main function of mitochondria is oxidative phosphorylation.
- **b.** True mitochondria are 'semi-autonomous', having their own organelles, including ribosomes, and DNA.
- **c.** True as do all eukaryotic membrane-bound organelles. In addition, mitochondria have a second, inner, phospholipid membrane, with very different properties to that of the outer membrane.
- **d.** True mitochondria are involved in programmed cell death, with many anti-apoptotic proteins being located in the outer mitochondrial membrane.
- **e.** False lysosomes contain acid hydrolases.

14. On vesicles:

- **a.** False secretory vesicles contain hormones and neurotransmitters. Lysosomes contain hydrolytic enzymes.
- **b.** True macrophages have abundant lysosomes, which become fused to pathogen containing phagosomes to create phagolysosomes.
- **c.** True other peroxisomal biogenic reactions include cholesterol and plasmalogen biosynthesis.
- **d.** False Zellweger syndrome results from a deficiency in peroxisome biogenesis.
- **e.** False insulin release is regulated by blood glucose concentration.

15. On the cytoskeleton:

- **a.** True during cell division, the mitotic spindle (microtubules) forms between centrioles.
- **b.** False microfilaments are made from actin, microtubules are made from tubulin.
- **c.** False microvilli are non-motile extensions of plasma membrane. Centrioles are the site of spindle assembly in cell division.
- **d.** False primary cilia have a 9+0 arrangement. Motile cilia have a 9+2 arrangement.
- **e.** False spermatozoa use flagella for motility.

16. On cell differentiation:

- **a.** False there are four tissue types: epithelia, connective tissue, nervous tissue and muscle.
- **b.** True in order to function as a collective unit, it is essential that specialized cells have the ability to interact and communicate with one another.
- **c.** False although all cell types are derived from a single cell type, this cell is the zygote.
- **d.** False the mature erythrocyte is anuclear.
- **e.** True even when removed from its normal environment the differentiated cell and its progeny will retain many of its functional characteristics.

17. On the fluid mosaic model:

- **a.** False the key concept relates to a phospholipid bilayer
- **b.** True stabilization is achieved through the hydrophobic interactions of the hydrophobic fatty acid molecules in the interior layer and by the presence of cholesterol.
- **c.** True the foundation of the plasma membrane is a phospholipid bilayer.
- **d.** False integral proteins span the entire phospholipid bilayer either once (monotopic), twice (bitopic) or many times (polytopic).
- **e.** False biological membranes are dynamic and fluid structures, hence the name the 'fluid mosaic model'.

18. On cell membrane lipids:

- **a.** True by definition lipids are insoluble in water but soluble in non-polar organic solvents.
- **b.** False triglycerides consist of three fatty acids attached to a glycerol backbone, while phospholipids are triesters of glycerol with two fatty acids and one phosphate ion.
- **c.** False sphingomyelin has a ceramide backbone.
- **d.** True the location of the different phospholipids within the lipid bilayer structure give the membrane asymmetry.

e. False cholesterol stabilize the plasma membrane by decreasing fluidity at physiological pH and increasing fluidity at low temperatures.

19. On cell membrane proteins:

a. False peripheral proteins adhere to the membrane, they do not pass through it.
b. False they are rich in polar amino acids.
c. True peripheral proteins are associated with the outer leaf of the bilayer.
d. False peripheral proteins may be attached by means of electrostatic attachment to integral proteins or covalent attachment to membrane consituents.
e. False glycoproteins are conjugated proteins in which the non-protein group is a carbohydrate.

20. On properties of biological membranes:

a. False membranes are more fluid if they have a high proportion of unsaturated fatty acids.
b. True fluidity is linked to the kinetic energy of phospholipid molecules.
c. True axial rotational movements are one of the four different modes of movement exhibited by membrane phospholipids.
d. True membrane asymmetry is functionally important, linking proteins with effector molecules.
e. True covalent attachment to lipid molecules also limit lateral movement.

21. On transport across the cell membrane:

a. False it is the movement of particles from a region of high concentration to a region of low concentration.
b. True this continues until equilibrium.
c. False One osmol/L of NaCl contains 0.5 mol of the ion Na^+ and 0.5 mol of the ion of Cl^-.
d. False it is 280–295 mosmol/L.
e. False it is expressed in terms of osmoles of solute per litre of solvent.

22. On ionic distributions across the membrane:

a. True distribution is controlled by the semipermeable membrane concept, electrochemical gradient and pumps.
b. False they cannot diffuse through the lipid bilayer, and they rely on transport proteins.
c. True these proteins are generally specific for particular molecules.
d. True energy is required to drive ions against their concentration gradient.
e. False Sodium is actively pumped out of the cell, maintaining a lower concentration intracellularly.

23. On transport across the membrane:

a. True passive diffusion occurs down a concentration gradient.
b. False sodium is actively pumped out of the cell.
c. True carrier proteins can become saturated.
d. False three Na^+ ions are pumped out and two K^+ ions move in to the cell.
e. True antiports transport ions in opposite directions.

24. On transport proteins:

a. True one type of which is permanently open leading to constant leakage of potassium through these channels, which is critical to the membrane potential.
b. True carrier proteins are implicated in active and passive processes.
c. True cells with lots of Ca^{2+} channels will be more permeable to Ca^{2+} than one with few.
d. False the sodium pump is an example of an antiport transporter.
e. False the sodium pump is composed of an α-subunit and a β-subunit.

25. On membrane potential:

a. True the opening of voltage-gated sodium ion channels leads to depolarization of the cell's membrane potential (from −70 mV towards 0).
b. False this is the Nernst equation.
c. True this is also known as the equilibrium potential.
d. True this is the basis of Gibbs–Donnan equilibrium.
e. False proteins are not ions! The third is Cl^-.

26. On receptors:

a. True they are ion channels controlled by a ligand activated gate.
b. False they are metabotrophic or G-protein coupled.
c. False metabotrophic receptors are seven pass transmembrane proteins.
d. True HREs are found in the promoters of hormone-responsive genes.
e. True receptor binding triggers a conformational change in the G-protein.

27. On signal transduction:

a. False IP_3 is a second messenger.
b. True activation of G_s containing G-proteins increase intracellular cAMP.
c. False first messengers attach themselves to the receptor, activating the intracellular second messenger.
d. False PKA is a cAMP-dependent protein kinase that phosphorylates target proteins. PKG is a cGMP-dependent protein kinase.
e. True the binding of a single molecule to a receptor can lead to a massive up or down regulation in end molecule production.

28. The Na^+/K^+ ATPase pump:

a. False three Na^+ ions are pumped outward and two K^+ ions inward.

b. False the Na^+/K^+ ATPase pump uses ATP and is, therefore, an example of primary active transport.
c. True the α-subunit is the catalytic unit, and the β-subunit is important for the assembly and localization of the pump.
d. False binding of Na^+ leads to phosphorylation of the pump. Binding of K^+ causes dephosphorylation.
e. True the ATPase couples the movement of Na^+ down its electrochemical gradient to the movement of a second ion against its gradient.

29. Receptors and drugs:

a. False agonists activate receptors.
b. True much like competitive inhibitors of enzymes, increasing the concentration of the agonist reduces competitive antagonism.
c. True propranolol is a non-specific β-antagonist.
d. True this describes a non-specific interaction.
e. True remember that all drugs have side effects, even those that bind their intended target.

30. Concerning microfilaments:

a. False they are polymers of G-actin.
b. True the terminal web is a specialized area of cortical cytoskeleton, which lies beneath microvilli and desmosomes.
c. False neurofilaments, a type of intermediate filament, determine the diameter of axons.
d. True gel–sol transition of the actin network facilitates movement in macrophages.
e. True α-actinin binds actin into tight arrays of parallel strands.

31. Concerning microtubules:

a. False all cells except mature erythrocytes contain microtubules.
b. True microtubules aid movement via cycles of ATP-powered dynein arm linkage.
c. True microtubules are abundant in neurons.
d. False lamins form the nucleoplasmic veil.
e. True microtubules extend from microtubule organizing centres, such as centrosomes.

32. Concerning the cytoskeleton:

a. True mechanical strength is one of the functions of the cytoskeleton.
b. True the red blood cell must squeeze through narrow blood capillaries without rupturing its membrane, this is achieved by the thin spectrin–actin cytoskeleton.
c. True nine fused pairs of microtubules ('doublets') form the cylinder, and one unfused pair sit in the centre.
d. True the cytoskeleton has a role in separating duplicated chromatids into separate cells during mitosis.
e. False actin gel–sol transition at the tip of phagocyte pseudopodia facilitates motility.

33. Concerning lysosomes:

a. False lysosome vesicles contain acid hydrolases.
b. True primary lysosomes bud off the trans-face of the Golgi.
c. False erythrocytes do not contain lysosomes.
d. False they are capable of both autophagy and heterophagy.
e. False their contents are glycosylated in the rough endoplasmic reticulum. Covalent modification occurs in the Golgi.

34. Concerning lysosomal storage diseases:

a. False all lysosomal storage diseases identified to date are single gene disorders.
b. True the incidence of these disorders is higher Ashkenazi Jews than that of the general population.
c. False lysosomal storage diseases result in macromolecules getting trapped within the lysosome, giving a characteristic enlargement of both lysosome and tissue.
d. True Gaucher's disease has an incidence of 1 in 25 000 live births.
e. False Gaucher's disease results from β-glucosidase deficiency. Tay–Sachs results from a deficiency in the hexosaminidase A α-chain.

35. Concerning cell junctions:

a. True cell–cell contact tight junctions are mediated by the protein occludin.
b. False tight junctions are responsible for bladder impermeability.
c. True adherens junctions attach the actin cytoskeletons of adjacent cells together.
d. True each gap junction is composed of two connexons, or hemi-channels.
e. False pemphigus results from autoantibodies generated against the cadherin desmoglein.

36. Concerning cadherins:

a. True they are calcium dependent single-pass glycoproteins.
b. False the N-terminal amino acids are important in ligand binding.
c. False N-cadherin, also known as CDH2, is found in nervous tissue and skeletal muscle. E-cadherin is found in epithelia and early nervous tissue.
d. True loss of cadherin expression has been shown to drive invasion and metastasis.
e. True catenins attach cadherins both to actin and to the cytoplasm of a cell.

37. Concerning the Immunoglobulin (Ig) superfamily:

a. True β-sheets roll up completely to join edges and form a cylinder or closed 'barrel', in which the first strand is hydrogen bonded to the last.
b. True they are Ca^{2+}-independent.
c. False selectins initiate leukocyte–endothelial interactions.

d. True members of this family include ICAM, VCAM and PECAM.
e. False they are formed by the association of two β-pleated sheets joined by cysteine–cysteine disulphide bonds.

38. Concerning basement membranes:

a. True the predominant collagen is type IV.
b. True the glomerular basement membrane forms the boundary between blood and urine, across which water and other small molecules from the blood are filtered.
c. True hemidesmosomes (intermedicate filaments) and focal contacts (microfilaments) anchor epithelial cells to the basement membrane.
d. True all endothelial cells sit on a basement membrane.
e. False integral proteins are a feature or the plasma membrane.

39. Concerning the extracellular matrix:

a. False glycosaminoglycans are very hydrophilic.
b. False chondroitin sulphate is found in cartilage.
c. False collagen is a triple helix.
d. True the major function of fibroblasts is the synthesis of collagen.
e. False elastin provides elasticity and flexibility to the ECM.

40. Concerning intermediate filaments:

a. False their stability comes, in part, from the fact that they do not dissociate under physiological conditions.
b. False lamins are found exclusively in the nucleus.
c. True desmin fibres anchor and orientate the Z bands in myofibrils, generating the striated pattern.
d. True as well as mesenchymal cells it is also found in endothelial cells.
e. True GFAP is found in glial cells surrounding neurons.

41. Concerning selectins:

a. True the carbohydrate ligands of selectins are called lectins.
b. True remember L is for leukocyte.
c. False it is recognized by E-selectin.
d. False P-selectin is also known as GMP-140.
e. True it is also rapidly activated by thrombin, platelet activating factor and phorbol esters.

42. Concerning integrins:

a. True they are involved in the transduction of signals across the cell membrane.
b. False β-2 integrins are found exclusively on leukocytes.
c. False the sequence is RGD.
d. True VLA stands for very late acting.
e. True this leads to an immunosuppressive phenotype and recurrent infection.

43. On amino acids:

a. False they possess carboxyl groups (COOH), not ketone group.
b. False Glycine is the only non-optically active amino acid.
c. True the reaction releases a molecule of water.
d. True as such they can be synthesized by the human body.
e. True essential amino acids must be must be supplied exogenously.

44. On amino acid structure and properties:

a. True hydrophobic character increases with increasing side chain size.
b. True an effect known as steric hindrance.
c. False they favour an α-helical secondary structure.
d. True they link thiol ($^-$SH) groups of cysteine residues.
e. False hydrophilic residues will be found on the outside, hydrophobic on the inside.

45. On amino acids as buffers:

a. True as such they have the ability to act as both donors and acceptors of protons.
b. True at this point the molecule would not move in an electric field.
c. False $pH = pK_a + \log [A^-]/[HA]$.
d. False both the amino and the carboxyl groups may act as buffers.
e. True at a pH of 9.8 the amino group of glycine functions as a buffer.

46. On functions and structures of proteins:

a. True Protein conformation is defined by the sequence of its amino acid residues, including those lost during processing.
b. True for example, myosin expressed in heart tissue has a different primary sequence to myosin expressed in fast muscle fibres.
c. False α-helices and β-pleated sheets are examples of secondary structure.
d. True these include including disulphide bridges and electrostatic interactions.
e. True the subunits are held together by the same types of bond that stabilize tertiary structure.

47. On forces and bonds of protein structure:

a. True it also has a length of 0.132 nm.
b. True hydrogen bonds occur between peptide bond atoms and polar side groups where a hydrogen atom is shared between two electronegative atoms.
c. False they result from strong attractions between positive and negative atoms.
d. True glycine and proline are usually found at the α-helix bends.
e. False antiparallel β-sheets are the more stable.

48. On monosaccharides:

a. True those with aldehyde carbonyl group are known as aldose sugars.
b. False most naturally occurring sugars are D isomers.
c. False they are called furanose rings.
d. True they have a high proportion of hydroxyl groups.
e. True each ring structure has one more optically active carbon than the straight-chain form, so can exist in two forms.

49. On disaccharides:

a. False disaccharides are formed by the joining of two monomers with the elimination of water.
b. True the α-glycosidic bond falls below the plane of the sugar ring.
c. False sucrose is a dimer of glucose and fructose.
d. True maltose is formed from the condensation of two glucose molecules and the formation of an α 1–6 link.
e. True cellibiose is formed from the condensation of two glucose molecules and the formation of a β 1–4 link.

50. On sugar derivatives:

a. True examples include ascorbic acid and glucuronic acid.
b. True examples include mannitol and ribitol.
c. True examples include glucosamine.
d. True examples include adenosine, deoxyadenosine, thymidine and deoxythymidine.
e. True examples include deoxyribose and fucose.

51. On polysaccharides:

a. False they have the general formula $C_n(H_2O)_{n-1}$.
b. False glycogen is highly branched – this allows it to be metabolized rapidly in times of need.
c. False in cellulose, glucose monomers are joined by β 1–4 glycosidic bonds.
d. True the most abundant heteropolysaccharides in the body are the glycosaminoglycans.
e. False they consist of a repeating disaccharide unit of an amino sugar and either a sugar or a sugar acid.

52. On properties of enzymes:

a. True they do so, greatly increasing the speed of the reaction.
b. True as a result they may work under different optimum conditions.
c. False it is the difference in free energy of the reactants and of the unstable intermediates.
d. True they remain in the same chemical state at the end of the reaction as at the beginning, so can be reused.
e. True examples of prosthetic groups include metal ions, for example the haem of haemoglobin.

53. On regulation of enzyme activity:

a. False it is the slowest enzyme. This is known as the rate limiting step.
b. True this shape adjustment helps to explain why enzymes only catalyse specific reactions.
c. True this type of modification and, therefore, change in enzyme activity, is usually reversible.
d. False binding of a small molecule to a site distant from the active site alters the conformation of the active site.
e. True enzymes may be irreversibly activated or deactivated in this manner.

54. On enzyme kinetics:

a. True rate of reaction is proportional to the enzyme concentration.
b. True as the enzyme becomes saturated the reaction rate reaches a limit.
c. False it is the substrate concentration at which the reaction rate is half of V_{max}.
d. False they increase K_m leading to decreased enzyme affinity.
e. False once plateau has occurred the reaction follows zero-order kinetics.

55. On structures in the nucleus:

a. False it is a bilayer, the outer of which is continuous with the endoplasmic reticulum.
b. False the periplasm inhabits the space between the two layers of the nuclear membrane.
c. False molecules of up to 60 kDa pass freely through the nuclear pores.
d. True lamins are associated with the inner nuclear membrane, and form a thin shell beneath it.
e. False genes that are active have less bound histone.

56. On nucleic acids:

a. True polymers of nucleotides make up the nucleic acids DNA and RNA.
b. False adenine is a purine, while thymine is a pyrimidine. Together they form a base pair.
c. True uracil is an RNA base; however, deamination of cytosine can yield uracil in DNA.
d. False the rate-limiting step of purine biosynthesis is controlled by PRPP synthetase.
e. True carbamoyl phosphate is the starting point for the biosynthesis of pyrimidines.

57. On DNA packaging:

a. True chromatin is the collective name for the long strands of DNA, RNA and their associated nucleoproteins.
b. False euchromatin is transcriptionally active, while heterochromatin is not active in RNA synthesis.
c. False nucleosomes consist of 146 bp of DNA wrapped twice around an octamer of histone proteins.

d. False nucleosomes are sometimes referred to as 'beads on a string'.
e. False Metacentric chromosomes have central centromeres, yielding p and q arms of similar lengths.

58. On gene structure:

a. True promoters are found in the upstream flanking region, which is found 5′ of the gene coding sequence.
b. False TATAAT boxes are a feature of prokaryotic genes.
c. False although exons are generally represented in mature mRNAs, one hnRNA may have many different polypeptide products, based on alternative splicing patterns.
d. False stop codons are found 3′, genes are transcribed 5′ to 3′.
e. True although there is some ongoing debate as to whether genes that are transcribed but produce a non-functional RNA should be classified as pseudogenes.

59. On DNA replication:

a. True the leading strand is produced as a continuous polymer, whereas the lagging strand is synthesized in short stretches, which are then ligated together.
b. True the segment of DNA that is copied starting from each unique replication origin is called a replicon.
c. True polymerase α adds the RNA primer and the first few hundred nucleotides of a newly synthesized stretch of DNA.
d. True it acts as a polymerase δ cofactor during DNA replication as well as during DNA synthesis associated with DNA damage repair.
e. False Okazaki fragments are joined together by a DNA ligase, endonucleases cleave nucleic acids at specific sequences.

60. On eukaryotic transcription and post–translational modification:

a. True DNA-dependent RNA polymerases unwind dsDNA to expose the template strand.
b. False this is the classical description of *cis* acting factors.
c. True RNA polymerase II contains at least six basal transcription factors (TF II) A, B, D, E, F and H.
d. False addition of a 5′ cap regulates nuclear export via the cap binding complex.
e. True there are four major classes or snRNP involved in splicing.

61. On protein synthesis:

a. True one mature mRNA produces one specific protein product.
b. True each amino acid is encoded by multiple codons.
c. False eukaryotic translation is more commonly cap-dependent.
d. True they yield products that are typically needed for maintenance of the cell.
e. False it is a post-translational modification.

62. On the cell cycle:

a. False the restriction point can be found in G_1.
b. False quiescence, or G_0, occurs after mitosis, but before the restriction point is passed.
c. True phosphorylation of Rb leads to the transcription of genes required for the transition from G1 into S phase.
d. False INK4 is a cyclin-dependent kinase inhibitor.
e. False the main site of action of p53 on the cell cycle is the G_1–S DNA integrity checkpoint.

63. On mitosis and meiosis:

a. False meiosis is reduction division. In human primordial germ cells it results in a reduction of chromosome number from 46 to 23.
b. False they are maximally visible at metaphase.
c. True diplotene is a phase of meiotic prophase I, and the point at which chiasmata separate.
d. False meiotic chromatids separate in anaphase II.
e. False they all have 23 as the first polar body is formed after meiosis I; that is after reduction division.

64. On DNA damage and repair:

a. False it is an example of endogenous mutagenesis.
b. True translocations can be balanced, without loss or gain of genetic material or unbalanced, and the gene at which the break occurs becomes altered.
c. False inability to repair DNA in healthy cells leads to a cell cycle arrest, and if the damage still cannot be repaired, the cell is apoptosed.
d. False HNPCC results from an inherited mutation in once copy of a *MMR* gene.
e. False NHEJ is a method for repairing double-stranded DNA damage and is highly error prone due to the loss of genetic material in the processing of broken, damaged chromosomal ends.

65. On eukaryotic RNA molecules:

a. True hnRNA is the precursor to mRNA, and contains both introns and exons.
b. False rRNA is required for ribosomal structure.
c. False introns are spliced out of the hnRNA to leave mRNA.
d. True post-translational modification of these bases is thought to be required for tRNA–protein interactions or in stabilizing the tRNA molecule.
e. False rRNA is synthesized in the nucleolus.

66. On telomeres and telomerase:

a. False they are repeats of the hexameric sequence 'TTAGGG'.
b. True T-loop formation prevents inappropriate telomerase activity.

- c. True they function to prevent end-on-end chromosome fusion, and protect the non-telomere ends of chromosomes from degradation.
- d. False telomerase is not normally active in somatic cells.
- e. True in humans they are known as TERT (TElomerase Reverse Transcriptases).

67. On isolation and preparation of nucleic acid:

- a. False depending on the restriction enzyme, sticky or blunt ends may be produced.
- b. True a sequence of eight specific base pairs will occur much less frequently than a sequence of four base pairs.
- c. False DNA carries a negative charge and moves towards the positive electrode.
- d. True it fluoresces with an orange-violet colour when exposed to ultraviolet light.
- e. True PFGE allows fragments well over 10 Mb to be separated out.

68. On PCR:

- a. False PCR can typically be used to amplify DNA segments less than 3 kb in length.
- b. True if a non-heat stable polymerase is used, it will be denatured during the first cycle, terminating amplification.
- c. True like any form of DNA replication, the direction of synthesis in 5′–3′.
- d. True this is why 72°C is taken as the extension temperature.
- e. False it uses RNA as a template to produce DNA.

69. On nucleic acid and protein hybridization techniques:

- a. True this allows many consecutive hybridization experiments to be conducted from one gel.
- b. False PCR takes around 12 h–2 days, while Southern blotting typically takes 3–5 days.
- c. False this is now largely achieved by PCR.
- d. False remember SNOW DROP. Western blotting is a protein technique.
- e. True RNA is the product of gene expression.

70. On DNA sequencing and polymorphisms:

- a. True For this reason, Sanger sequencing is also known as chain terminator sequencing.
- b. True as each dNTP can have its own coloured fluorophore, a single reaction can be used, rather than the four of traditional Sanger sequencing.
- c. False SNPs are more common in intronic DNA.
- d. False the level is taken to be 1%.
- e. True other features include a repeat unit of 15–100 bp running over a length of 0.5–30 kb.

71. On cytogenetics:

- a. False it is named after the stain used; Giemsa.
- b. False G-banding is generally regarded as too crude to detect small changes in chromosome composition.
- c. True the probe is then localized via fluorescent microscopy.
- d. False the resolution of CGH is 5–10 kb.
- e. True it can resolve single exon deletions.

72. On cloning and cloning techniques:

- a. True the sequence of DNA is replicated alongside the microbes own complement of DNA, leading to amplification.
- b. False BACs have an insert size limit of around 100–300 kb.
- c. False phages are viruses that infect bacteria.
- d. True only colonies that contain the vector plus insert turn blue.
- e. True often, if the PCR product is large, it will be digested with a restriction enzyme to break it into smaller pieces.

73. On maps:

- a. False genetic maps are measured in centimorgans, physical maps are measured in bases.
- b. False it is equivalent to a 1% chance of recombination.
- c. False as recombination occurs more frequently in female meiosis, there are separate maps for males and females.
- d. True the most recent high-resolution genetic maps have polymorphic markers spaced at intervals of less than 1 cM.
- e. True using the human genome was sequenced using overlapping clones spanning the whole human genome.

74. On the human genome project:

- a. True correction of these errors is expected to continue for the foreseeable future.
- b. True the human genome is being compared with the genomes of model organisms, such as *E. coli* and *S. cerevisiae*.
- c. False estimates lie at 20 000–25 000.
- d. False it involves sequence analysis, open reading frame (ORF) analysis and gene scanning.
- e. False it refers to all the metabolites produced by a genome, which includes the whole protein complement.

75. On genetic linkage analysis:

- a. True this is the basic principle of linkage analysis.
- b. False chiasmata form during meiosis.
- c. True a recombination fraction cannot exceed 50%.
- d. True therefore, also equivalent to a 1% chance of recombination.
- e. False a LOD of 3.0 suggests linkage.

76. On gene therapy:

- a. False germ-line strategies are considered unethical on the grounds that genetic changes would be transmitted to future generations.
- b. False acquired conditions can also, in theory, be targeted.

c. True a gene therapy trial cannot be conducted until this criterion is satisfied.
d. False unless the vector is inert it, and the gene insert, will be destroyed.
e. False they are artificial lipid vesicles that fuse with the cell membrane to deliver insert DNA in to the cell.

77. On mechanisms of mutation:

a. False mutations by definition are permanent changes to the genetic material.
b. False transition substitutions are an interchange between a purine and a purine or a pyrimidine and a pyrimidine.
c. True such sequences are often similar, and predispose to recombination errors.
d. False frameshift mutations destroy open reading frames and as such are never multiples of three nucleotides.
e. True most commonly: CGG, CCG, CAG or CTG.

78. On structural and functional effects of mutation:

a. False point mutation may lead to a silent mutation, with no overall alteration of amino acid sequence in the gene product.
b. False missense mutations change the codon to another, which in turn results in a different amino acid being inserted into the growing peptide chain.
c. True nonsense mutations see the insertion of a stop codon, terminating the protein product.
d. True leading to partial loss of function.
e. True this usually arises by dimerization of the product leading to inactivation.

79. On dominant monogenic disorders:

a. True older texts may refer to this as 'vertical inheritance' to represent the pedigree pattern observed.
b. True such mutations, when present in the homozygote, are embryonically lethal.
c. True loss of half of normal protein activity is sufficient for disease to occur.
d. True sons receive their X chromosome from their mother, so are not affected.
e. False it is X–linked recessive.

80. On recessive monogenic disorders:

a. True generally, two mutant copies of an allele are required for autosomal recessive conditions to manifest.
b. True autosomal recessive disease does not show any inherent sex selection.
c. False females can be affected, for example if she is the daughter of an affected male and a carrier female, by skewed Lyonization or as a result of having Turner's syndrome.
d. True all offspring will receive one X chromosome from their mother.
e. True as males can only receive a Y chromosome from their father, there is no male to male transmission.

81. On non-Mendelian inheritance of single-gene disorders:

a. True anticipation is commonly seen in trinucleotide repeat disorders, such as myotonic dystrophy.
b. False progressive external ophthalmoplegia is a mitochondrially inherited disorder, and its expression results from the relative proportion of normal to mutant mitochondrial DNA
c. False this is imprinting.
d. False Angelman syndrome corresponds to a switching off of the critical region on the paternally derived chromosome.
e. True mitochondrial DNA is maternally inherited.

82. On chromosome disorders:

a. True monosomies result in haploinsufficiency, while polyploidy and trisomy result in simple gain of function.
b. True a delay in the movement of one chromosome from the metaphase plate during anaphase may result in its loss, if it does not reach the pole before the cell membrane is reformed.
c. False trisomy 18 is Edwards syndrome. Patau syndrome is trisomy 13.
d. False this would be in keeping with monosomy X, Turner's syndrome.
e. True it typically arises as a result of unequal crossing over between X and Y chromosomes during meiosis.

83. On multifactoral disorder inheritance:

a. False multifactorial disorders arise from interactions between genetic and environmental factors, so such conditions are not inherited in a Mendelian fashion.
b. True abnormalities do not usually have a distinct phenotype, but are extremes of the curve.
c. True sets of individuals commonly compared include monozygotic and dizygotic twins.
d. True the threshold model states that a condition only manifests when liability for a disease exceeds the threshold for that condition.
e. True if heritability is high, there is a high correlation in relatives.

84. On multifactoral disorders:

a. True 95% of affected individuals have human leukocyte antigens (HLA) DR3 and/or DR4.
b. True other factors include advancing age, decreased physical activity and body weight.
c. False amyloid precursor protein is associated with autosomal dominant inheritance patterns.
d. True other genetic factors identified include laeverin and 11β-hydroxysteroid dehydrogenase type 2.
e. False the hereditability of asthma is currently thought to lie at around 60%.

85. On Down syndrome:

a. False Down syndrome can also result from a Robertsonian translocation or from genetic mosaicism.
b. True the most common leukaemia seen in DS patients is acute lymphoblastic leukaemia.
c. False there is a strong correlation between maternal age and incidence of DS.
d. False aplasia cutis is usually more associated with trisomy 13.
e. False NEVER use terms such as mongolism or mentally retarded in relation to ANY patient, either in conversation or in written form.

86. On cancer genetics:

a. False only around 5% of cancers are familial.
b. True carcinogenesis generally results from an accumulation of mutations leading to genetic instability.
c. False proto-oncogenes are normal genes that act to promote normal cell growth. Oncogenes are pro-carcinogenic.
d. False tumour suppressor genes are also known as anti-oncogenes.
e. True most mutations are gain of function mutations that result in increased expression of the gene.

87. On cancer syndromes:

a. True this is one of a number of features suggestive of inherited cancer susceptibility in a family.
b. True cell cycle arrest is critical for the ability to repair DNA.
c. True although better known for their role in breast cancer, *BRCA1* and *BRCA2* mutations are also seen in ovarian cancer.
d. False *APC* is a tumour suppressor gene, and is mutated in familial adenomatous polyposis coli.
e. False they develop basal cell carcinomas.

88. On chromosomal disorder nomenclature:

a. True.
b. False it describes an inversion of a segment of chromosome 9 between position 12 of the short arm and position 14 of the long arm.
c. True it is Turner's syndrome.
d. False duplications are denoted by '(dup)', so as not to confuse them with deletions '(del)'.
e. True for example 46,XX,r(X)(p12,q14), describes a ring chromosome X syndrome, where p12 is fused to q14.

89. On the Hardy–Weinberg equilibrium:

a. True only then can the law be extended to incorporate the Hardy–Weinberg equilibrium.
b. False the Hardy–Weinberg equilibrium relies on an absence of selection.
c. False as the likelihood of producing a heterozygote is $(p \times q) + (p \times q) = 2pq$, Hardy–Weinberg equilibrium is expressed as $p^2 + 2pq + q^2 = 1$.
d. False the heterozygote frequency would be approximately 1 in 20.
e. True nearly all those affected by autosomal dominant conditions are heterozygotes.

90. On factors which disturb Hardy–Weinberg equilibrium:

a. False assortative mating is non-random and describes the selection of a mate based on a certain desirable character.
b. False high mutation rates are balanced against a reduction in reproductive fitness in affected individuals, preventing propagation.
c. True the selection of heterozygosity for its ability to confer a specific ability is a positive selective force.
d. True immigration into a population may lead to the introduction of new alleles, while emigration may result in the loss of alleles.
e. True the founder effect can lead to a larger than expected frequency of an allele within a population as the colony expands.

91. On screening:

a. False population screening may be offered to all, regardless of family history.
b. False this is the Wilson–Jungner criteria.
c. True this is one of the Wilson–Jungner criteria.
d. True multiple of the median calculations compare the marker in question with the normal median calculation, which is taken as being 1.0.
e. True it allows more accurate estimation of risk to be made. It is controversial, however, as it may put pressure on relatives to test, therefore undermining their autonomy.

92. On methods used for detection of carriers and sufferers:

a. False although linked markers are usually transmitted together and give an indication of risk, they do have the potential for independent segregation.
b. False the newborn blood spot test can detect: phenylketonuria, congenital hypothyroidism, cystic fibrosis and sickle-cell disorders.
c. True if the mutation(s) that cause a specific disease has been defined, PCR can identify both sufferers and carriers.
d. True people with a high risk of having a child with cystic fibrosis may be offered antenatal diagnostic tests.
e. False although Sickledex is used for adult diagnoses, sickle-cell anaemia can also be detected using the Guthrie card method.

93. On taking a genetic history:

a. False in families with a high incidence of an autosomal recessive disease, consanguinity increases the likelihood of two carriers coming together. This explains a large part of the high incidence of certain conditions in certain ethnic populations.
b. False the symbol is a filled in square.
c. False the index is known as the proband.
d. True not all genetic disorders are monogenic.
e. False many genetic diseases arise spontaneously, through mutation.

94. On risk assessment:

a. False this describes an autosomal recessive condition and as such, the risk to the child is 25%, or 1 in 4.
b. False in very rare cases a parent can be homozygous for an autosomal dominant condition, in which case the risk to the child would be 100%. Homozygosity for autosomal dominant conditions is usually embryonically lethal. If the affected parent is heterozygous, as is usually the case, the risk to the child would indeed be 1/2 or 50%.
c. False they will be carriers, unless mother is a carrier too then it is 50%.
d. True it is so-called as it is the probability that one event or another event will occur.
e. False risk assessment gives a probability that an individual will be affected by a specific condition.

95. On Bayes' theorem:

a. True it corrects for additional information, such as test data.
b. False it gives a better estimation of risk, but not an absolute value.
c. False family history is crucial for determining the prior probability to be modified by test data.
d. False prior probability is determined using classical Mendelian inheritance patterns.
e. True dividing the 'joint' probability by the sums of the 'joint' probabilities yields the 'relative' probability.

96. On genetic counselling and consultation:

a. False it should be non-judgemental and non-directive towards the consultands.
b. True and the risk to children born of first cousins is 1/20.
c. True the mother will possess 50% of the child's VNTRs and the father a further 50%.
d. False all reported dysmorphic features should be investigated and each system carefully examined for non-reported features.
e. True certain geographical regions have a higher incidence of specific genetic diseases, for example β-thalassaemia and the Mediterranean.

97. On common presentations of genetic disease:

a. True as such careful measurements of facial landmarks are crucial if an abnormality is suspected.
b. False hypotelorism is a decreased interpupillary distance.
c. True cataracts are seen in around 60% of patients.
d. False only 10% of genetic causes of sensorineural deafness are attributable to the X chromosome.
e. False osteopetrosis is a rare genetic disease in which bones become harder and more dense, causing them to break more easily

98. On current treatment techniques for genetic disease:

a. True it is concerned with relieving symptoms and reducing complications, and concerns medics, surgeons, physiotherapists and occupational therapists among others.
b. True patients with α_1-antitrypsin deficiency may delay the rate and severity of emphysema by avoiding tobacco smoke.
c. False phenotype may also be corrected for functional reasons.
d. False phenylketonuria is treated with dietary limitation of phenylalanine.
e. False it targets it to macrophage lysosomes.

99. On future treatment techniques for genetic disease:

a. False ADA-SCID may already be treated using somatic gene therapy.
b. True the host immune system detects 'foreign' DNA and targets it for destruction.
c. False viral vectors are modified to remove their inherent disease causing capability.
d. True thus silencing the expression of target genes.
e. True bone marrow is a rich source of haematological stem cells.

100. On genetic ethics, its theories and principles:

a. False it is the ethics of duty and judges an action independently of its consequences.
b. False non-maleficence is the principle of seeking, overall, to do no harm.
c. False they are integral to autonomy, fairness and even handedness are integral to justice.
d. False presymptomatic diagnosis can only influence management if management options are available.
e. False other indications for abortion post-24 weeks are: a substantial risk that the resulting child would be born with serious handicap or if there is a risk to the mothers' life, physical or mental health.

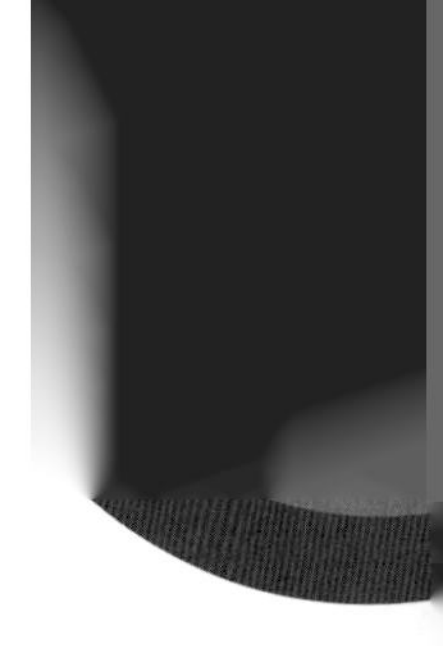

1. Salbutamol is a short-acting β-agonist (a 'SABA'). It acts on β-adrenergic receptors, which are members of the metabotropic receptor family. Metabotropic receptors exert their effects by activating GTP-binding proteins. Activated G-proteins then activate intracellular second-messenger pathways. See Chapter 3.

2. The gene involved should be cloned and characterized; the specific tissue to be targeted should be accessible and identified; a safe and efficient vector system for the gene should be defined; the scientific rationale for the gene therapy approach should be sound and the perceived risks commensurate with the potential benefits. See Chapter 7.

3. Bacteria may acquire genetic material by three mechanisms: transformation, transduction and conjugation. These mechanisms are described in Chapter 1.

4. Ms S is experiencing the prodromal symptoms of herpes simplex virus (HSV) infection, leading to herpes labialis (a 'cold sore'). Aciclovir can reduce the duration of active HSV. Aciclovir is a prodrug and is converted to aciclovir monophosphate by virally encoded thymidine kinase, and to aciclovir triphosphate by the host cell. Aciclovir triphosphate is a chain terminating nucleotide analogue. See Chapter 1.

5. Peroxisomes are derived from the endoplasmic reticulum and are involved in both biogenesis (i.e. cholesterol and bile) and degradation (fatty acids) reactions and in the breakdown of toxic compounds, for instance breaking down excess purines to urea. See Chapter 2.

6. Familial adenomatous polyposis coli is a rare autosomal dominant condition caused by a deletion in 5q21 resulting in a loss of function of the tumour suppressor gene, *APC*. Factors that might make you suspect an inherited cancer susceptibility syndrome include: several close (first- or second-degree) relatives with a common cancer; several close relatives with genetically associated cancers; two family members with the same rare cancer; an unusually early age of onset; bilateral tumours in paired organs; synchronous or successive tumours; and tumours in two different organ systems in one individual. See Chapter 8.

7. Gaucher's disease is defined by age of onset and degree of neurological involvement. Three types are commonly described: type I – adult type, non-neuronopathic, lifespan is shortened but not markedly; type II – severe infantile, rare, neurological signs seen at 3 months, death usually by 2 years of age; and type III – 'juvenile' subacute, neuronopathic, variable presentation from childhood to 70 years of age. See Chapter 4.

8. DNA, RNA and their associated nucleoproteins are known as 'chromatin', DNA winds onto nucleosome spools, the nucleosome chain coils into a structure known as the solenoid. The solenoid forms loops and the loops attach to a central scaffold. The scaffold plus loops arrange themselves into a giant supercoil. During mitosis, the giant supercoils are organized into chromosomes. See Chapter 6.

9. Proto-oncogenes are normal cellular genes that drive cell growth and division. When mutated or up regulated (oncogenes), control of cell division is lost and as such cancer is promoted. Tumour suppressor genes are genes which encode proteins involved in the regulation of cell division and differentiation. Their loss may result in uninhibited growth, driving the carcinogenesis process. See Chapters 3,6 and 9.

10. The six main aims of the HGP are: mapping human genes and markers; sequencing the human genome, functional analysis and post-genomic genetics, comparing the human genome with the genomes of model organisms; developing new DNA technologies and developing bioinformatics. See Chapter 7.

11. CF is autosomal recessive, so a mutated gene must be inherited from each parent for the disease to manifest. Unfortunately it is likely that the foetus will have CF because it has the only possible combination of alleles not seen in the healthy siblings. However, the test can not be considered conclusive because recombination between the microsatellite marker and the causal mutation is possible (albeit unlikely). See Chapter 8.

12. Prokaryotic ribosomes are composed of a small 30S and a large 50S subunit, the so called '70S' ribosome). Eukaryotic cells consist of a small 40S subunit and a large 60S subunit, the '80S' ribosome. This difference is important, and the basis of many antimicrobial therapies. See Chapters 1 and 2

13. Angelman syndrome results from the loss of imprinted genomic material within a specific locus on chromo-

some 15 (the so-called 'critical region' between 15q11 and 13). This results in the loss of the only active copy of the *UBE3A* gene. Features seen in Angelman syndrome include: hypertonia; ataxic, broad-based gait; characteristic arm posture (arms up and flexed); prominent jaw; deep set eyes; an absence of speech; smiling face and inappropriate laughter; flat occiput; learning difficulties and epilepsy. See Chapter 8.

---

14. Hardy–Weinberg equilibrium states states that allele frequencies remain constant from one generation to the next, providing there is an absence of selection. It can be denoted as: $p^2 + 2pq + q^2 = 1$, where p is defined as the frequency of the dominant allele and q as the frequency of the recessive allele (for a trait controlled by a pair of alleles). In order for Hardy–Weinberg to apply an organism has to be: diploid; sexually reproducing and have discrete generations. In addition, the following criteria have to be met: the population is large, to minimize genetic drift; mating is random; the mutation rate remains constant; alleles are not selected for (i.e. they confer no survival or reproductive advantage); and there is no migration into or out of the population (gene flow). See Chapter 9.

---

15. See Fig 5.4. You diagram should show an amino group, a carboxyl group on the right, and an R/'side' group'. See Chapter 5.

---

16. In cases of dehydration, serum osmolality will increase (as a result of a relative increase in serum solute load). This change is detected in the hypothalamus, triggering the release of anti-diuretic hormone (ADH) from the anterior pituitary. ADH acts to increase the permeability of the distal tubules and collecting ducts of the kidney, reducing the amount of water excreted and leading to a relative decrease in serum solute load and a lowering of serum osmolality. See Chapter 3.

---

17. Situs inversus is a congenital condition in which the major visceral organs are reversed from left to right. The heart and stomach are found on the right, and the liver on the left. It is thought to arise as a result of cilia dysfunction as interplay between motile and sensory cilia is required for determination of left–right axis in early vertebrate development. See Chapter 2.

---

18. See Fig 1.2. Your diagram should show that the Gram-positive bacterial cell wall is made mostly of peptidoglycan and teichoic acids, while Gram-negative walls contains a thin peptidoglycan surrounded by an outer membrane, which contains antigenic lipopolysaccharides. See Chapter 1.

---

19. You will need to take a careful history. This should include a careful family history, and a family tree should be constructed to show how any conditions have been passed on through the family – Fig 8.4 will help you with this. Enquire specifically about infant deaths, stillbirths and abortions as this may alter recurrence risks; degree of consanguinity; and ethnicity and country of descent. The risk of disease or serious congenital malformation in a child born to first cousins is taken to be 1 in 20. This is because such second-degree relatives share one-quarter of their genetic information. The couple should be advised as to this increased risk, the true value of which lies in the pedigree you have constructed, and if/when they do get pregnant, detailed anomaly scanning is indicated during the pregnancy as well as careful monitoring during and after the pregnancy. See Chapter 9.

---

20. You could have listed any of the following: progressive neurological degeneration; hepato(spleno)megaly; skeletal dysplasia; coarse facies; specific eye changes (e.g. cherry red spot or corneal clouding) or angiokeratoma. Examples of lysosomal storage disorders include: Gaucher's disease, Tay–Sachs disease, Pompe's disease, Fabry's disease, Hunter's disease and Danon's disease. See Chapter 4.

1. **Theme: prokaryotic structure and function.**
   1. g – Transformation
   2. j – Filament
   3. d – FtsZ
   4. i – 70S
   5. k – Gram-negative

2. **Theme: anti-microbial chemotherapy.**
   1. d – RNA polymerase activity
   2. i – Aciclovir
   3. j – Teicoplanin
   4. h – Cefuroxime
   5. f – Ciprofloxacin

3. **Theme: receptors and signalling.**
   1. h – Tyrosine-kinase associated receptor
   2. c – Ionotropic receptor
   3. m – G_q
   4. j – Hormone response elements
   5. k – Autocrine

4. **Theme: cellular proteins.**
   1. l – Chondroitin sulphate
   2. e – Lamins
   3. d – Keratin
   4. h – Cathepsins
   5. k – β-2 integrin

5. **Theme: mechanisms of genetic disease.**
   1. e – Haemochromatosis
   2. f – Developmental dysplasia of the hip
   3. j – Leber hereditary optic neuropathy
   4. c – Vitamin D resistant rickets
   5. b – Prader–Willi syndrome

6. **Theme: genetic disease.**
   1. k – Tay–Sachs disease
   2. i – Phenylketonuria
   3. j – Hereditary nonpolyposis colon cancer
   4. f – Ataxia telangiectasia
   5. c – Cystic fibrosis

7. **Theme: eukaryotic organelles.**
   1. c – Peroxisomes
   2. i – Microtubules
   3. b – Mitochondria
   4. g – Golgi apparatus
   5. h – Nucleoli

8. **Theme: chromosomal disease.**
   1. a – Down syndrome
   2. i – Klinefelter syndrome
   3. h – Turner's syndrome
   4. a – Down syndrome
   5. j – de la Chapelle syndrome

9. **Screening and risk assessment.**
   1. l – Large population size
   2. f – The natural history of the condition should be understood
   3. b – VNTR
   4. j – Maternal serum α-fetoprotein
   5. a – Non–random mating

10. **Features associated with genetic disease.**
    1. e – Blue sclerae
    2. a – Mandibular hypoplasia
    3. m – Basal cell carcinoma
    4. g – Macro-orchidism
    5. j – Imperforate anus

A

D

E

F

G

H

I

J

K

L

P

Q

R

S